PHANEROZOIC DIVERSITY PATTERNS

PRINCETON SERIES IN GEOLOGY AND PALEONTOLOGY
Edited by Alfred G. Fischer

PHANEROZOIC
DIVERSITY PATTERNS

Profiles in Macroevolution

JAMES W. VALENTINE, EDITOR

1985

PRINCETON UNIVERSITY PRESS
and
PACIFIC DIVISION, AMERICAN ASSOCIATION
FOR THE ADVANCEMENT OF SCIENCE

PRINCETON, NEW JERSEY and
SAN FRANCISCO, CALIFORNIA

Copyright © 1985 by Princeton University Press
Published by Princeton University Press,
41 William Street, Princeton, New Jersey 08540
In the United Kingdom: Princeton University Press,
Guilford, Surrey

All Rights Reserved
Library of Congress Cataloging in Publication Data
will be found on the last printed page of this book

ISBN 0-691-08374-6
ISBN 0-691-08375-4 (pbk.)

This book has been composed in IBM Press Roman
Clothbound editions of Princeton University Press books
are printed on acid-free paper, and binding materials
are chosen for strength and durability.
Paperbacks, although satisfactory for personal collections,
are not usually suitable for library rebinding.

Typeset in the United States of America by
Pacific Division, American Association for the
Advancement of Science, San Francisco, California

Printed in the United States of America by
Princeton University Press, Princeton, New Jersey

CONTENTS

PREFACE

This book contains the proceedings of a symposium held on June 21, 1982 during the sixty-third annual meeting of the Pacific Division, American Association for the Advancement of Science, at the University of California, Santa Barbara. The symposium was part of the program of Section E (Geology and Geography), coordinated by Dr. Peter U. Rodda, Department of Geology, California Academy of Sciences. The meetings as a whole were coordinated by Dr. Alan E. Leviton, Executive Director, Pacific Division, AAAS, of the Department of Herpetology, California Academy of Sciences; without his enthusiastic support this volume could not have been developed.

The continued improvement of the data on Phanerozoic diversity and the rising intensity of effort aimed at explaining the major questions raised by the diversity patterns provided a strong basis for a meeting to exchange views and to produce a collection of work representative of present research in this field. Considering the plethora of symposia in paleontology, it is remarkable that this is the first treatment of Phanerozoic diversity as such; one must turn to such symposia as that of Harland et al. (*The Fossil Record*, Geol. Soc. London, 1967) or the early effort organized by Newell (J. Paleontology 26, 1952) to find fairly close though collateral relations. The differences between the present volume and those earlier works are illuminating. The earlier works are chiefly concerned with establishing whether there *are* patterns and what they are like, with comments on patterns associated with some major events such as the Permian-Triassic extinction; most of the present papers are routinely involved or explicitly concerned with theory or at least with explanation, and should be of interest to biologists as well as paleobiologists. Indeed it is a primary aim of this volume that it serve as a source for biologists, bringing together papers that would otherwise be scattered, as an introduction to what the fossil record may contribute to understanding some macroevolutionary processes.

Special thanks are due Dr. Edward Tenner, Princeton University Press, for his interest and patience. Dr. Alfred G. Fischer, Department of Geology, Princeton University and an anonymous reviewer offered encouragement and constructive suggestions. Much of the editorial work was accomplished while I was in residence at the Department of Geology, Field Museum of Natural History, Chicago, under the Visiting Scientist Program.

James W Valentine

PHANEROZOIC DIVERSITY PATTERNS

INTRODUCTION

DIVERSITY AS DATA

JAMES W. VALENTINE

Department of Geological Sciences, University of California, Santa Barbara

Charles Darwin had a lot of trouble with the fossil record. He was able to use it as strong evidence that evolution had occurred, observing that there had been a long history of ancient life, with biotas succeeding each other through time. However, when it came to using fossil data to support the idea that natural selection was responsible for evolution, the fossil record was singularly unhelpful. Darwin was forced to argue that the record was good enough to establish evolution as a fact but poor enough that evidence of change via natural selection could not be found (Rudwick, 1976).

Within a year of the publication of the *Origin of Species*, a number of leading paleontologists had pointed out that the record was not really neutral with regard to Darwin's expectations, but was best interpreted as being at odds with them (for example Owen, Philipps, and Pictet; see Rudwick, 1976). Despite such opinions, or perhaps because of them, it has been a common notion since Darwin's time that data from the fossil record are inappropriate for the study of evolutionary processes, although it is usually conceded that the data may be used in tracing the broad course of evolutionary history (see Valentine, 1981). Similar sentiments can be heard when the fossil record is studied from an ecological standpoint. For example, some believe that the record may provide evidence of the associations of fossilizable species in

©1985 by Princeton University Press
Phanerozoic Diversity Patterns:
Profiles in Macroevolution

empirically established "fossil communities" and of community distributions in space and time, but that it cannot be used to study the ecological processes associated with those communities.

There is wide agreement that the contribution to geological science of both evolutionary and ecological aspects of the fossil record has been quite significant, but these successes have been almost entirely empirical. The patterns in space and time of fossil taxa or of fossil associations were invaluable in developing the very principles of stratigraphy and more recently in establishing sophisticated interpretations of ancient environmental frameworks of deposition. Yet until quite recently there has been relatively little theoretical feedback from paleontology into biology. A major exception has been the work of George Gaylord Simpson (see Gould, 1980), which however assumed the correctness of a certain body of evolutionary opinion (which became the "synthetic theory") and then rationalized the fossil patterns, albeit brilliantly, in an attempt to show that evolutionary processes as then understood were essentially sufficient to account for the fossil patterns.

As continuing work on the fossil record has improved our understanding of its quality and limitations, it has become clear that the pessimistic assessments of its potential contributions to biological theory have been largely incorrect. Hypotheses concerning evolutionary or ecological processes commonly entail predictions as to what fossil patterns must be like, should a given hypothesis be correct. It has turned out that many patterns have been different from the predictions, sometimes strikingly so, and that these patterns cannot be argued away as artifacts of an incomplete fossil record. The gradualistic pattern of evolution predicted by Darwin is simply not found—simply did not occur—in many cases. The circumstances acting to produce the patterns that we do find—the punctuational pattern of species lineages (Eldredge, 1971; Eldredge and Gould, 1972) and the sudden introductions of major new body plans—are still not understood.

When paleontologists discover patterns that differ from predicted ones, or when they find patterns that are simply not anticipated by prevailing theories, they have the obligation to create new hypotheses to explain the observations. Such hypotheses can then be tested, perhaps among the living biota, in the usual manner. The source of an hypothesis—whether from the fossil record or the genetic laboratory—is not in itself relevant to its success.

Among the more interesting sorts of patterns that are displayed in the fossil record are patterns of diversity trends. Diversity varies spatially in today's world, in patterns which are still under intense study, and by causes which are still under hot debate. The spatial patterns have varied through geological time, and diversity levels have varied as well, locally and globally. It is plausible that strong evidence can be found in the varied fossil patterns to falsify or to corroborate extant hypotheses that deal with the processes that effect diversity regulation and change; fossil evidence may also lead to additional hypotheses as required.

DIVERSITY IN THE FOSSIL RECORD

To paleontologists, the term *diversity* usually means simply the numbers of fossil taxa associated with a certain place and time. The taxa may be at any given taxonomic level, the place may vary from a microscopic rock sample to the entire planet, and the time may range from a theoretical instant to the entire duration of earthly life.

TAXA

The familiar nested hierarchy of taxonomic units permits us to study diversity at a series of levels that vary in inclusiveness and in the quality of their fossil records. A basic problem is that many of the processes in which we are interested occur at the species level or below, while owing to the nature of the fossil record our data are best at higher taxonomic levels. Phanerozoic marine species diversity cannot be read directly from the fossil record via counts of fossil species; the record is simply too fragmentary. The record of families should be much better, and therefore many paleodiversity studies are conducted at the family level. However, a couple of major problems remain. First, soft-bodied or thin-shelled forms are unlikely to leave records at any taxanomic level, or if they do become fossilized, to leave records that are comparable to those of the well-skeletonized groups. And second, family diversity is not a good predictor of the diversity in, say, classes (Valentine, 1969) and there is no special reason to expect that it predicts species diversity closely either. In fact, the species/family ratio varies by over a factor of four between the tropics and the arctic in the eastern Pacific today.

Fortunately, among the living fauna, faunal diversity *trends* exhibited by the living species distribution of the easily fossilizable fraction of the fauna are reflected, though muted, at the family level (Stehli et al., 1967; Campbell and Valentine, 1977), and gross temporal *trends* of marine species numbers also seem to be reflected on the family level (Sepkoski et al., 1980). Cautious use of family-level fossil data as a basis for the interpretation of diversity patterns and trends therefore seems warranted.

A common practice is to study the diversity trend within a given clade rather than within entire faunal associations, giving rise to the familiar spindle diagrams (or other curves) which depict the changing fortunes of, say, trilobites or ammonites through time. With such data, an important concern has been whether the variations evident from the fossil record reflected the workings of deterministic processes, or whether they were chiefly the products of chance events and therefore subject to overinterpretation. Studies by Raup et al. (1973), Stanley et al. (1981) and Raup (1981) have had the final effect of reinforcing the belief, based upon previous empirical and theoretical findings, that most of the abrupt, fairly large, or long-term changes in clade size can be interpreted in terms of ecological and evolutionary processes. For each individual case, however, caution is certainly indicated.

PLACES

The biosphere can also be regarded as a nested hierarchy, of ecological rather than taxonomic units, with levels composed of individuals, populations, communities, provinces and the biosphere. Diversity is accommodated on more than one level (Valentine, 1968, 1969); taxonomic richness may vary with the packing of populations within communities, with the packing of communities within biotic provinces, and with the packing of provinces within the biosphere. Thus the number of taxa in the biosphere may vary greatly if, say, the numbers of communities change or if the numbers of provinces vary, even if packing within communities remains constant. The fossil record provides examples of change on all these levels: species packing within communities has on balance increased during the Phanerozoic (Bambach, 1977); community packing has varied with the appearance and disappearance of reef community complexes (Newell, 1971); and the number of marine provinces has varied from a few to over 30 (Valentine et al., 1978), for example. To study ecological aspects of diversity at whatever taxonomic level, fossil samples should be amenable to interpretation in terms of the communities and/or provinces in which the biota lived, or should be representative of an entire ecological realm, such as the marine biosphere. These considerations permit us to scale our sampling efforts accordingly.

TIMES

The fossil record is a notoriously poor place to study paleobiological events which occurred within relatively short time spans. The problems of time resolution in the geological record are complex (see for example the symposium "Time Resolution in Evolutionary Paleobiology" summarized by Behrensmeyer and Schindel, 1983, and references therein). Processes such as speciation are not routinely amenable to paleontological investigation. Diversity fluctuations involving thousands to tens of thousands of years may be "ephemeral" so far as the actual record of the change is concerned. Nevertheless such events leave behind indications that they have occurred. New species appear following speciations, and diversity fluctuations require either extirpations from or additions to the biota, and these are then indicated by compositional changes. Thus paleobiologists can study the long-term results of short-term events.

Because time resolution is often coarse, the major diversity patterns are time-averaged, with the loss or masking of short-term dynamics of fossil change. Actually we are still short of achieving the scope and detail in correlation and dating of fossiliferous rocks that is technically possible, so that as work goes on important improvements occur steadily; as this continues more of the shorter-term patterns will be elucidated. It is not yet clear that this will in fact help to interpret the gross patterns of Phanerozoic diversity. At any rate we do have marvelous patterns now at the resolutions presently available and there is no reason not to proceed in their interpretation, so long as we take due account of the scales at which we operate.

SWEET ARE THE USES OF DIVERSITY

For the study of paleodiversity, the final goals involve synthesis of the processes connecting the ecological structure of the environment, the taxonomic structure of the biota over all categories, and the history of the individual clades at each taxonomic level. We are far from such goals, which are tantamount to understanding macroevolution, but nevertheless they appear to be within eventual reach. Paleontologists have been piling up data on biotic history for many decades now at an accelerating pace. Perhaps within some of these messy piles of data there are elegant principles struggling to get out? If so, it may well be diversity studies that eventually set them free.

REFERENCES

Bambach, R. K., 1977. Species richness in marine benthic habitats through the Phanerozoic. Paleobiology 3:152-167.

Behrensmeyer, A. K., and Schindel, D., 1983. Resolving time in paleobiology. Paleobiology 9:1-8.

Campbell, C. A., and Valentine, J. W., 1977. Comparability of modern and ancient marine faunal provinces. Paleobiology 3:49-57.

Eldredge, N., 1971. The allopatric model and phylogeny in Paleozoic invertebrates. Evolution 25:156-167.

Eldredge, N., and Gould, S. J., 1972. Punctuated equilibria: An alternative to phyletic gradualism. In Schopf, T.J.M. (ed.), Models in Paleobiology, San Francisco: Freeman and Cooper, 82-115.

Gould, S. J., 1980. G. G. Simpson, paleontology and the modern synthesis. In Mayr, E., and Provine, W. B. (eds.), The evolutionary synthesis—

Perspectives on the unification of biology, Cambridge: Harvard Univ. Press, 152-172.

Newell, N. D., 1971. An outline history of tropical organic reefs. Am. Mus. Novitates, no. 2465.

Raup, D. M., 1981. Extinction: Bad genes or bad luck? Acta Geol. Hispanica 16:25-33.

Raup, D. M., Gould, S. J., Schopf, T. J. M., and Simberloff, D. S., 1973. Stochastic models of phylogeny and the evolution of diversity. J. Geol. 81: 525-542.

Rudwick, M. J. S., 1976. The meaning of fossils (rev. ed.). New York: Science History Pubs.

Sepkoski, J. J., Jr., Bambach, R. K., Raup, D. M., and Valentine, J. W., 1981. Phanerozoic marine diversity and the fossil record. Nature 293:435-437.

Stehli, F. G., McAlester, A. L., and Helsley, C. E., 1967. Taxonomic diversity of recent bivalves and some implications for geology. Geol. Soc. Amer. Bull. 78:455-466.

Stanley, S. M., Signor, P. W., III, Lidgard, S., and Karr, A. F., 1981. Natural clades differ from "random" clades: Simulations and analyses. Paleobiology 7:115-127.

Valentine, J. W., 1968. The evolution of ecological units above the population level. Paleontology J. 82:253-267.

Valentine, J. W., 1969. Patterns of taxonomic and ecological structure of the shelf benthos during Phanerozoic time. Palaeontology 12:684-709.

Valentine, J. W., 1981. Darwin's impact on paleontology. BioScience 32: 513-518.

Valentine, J. W., Foin, T. C., and Peart, D., 1978. A provincial model of Phanerozoic marine diversity. Paleobiology 4:55-66.

Part I

PHANEROZOIC DIVERSITY TRENDS

This first section leads off with an atlas of Phanerozoic diversity patterns by Sepkoski and Hulver, which provides up-to-date information on diversity trends on a global scale at the family level across the whole span of multi-cellular life. In a very real sense, the remaining papers in this volume attempt to analyze and explain aspects of these patterns. The atlas includes a revised version of the spindle diagrams of Phanerozoic marine family diversity, chiefly by phyla, published previously by Sepkoski (1981), together with new detailed breakdowns of family diversity within orders or other appropriate mid-level taxonomic categories. Similar diagrams for taxa inhabiting the terrestrial biosphere are published for the first time.

Each of the following three papers in this section examines the Phanero-zoic diversity of a major biotic component. Terrestrial vertebrate diversity is not as well represented in the fossil record as is marine invertebrate diversity. Padian and Clemens discuss the problems and promises of these vertebrate data and show that whatever biases are present, the changing patterns that can be discerned do provide important information on extinctions and on diversification processes. The record of vascular plants is even more spotty, and the fossil species diversity pattern is expected to be little more than a sketchy cartoon of historical conditions. Niklas, Tiffney, and Knoll neverthe-less show that plant species data do indicate the rise and fall of major floral assemblages and also display some intriguing regularities. Signor attempts to infer the species diversity record of marine invertebrates, for which the fossil record is believed to be inadequate, from the relatively good family record. This adds fresh fuel to the fire of an old debate and may bring us closer to understanding the history of diversity at the level where most biotic processes actually operate.

Chapter 1

AN ATLAS OF PHANEROZOIC
CLADE DIVERSITY DIAGRAMS

J JOHN SEPKOSKI, Jr. and MICHAEL L. HULVER

Department of the Geophysical Sciences, University of Chicago

Clade diversity diagrams are spindle-shaped graphs that summarize patterns of taxonomic evolution within higher taxa through geologic time. Most clade diversity diagrams are constructed about a central axis that represents time (scaled either metrically or ordinally, by stratigraphic interval). Some measure or estimate of taxonomic diversity (or "richness") is then plotted symmetrically about the axis to give the diagram an overall spindle shape (e.g., Figure 1).

Diversity diagrams for individual clades convey information about their size, shape, and variability in the fossil record (cf. Gould et al., 1977). Such "morphologic" information is valuable for assessing how evolutionary rates (that is, rates of origination and extinction) vary within the taxa through geologic time. Clade diversity diagrams for groups of higher taxa hypothesized to be related by phylogeny or by function are useful for comparisons of the histories of the taxa. Common patterns of expansion or contraction may relate to general factors governing all taxa, whereas reciprocal patterns may be interpretable as negative interactions between pairs of ecologically similar taxa (e.g., Simpson, 1953; Bambach, this volume). Sets of clade diversity diagrams also are useful for summarizing variation among large numbers of clades for the purpose of testing general macroevolutionary models (e.g., Raup et al., 1973; Gould et al., 1977).

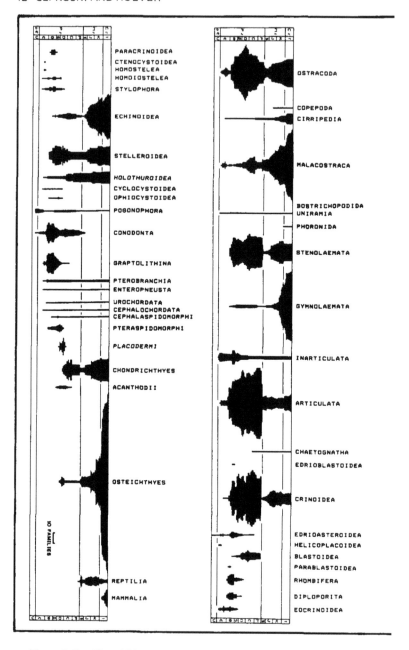

Figure 1. Families within classes of fossil marine mammals.

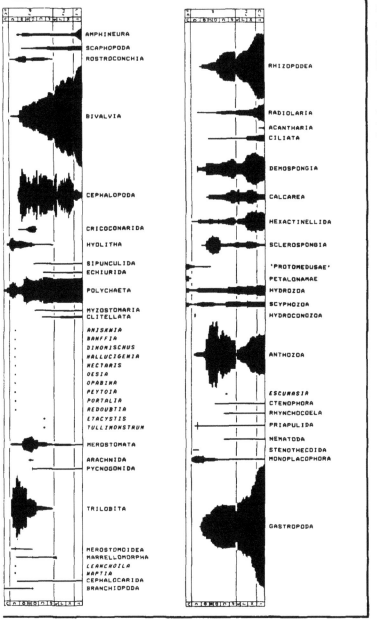

AMPHINEURA
SCAPHOPODA
ROSTROCONCHIA

BIVALVIA

CEPHALOPODA

CRICOCONARIDA

HYOLITHA

SIPUNCULIDA
ECHIURIDA

POLYCHAETA

MYZOSTOMARIA
CLITELLATA

ANISKNIA
BANFFIA
DINOMISCHUS
HALLUCIGENIA
NECTARIS
OESIA
OPABINA
PEYTOIA
PORTALIA
REDOUBTIA
ETACYSTIS
TULLIMONSTRUM

MEROSTOMATA

ARACHNIDA
PYCNOGONIDA

TRILOBITA

MEROSTOMOIDEA
MARRELLOMORPHA
LEANCHOILA
NAPTIA
CEPHALOCARIDA
BRANCHIOPODA

RHIZOPODEA

RADIOLARIA
ACANTHARIA
CILIATA

DEMOSPONGIA

CALCAREA

HEXACTINELLIDA

SCLEROSPONGIA

'PROTOMEDUSAE'
PETALONAMAE
HYDROZOA
SCYPHOZOA
HYDROCONOZOA

ANTHOZOA

ESCUMASIA
CTENOPHORA
RHYNCHOCOELA
PRIAPULIDA

NEMATODA
STENOTHECOIDA
MONOPLACOPHORA

GASTROPODA

This chapter presents a collection of clade diversity diagrams which we hope will be useful for examining the general histories of a wide variety of animal taxa. The main body of the chapter is a series of 12 figures displaying spindle diagrams for orders, classes, and phyla of both marine and nonmarine (or "continental") animals for the whole of the Phanerozoic (including the Vendian). Nearly all of the diagrams are plotted at a uniform taxonomic and temporal resolution, specifically that of familial diversity per stratigraphic stage. The taxonomic rank of family is used simply because comprehensive data with good stratigraphic resolution can be obtained for all animal groups at this level. Although families do not display all of the detail of the fossil record, they should be sufficiently sensitive to show major evolutionary trends and patterns with characteristic timescales of fives to tens of million years (see also Sepkoski, 1979, 1982a; Raup and Sepkoski, 1982).

The clade diversity diagrams in most of the figures are formatted in strips that have time in the vertical dimension. Most of the strips are scaled from 625 myr at their bottoms to approximately 1 myr BP at their tops. (No data on Recent diversity are directly included in the diagrams.) Geologic eras and systems are indicated at the lefthand ends of the strips, with eras denoted by *Cz* = Cenozoic, *Mz* = Mesozoic, *Pz* = Paleozoic, and *pε* = latest Precambrian; systems are denoted by standard symbols, with *V* = Vendian. The widths of the clade diversity diagrams in each strip indicate the numbers of families known from direct fossil evidence or from interpolation between known occurrences to be present in the "clades" in each of 80 stratigraphic stages or comparable intervals (see Table 1 in Sepkoski, 1982b for a listing of the stages used). A scale for the familial diversities appears in the lower righthand part of most of the figures. All of the diagrams were produced with an IBM Personal Computer and Epson dot-matrix printer.

The first two figures in this chapter contain class-level summaries of the entire Phanerozoic fossil record. Figure 1 displays clade diversity diagrams for the 87 classes and 15 unique, problematic genera that have representatives in the marine fossil record. This illustration is an updated version of Figure 1 in Sepkoski (1981) with corrections based on new data in Sepkoski (1982b). The second figure in this chapter summarizes the continental fossil record. The diversity diagrams display numbers of freshwater and terrestrial families within the 39 animal classes known from the nonmarine fossil record; data on the classes were compiled from the literature sources listed in Table 1. Also shown at the bottom of Figure 2 are clade diversity diagrams for numbers of *species* within the 13 taxonomic divisions of the tracheophytes and bryophytes; the data for these diagrams were taken from Niklas, Tiffney and Knoll (this volume).

The next eight figures illustrate a breakdown of the class-level clades into their constituent orders. Time and diversity in all diagrams are plotted at the

TABLE 1. Principal literature sources of information on the taxonomy
and stratigraphy of continental animal families.

Taxon	References
MOLLUSCA:	Davies (1971), Henderson (1935), Moore, Teichert, and Robison (1953-1982), Orlov (1958-1964), Solem and Yochelson (1979), Taylor and Sohl (1962).
ARTHROPODA: (excl. insects)	Cooper (1964), Harland et al. (1967), Kukalová-Peck (1973), Moore, Teichert, and Robison (1953-1982), Morris (1979), Mundel (1979), Orlov (1958-1964), Piveteau (1952-1969), Rolfe et al. (1983), Schram (1969), Schram and Schram (1979).
INSECTA:	Barthel (1978), Bode (1953), Burnham (1978), Carpenter (1976, 1979, 1980), Evans (1956), Grande (1980), Harland et al. (1967), Hoganson and Ashworth (1982), Jarzembowski (1980), Kukalová (1966, 1969), Kukalová-Peck (1973, 1975), MacLeod (1970), Müller (1963-1970), Orlov (1958-1964), Piveteau (1952-1969), Rodendorf (1968), Rolfe et al. (1983), Whalley (1980), Wighton (1982), Wilson (1978).
CHORDATA:	Brodkorb (1967, 1971, 1978), Carroll (1977), Charig et al. (1976), Denison (1978, 1979), Eisenberg (1981), Estes (1981), Grande (1980), Harland et al. (1967), Kuhn (1969), Lillegraven et al. (1979), Mlynarski (1976), Moy-Thomas and Miles (1971), Nelson (1976), Olsen and Galton (1977), Orlov (1958-1964), Romer (1966), Russell (1975), Steel (1970, 1973), Zangerl (1981).
OTHERS:	Clark (1969), Conway Morris (1981), Conway Morris et al. (1982), Harland et al. (1967), Kukalová-Peck (1973), Moore, Teichert and Robison (1953-1982), Schram (1979), Southcott and Lange (1971), Thompson and Jones (1980).

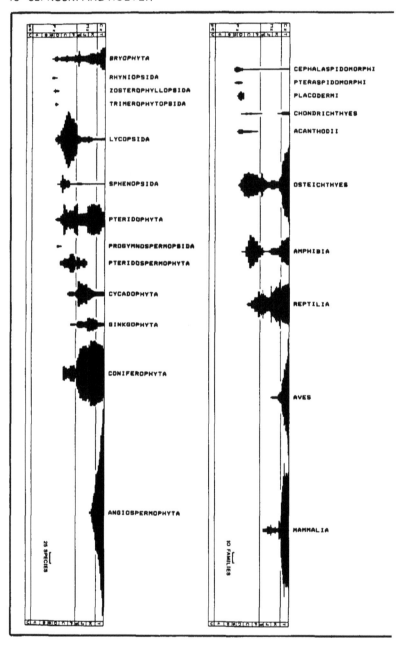

Figure 2. Families within classes of continental (i.e. terrestrial and freshwater) fossil

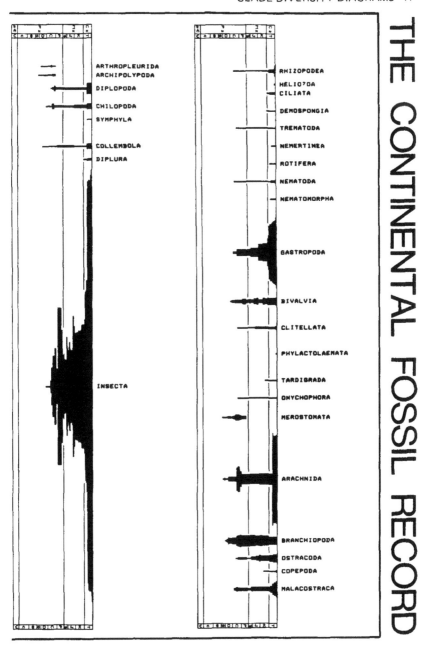

THE CONTINENTAL FOSSIL RECORD

animals and species within divisions of continental plants.

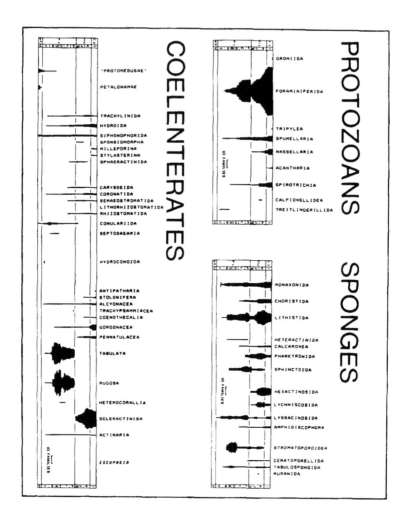

Figure 3. Families within orders of fossil "Protozoa," Porifera, and Coelenterata.

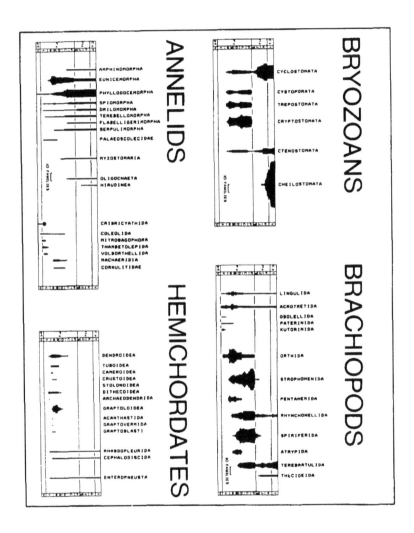

Figure 4. Families within orders of fossil Bryozoa, Brachiopoda, Annelida, and Hemichordata.

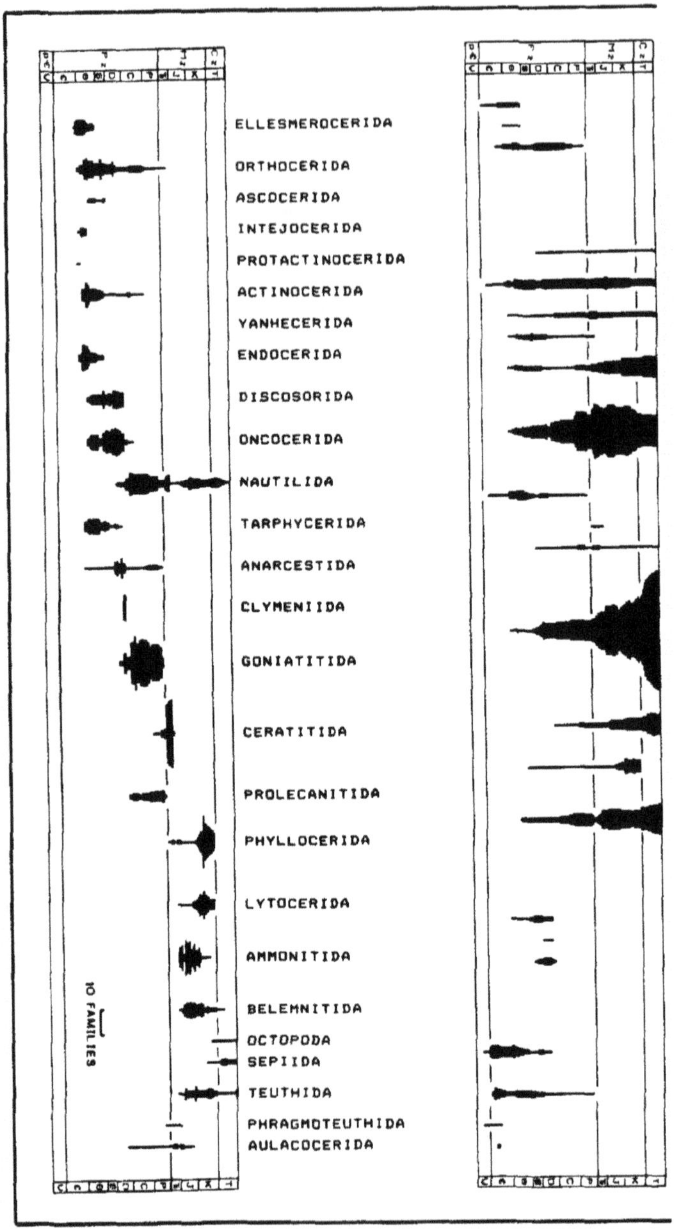

Figure 5. Families within orders of fossil marine Mollusca.

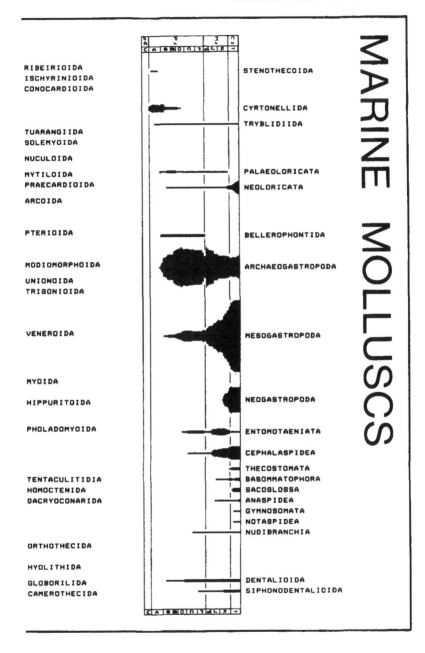

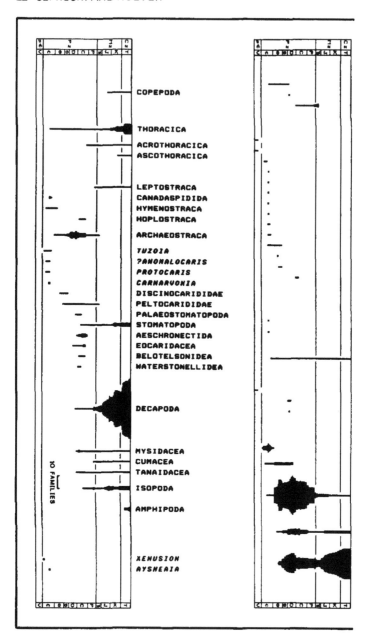

Figure 6. Families within orders of fossil marine Arthropoda.

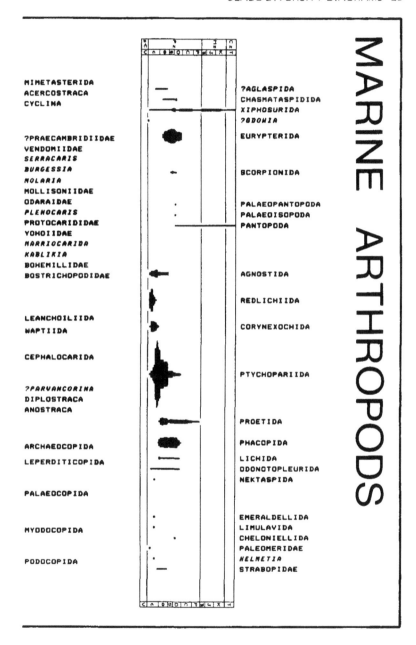

MARINE ARTHROPODS

MIMETASTERIDA
ACERCOSTRACA
CYCLINA

?PRAECAMBRIDIIDAE
VENDOMIIDAE
SERRACARIS
BURGESSIA
NOLARIA
MOLLISONIIDAE
ODARAIDAE
PLENOCARIS
PROTOCARIDIDAE
YOHOIIDAE
MARRIOCARIDA
KABLIKIA
BOHEMILLIDAE
BOSTRICHOPODIDAE

LEANCHOILIIDA

NAPTIIDA

CEPHALOCARIDA

?PARVANCORINA
DIPLOSTRACA
ANOSTRACA

ARCHAEOCOPIDA

LEPERDITICOPIDA

PALAEOCOPIDA

MYODOCOPIDA

PODOCOPIDA

?AGLASPIDA
CHASMATASPIDIDA
XIPHOSURIDA
?BDOWIA

EURYPTERIDA

SCORPIONIDA

PALAEOPANTOPODA
PALAEOISOPODA
PANTOPODA

AGNOSTIDA

REDLICHIIDA

CORYNEXOCHIDA

PTYCHOPARIIDA

PROETIDA

PHACOPIDA

LICHIDA
ODONOTOPLEURIDA
NEKTASPIDA

EMERALDELLIDA
LIMULAVIDA
CHELONIELLIDA
PALEOMERIDAE
NELMETIA
STRABOPIDAE

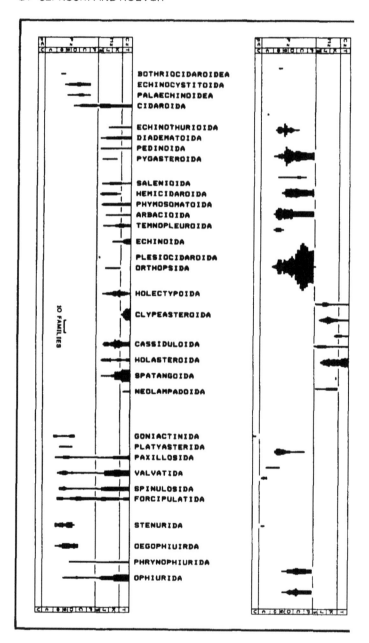

Figure 7. Families within orders of fossil Echinodermata.

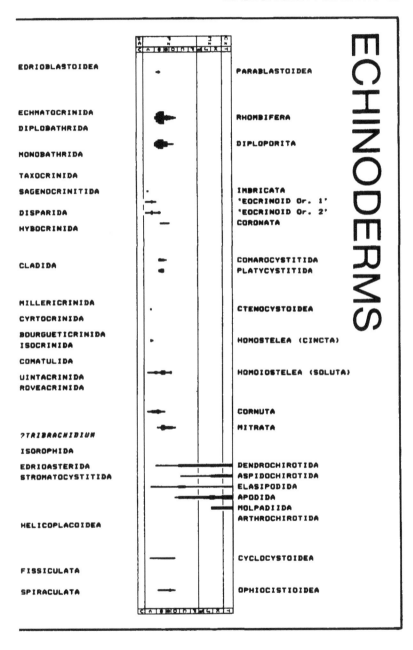

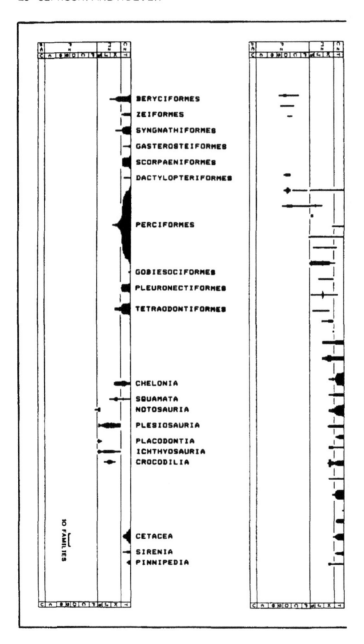

Figure 8. Families within orders of marine Vertebrata.

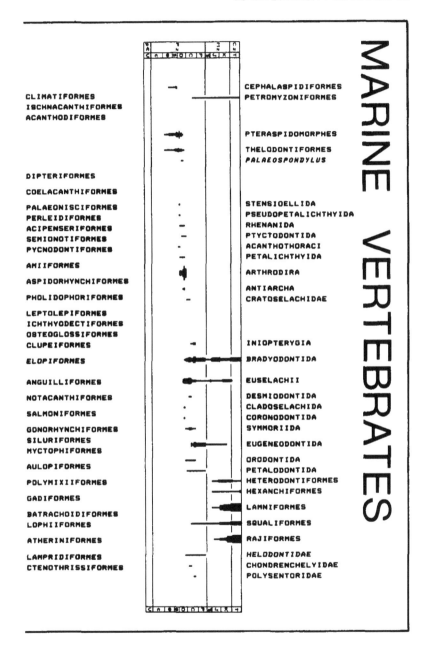

CLIMATIFORMES
ISCHNACANTHIFORMES
ACANTHODIFORMES

CEPHALASPIDIFORMES
PETROMYZONIFORMES

PTERASPIDOMORPHES

THELODONTIFORMES
PALAEOSPONDYLUS

DIPTERIFORMES

COELACANTHIFORMES

PALAEONISCIFORMES
PERLEIDIFORMES
ACIPENSERIFORMES
SEMIONOTIFORMES
PYCNODONTIFORMES

STENSIOELLIDA
PSEUDOPETALICHTHYIDA
RHENANIDA
PTYCTODONTIDA
ACANTHOTHORACI
PETALICHTHYIDA

AMIIFORMES

ARTHRODIRA

ASPIDORHYNCHIFORMES

ANTIARCHA

PHOLIDOPHORIFORMES

CRATOSELACHIDAE

LEPTOLEPIFORMES
ICHTHYODECTIFORMES
OSTEOGLOSSIFORMES
CLUPEIFORMES

INIOPTERYGIA

ELOPIFORMES

BRADYODONTIDA

ANGUILLIFORMES

EUSELACHII

NOTACANTHIFORMES

DESMIODONTIDA

SALMONIFORMES

CLADOSELACHIDA
CORONODONTIDA

GONORHYNCHIFORMES
SILURIFORMES
MYCTOPHIFORMES

SYMMORIIDA

EUGENEODONTIDA

AULOPIFORMES

ORODONTIDA
PETALODONTIDA

POLYMIXIIFORMES

HETERODONTIFORMES
HEXANCHIFORMES

GADIFORMES

BATRACHOIDIFORMES
LOPHIIFORMES

LAMNIFORMES

SQUALIFORMES

ATHERINIFORMES

RAJIFORMES

LAMPRIDIFORMES
CTENOTHRISSIFORMES

HELODONTIDAE
CHONDRENCHELYIDAE
POLYSENTORIDAE

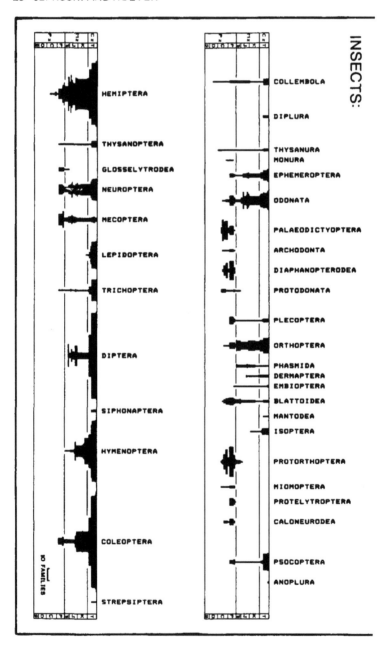

Figure 9. Families within orders of fossil nonmarine Arthropoda.

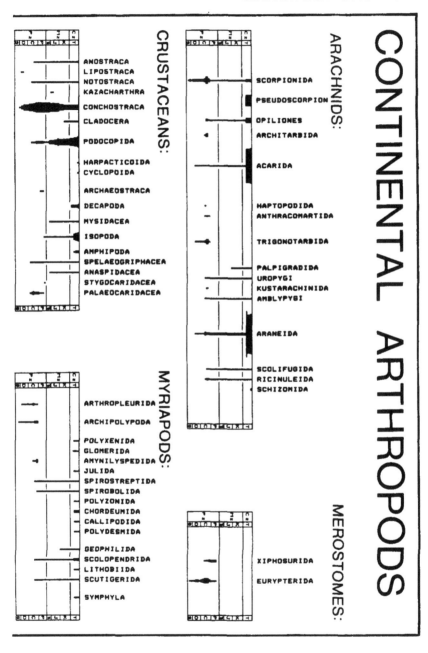

CONTINENTAL ARTHROPODS

CRUSTACEANS:

ANOSTRACA
LIPOSTRACA
NOTOSTRACA
KAZACHARTHRA
CONCHOSTRACA
CLADOCERA
PODOCOPIDA
HARPACTICOIDA
CYCLOPOIDA
ARCHAEOSTRACA
DECAPODA
MYSIDACEA
ISOPODA
AMPHIPODA
SPELAEOGRIPHACEA
ANASPIDACEA
STYGOCARIDACEA
PALAEOCARIDACEA

ARACHNIDS:

SCORPIONIDA
PSEUDOSCORPION
OPILIONES
ARCHITARBIDA
ACARIDA
HAPTOPODIDA
ANTHRACOMARTIDA
TRIGONOTARBIDA
PALPIGRADIDA
UROPYGI
KUSTARACHINIDA
AMBLYPYGI
ARANEIDA
SCOLIFUGIDA
RICINULEIDA
SCHIZOMIDA

MYRIAPODS:

ARTHROPLEURIDA
ARCHIPOLYPODA
POLYXENIDA
GLOMERIDA
AMYNILYSPEDIDA
JULIDA
SPIROSTREPTIDA
SPIROBOLIDA
POLYZONIDA
CHORDEUMIDA
CALLIPODIDA
POLYDESMIDA
GEOPHILIDA
SCOLOPENDRIDA
LITHOBIIDA
SCUTIGERIDA
SYMPHYLA

MEROSTOMES:

XIPHOSURIDA
EURYPTERIDA

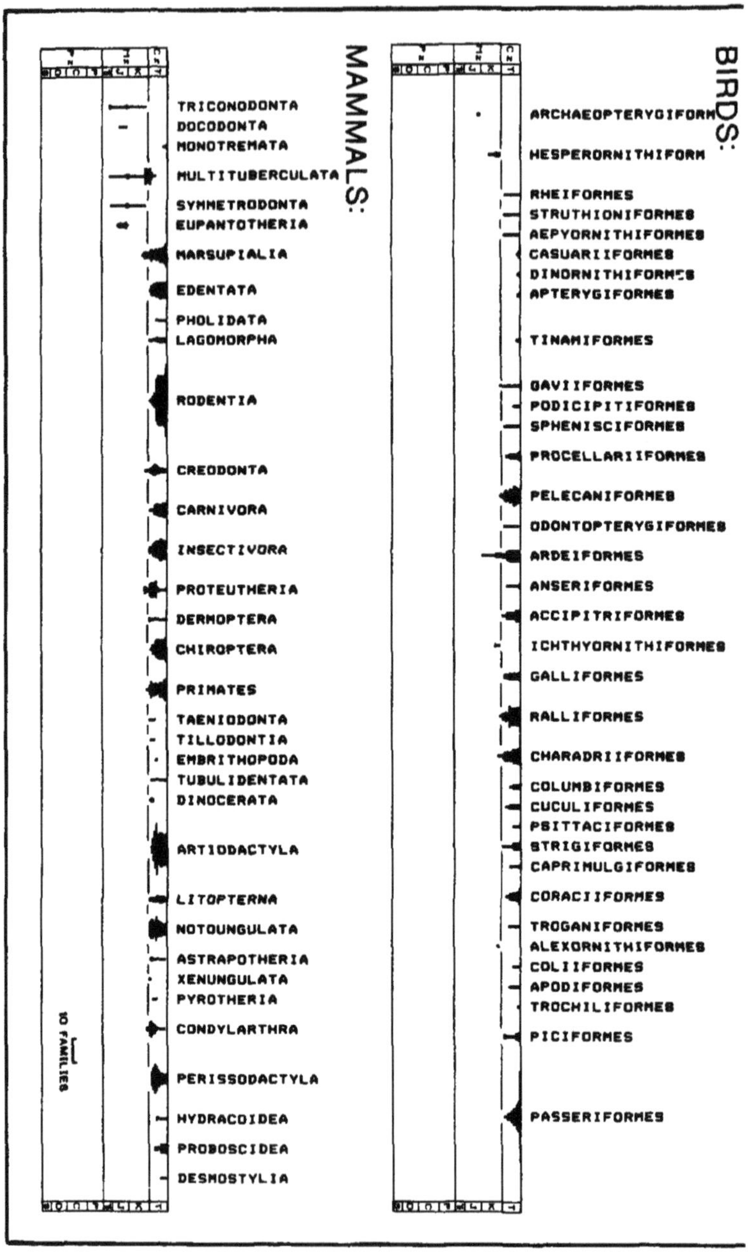

Figure 10. Families within orders of fossil nonmarine Vertebrata.

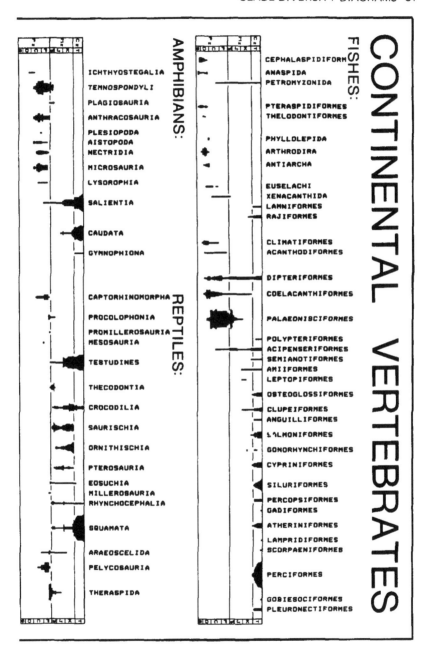

CONTINENTAL VERTEBRATES

FISHES:

CEPHALASPIDIFORM
ANASPIDA
PETROMYZONIDA

PTERASPIDIFORMES
THELODONTIFORMES

PHYLLOLEPIDA
ARTHRODIRA
ANTIARCHA

EUSELACHI
XENACANTHIDA
LAMNIFORMES
RAJIFORMES

CLIMATIFORMES
ACANTHODIFORMES

DIPTERIFORMES
COELACANTHIFORMES

PALAEONISCIFORMES

POLYPTERIFORMES
ACIPENSERIFORMES
SEMIANOTIFORMES
AMIIFORMES
LEPTOPIFORMES
OSTEOGLOSSIFORMES
CLUPEIFORMES
ANGUILLIFORMES
SALMONIFORMES
GONORHYNCHIFORMES
CYPRINIFORMES
SILURIFORMES
PERCOPSIFORMES
GADIFORMES
ATHERINIFORMES
LAMPRIDIFORMES
SCORPAENIFORMES
PERCIFORMES
GOBIESOCIFORMES
PLEURONECTIFORMES

AMPHIBIANS:

ICHTHYOSTEGALIA
TEMNOSPONDYLI
PLAGIOSAURIA
ANTHRACOSAURIA
PLESIOPODA
AISTOPODA
NECTRIDIA
MICROSAURIA
LYSOROPHIA
SALIENTIA
CAUDATA
GYMNOPHIONA

REPTILES:

CAPTORHINOMORPHA
PROCOLOPHONIA
PROMILLEROSAURIA
MESOSAURIA
TESTUDINES
THECODONTIA
CROCODILIA
SAURISCHIA
ORNITHISCHIA
PTEROSAURIA
EOSUCHIA
MILLEROSAURIA
RHYNCHOCEPHALIA
SQUAMATA
ARAEOSCELIDA
PELYCOSAURIA
THERASPIDA

same relative scale as in Figures 1 and 2 in order to facilitate comparison. Figures 3 and 4 display family-level clade diagrams for orders within the moderately diverse marine phyla: the Protozoa, Porifera, Coelenterata, Bryozoa, Brachiopoda, Annelida, and Hemichordata. (The set of clade diversity diagrams for the Annelida includes several taxa of questionable affinities which might best be considered *incertae sedis*; these are in the group of diagrams beginning with Cribricyathida and ending with Cornulitidae.) The more diverse marine phyla are represented in Figures 5 to 8. Figure 5 displays orders of marine molluscs; Figure 6 orders of marine arthropods; Figure 7 orders of echinoderms; and Figure 8 orders of marine vertebrates.

The two large phyla of continental animals, the nonmarine Arthropoda and Chordata, are featured in Figures 9 and 10. Nonmarine taxa have been segregated from their marine relatives because we believe that the land and sea are best treated as separate major arenas of evolution (see also Boucot, 1983). Despite the fact that some continental clades contain secondary species which alternate between marine and nonmarine habitats, and that all clades ultimately had their origins in the oceans, the great majority of continental animals evolved *in situ*, isolated from evolutionary activity in the seas. Thus, the segregation of marine and continental taxa enhances assessment of evolutionary patterns within the two arenas as well as comparisons between them. Note that the time axes for the continental clade diversity diagrams in Figures 9 and 10 have been truncated below the Silurian; this is because there is virtually no nonmarine fossil record prior to the mid Paleozoic (see Boucot and Janis, 1983).

The final pair of figures in this chapter (Figures 11 and 12) contains 14 diversity diagrams for families within entire phyla, split again between marine and continental. These diagrams are formatted somewhat differently than in the preceding figures. The spindles have been cut in half and rotated so that the time axis runs horizontally. This arrangement permits easier assessment of the times and magnitudes of diversity change but impedes comparison of changes between groups.

The use of a single level of taxonomic and stratigraphic resolution in all clade diversity diagrams is intended to aid interpretation and comparison of patterns among the various taxa. However, the constancy of resolution does not imply a uniformity of quality throughout the data. The accuracy of the taxonomic and stratigraphic information varies considerably among the taxonomic groups. In general, the quality is much better for marine taxa than for nonmarine taxa. Also, as should be expected, the fossil data are much better (and much more complete) for heavily skeletonized animals than for soft-bodied and lightly sclerotized animals. In fact, many of the diagrams for the latter groups reflect little more than the geologic distribution of Lagerstätten that preserve unusual fossils. This is particularly evident in the long, thin

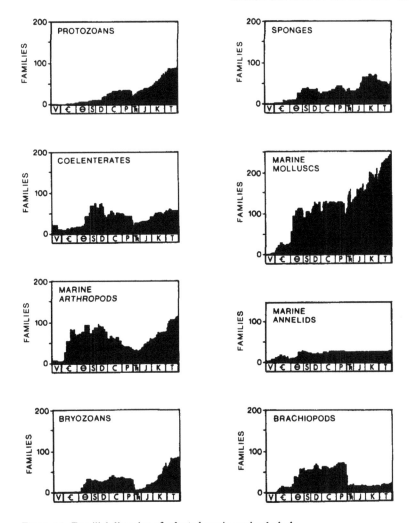

Figure 11. Familial diversity of selected marine animal phyla.

clade diagrams for such extant groups as the Nemertinea and Priapulida (Figure 1); these diagrams show only the extension of stratigraphic ranges from the Recent to the one or more Lagerstätten that happen to contain the groups' early members.

Much of the character of the diversity diagrams for some large clades, such as insects (Figures 2 and 9), also represents the effects of Lagerstatten. For the insects, the more important Lagerstätten include the Upper Carboniferous

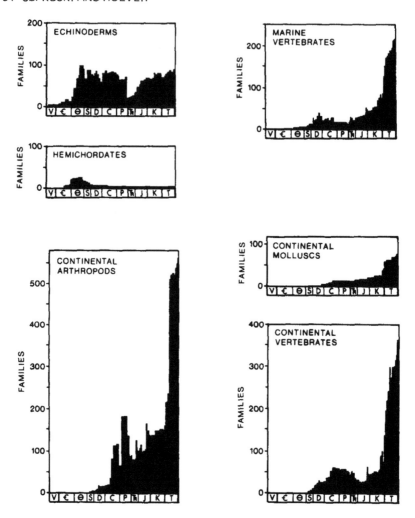

Figure 12. Familial diversity of selected marine and continental animal phyla.

siderite concretion deposits of North America and Europe, the mid-Permian lake deposits of Kansas and Kazakhstan, the Eocene Green River deposits of Wyoming, and especially the Oligocene Baltic Amber of northern Europe. The Baltic Amber alone contributes most of the Cenozoic bulge in the clade diversity diagrams for both insects and other lightly sclerotized terrestrial arthropods (Figures 2, 9, and 12). The effects of Lagerstatten, or of their non-occurrence, are even seen in some well-skeletonized groups with fairly extensive fossil records. The drop in the diversity of continental vertebrates in the Jurassic (Figure 12), for example, probably reflects largely a paucity of fossiliferous continental deposits between the Rhaetian and Tithonian (see also Carroll, 1977; Padian and Clemens, 1985, this volume).

These shortcomings of the fossil record, along with the problems associated with family-level data and 5 to 10 myr-long stages, do limit the value of the clade diversity diagrams presented here. However, we believe that a great deal still can be learned from them about the shape of evolution—about the success and failure of taxa and about the apparent order, or disorder, in their radiations and extinctions. Thus, we hope that this "atlas" will aid in the assessment and interpretation of evolutionary history as well as serve as a baseline for the compilation of more accurate and detailed diversity data.

ACKNOWLEDGEMENTS

We thank K. J. Niklas, B. H. Tiffney, and A. H. Knoll for permission to reproduce their data on continental plant diversity in Figure 2. We also thank J. Cracraft, P. Crane, and J. A. Hopson for help and advice during compilation of data on continental animals. Production of this paper received partial support from NSF Grant DEB81-08890 to J. J. S.

REFERENCES

Bambach, R. K., 1985. Classes and adaptive variety: The ecology of diversification in marine faunas through the Phanerozoic. In Valentine, J. W. (ed.), Phanerozoic diversity patterns: Profiles in macroevolution, Princeton, N.J.: Princeton Univ. Press and Amer. Assoc. Adv. Sci. (this volume).

Barthel, K. W., 1978. Solnhofen. Ein Blick in die Erdgeschichte. Thun, Switzerland: Ott Verlag, 393.

Bode, A., 1953. Die Insektenfauna des Ostnidersächsischen Oberen Lias. Palaeotographica Abt. A. 103:1-375.

Boucot, A. J., 1983. Does evolution take place in an ecological vacuum? II. J. Paleontol. 57:1-30.

Boucot, A. J., and Janis, C., 1983. Environment of the early Paleozoic vertebrates. Palaeogeogr., Palaeoclimat., Palaeoecol. 41:251-288.

Brodkorb, P., 1967. Catalogue of fossil birds, Part 3 (Ralliformes, Ichthyornithiformes, Charadriiformes). Florida State Mus. Bull., Biol. Sci. 11(3): 99-220.

Brodkorb, P., 1971. Catalogue of fossil birds, Part 4 (Columbiformes through Piciformes). Florida State Mus. Bull., Biol. Sci. 15(4):163-266.

Brodkorb, P., 1978. Catalogue of fossil birds, Part 5 (Passeriformes). Florida

State Mus. Bull., Biol. Sci. 23(3):139-228.

Burnham, L., 1978. Survey of social insects in the fossil record. Psyche 85: 85-133.

Carpenter, F. M., 1976. The Lower Permian insects of Kansas, Part 12. Protorthoptera (continued), Neuroptera, additional Palaeodictyoptera, and families of uncertain position. Psyche 83:336-378.

Carpenter, F. M., 1979. Lower Permian insects from Oklahoma, Part 2. Orders Ephemeroptera and Palaeodictyoptera. Psyche 86:261-290.

Carpenter, F. M., 1980. Studies on North American Carboniferous insects. 6. Upper Carboniferous insects from Pennsylvania. Psyche 87:107-119.

Carroll, R. L., 1977. Patterns of amphibian evolution: An extended example of the incompleteness of the fossil record. In Hallam, A. (ed.), Patterns of evolution, Amsterdam: Elsevier, 405-437.

Charig, A. J., Krebs, B., Sues, H. -D., and Westphal, F., 1976. Thecodontia. Encyclopedia of paleoherpetology, Pt. 13. Stuttgart: Gustav Fischer Verlag, 137.

Clark, R. B., 1969. Systematics and phylogeny: Annelida, Echiura, Sipuncula. In Florkin, M. and Schoer, B. T. (eds.), Chemical zoology, v. 4, New York: Academic Press, 1-68.

Conway Morris, S., 1981. Parasites and the fossil record. Parasitology 82:489-509.

Conway Morris, S., Pickerill, R. K., and Harland, T. L., 1982. A possible annelid from the Trenton Limestone (Ordovician) of Quebec, with a review of fossil oligochaetes and other annulate worms. Can. J. Earth Sci. 19: 2150-2157.

Cooper, K. W., 1964. The first fossil tardigrade: *Beorn leggi* Cooper, from Cretaceous amber. Psyche 71:41-48.

Davies, A. M., 1971. Tertiary faunas. v.1. The composition of Tertiary faunas. Amsterdam: Elsevier, 571.

Denison, R., 1978. Placodermi. Handbook of paleoichthyology, v.2. Stuttgart: Gustav Fischer Verlag, 128.

Eisenberg, J. F., 1981. The mammalian radiations. Chicago: Univ. Chicago Press, 610.

Estes, R., 1981. Gymnophiona, Caudata. Encyclopedia of paleoherpetology, Pt. 2. Stuttgart: Gustav Fischer Verlag, 115.

Evans, J. W., 1956. Palaeozoic and Mesozoic Hemiptera (insects). Aust. Zool. J. 4:165-258.

Gould, S. J., Raup, D. M., Sepkoski, J. J. Jr., Schopf, T. J. M., and Simberloff, D. S., 1977. The shape of evolution: A comparison of real and random clades. Paleobiol. 3:23-40.

Grande, L., 1980. Paleontology of the Green River Formation with a review of the fish fauna. Geol. Surv. Wyoming Bull. 63.

Harland, W. B., et al., eds., 1967. The fossil record. London: Geol. Soc. London, 828.

Henderson, J., 1935. Fossil non-marine Mollusca of North America. Geol.

Soc. Amer. Spec. Pap. 3, 313.

Hoganson, J. W. and Ashworth, A. C., 1982. The late-glacial climate of the Chilean Lake region implied by fossil beetles. Third N. Am. Paleont. Conv., Proc. 1:251-256.

Jarzembowski, E. A., 1980. Fossil insects from Bembridge Marls, Palaeogene of the Isle of Wright, southern England. Br. Mus. Nat. Hist. Bull. (Geol.) 33:237-293.

Kuhn, O., 1969. Cotylosauria. Encyclopedia of paleoherpetology, Pt. 6. Stuttgart: Gustav Fischer Verlag, 89.

Kukalová, J., 1966. Protelytroptera from the Upper Permian of Australia, with a discussion of Protocoleoptera and Paracoleoptera. Psyche 73: 89-111.

Kukalová, J., 1969. Revisional study of the order Palaeodictyptera in the Upper Carboniferous shales of Commentry, France. Part 1. Psyche 76: 163-215.

Kukalová-Peck, J., 1973. A phylogenetic tree of the animal kingdom (including orders and higher categories). Canada: Ottawa, Nat. Museums of Canada, Publ. in Zool. no. 8, 78.

Kukalová-Peck, J., 1975. Megasecoptera from the Lower Permian of Moravia. Psyche 82:1-19.

Lillegraven, J. A., Kielan-Jaworowska, Z., and Clemens, W. A., eds., 1979. Mesozoic mammals. Berkeley: Univ. Calif Press, 311.

MacLeod, E. G., 1970. The Neuroptera of the Baltic Amber. I. Ascalaphidae, Nymphidae, and Psychopsidae. Psyche 77:147-180.

Młynarski, M., 1976. Testudines. Encyclopedia of paleoherpetology, Pt. 7. Stuttgart: Gustav Fischer Verlag, 130.

Moore, R. C., Teichert, C., and Robison, R. A., eds., 1953-1982. Treatise on invertebrate paleontology. Lawrence, Kansas: Geol. Soc. Am. and Univ. Kansas Press.

Morris, S. F., 1979. A new fossil terrestrial isopod with implications for the East African Miocene land fauna. Br. Mus. Nat. Hist. Bull. 32:71-75.

Moy-Thomas, J. A., and Miles, R. S., 1971. Paleozoic fishes, 2nd ed. Philadelphia: Saunders, 259.

Muller, A. H., 1963-1970. Lehrbuch der Palaozoologie. Jena: Veb Gustav Fischer Verlag.

Mundel, P., 1979. The centipedes (Chilopoda) of the Mazon Creek. In Nitecki, M. H., (ed.), Mazon Creek fossils, New York: Academic Press, 361-368.

Nelson, J. S., 1976. Fishes of the world. New York: Wiley-Interscience, 416.

Niklas, K. J., Tiffney, B. H., and Knoll, A. H., 1985. Patterns in vascular land plant diversification: An analysis at the species level. In Valentine, J. W. (ed.), Phanerozoic diversity patterns: Profiles in macroevolution, Princeton: Princeton Univ. Press and Amer. Assoc. Adv. Sci. (this volume).

Olsen, P. E., and Galton, P. M., 1977. Triassic-Jurassic tetrapod extinctions: Are they real? Science 197:983-986.

Orlov, Yu. A., ed., 1958-1964. Osnovy Paleontologii. Moscow: Nedra.

Padian, K., and Clemens, W. A., 1985. Terrestrial vertebrate diversity: Episodes and insights. In Valentine, J. W. (ed.), Phanerozoic diversity patterns: Profiles in macroevolution, Princeton, N.J.: Princeton Univ. Press and Amer. Assoc. Adv. Sci. (this volume).

Piveteau, J., ed., 1952-1969. Traité de Paléontologie. Paris: Masson et Cie Éditeurs.

Raup, D. M., Gould, S. J., Schopf, T. J. M., and Simberloff, D. S., 1973. Stochastic models of phylogeny and the evolution of diversity. J. Geol. 81: 525-542.

Raup, D. M., and Sepkoski, J. J., Jr., 1982. Mass extinctions in the marine fossil record. Science 215:1501-1503.

Rodendorf, B. B., ed., 1968. Jurski Nasekomie Karatau. Moscow: "Nauka," 252.

Rolfe, W. D. I., Bonamo, P. M., Grierson, J. D. and Shear, W. A., 1983. The earliest land animals. Am. Assoc. Adv. Sci., 149th Ann. Mtg., Abst. of Papers, 38-39.

Romer, A. S., 1966. Vertebrate paleontology, 3rd ed. Chicago: Univ. Chicago Press, 468.

Russell, D. A., 1975. Reptilian diversity and the Cretaceous-Tertiary transition in North America. Geol. Assoc. Canada Spec. Pap. (13):119-136.

Schram, F. R., 1969. The stratigraphic distribution of the Paleozoic Eumalacostraca. Fieldiana Geol. 12:213-234.

Schram, F. R., 1979. Worms of the Mississippian Bear Gulch Limestone of central Montana, USA. San Diego Soc. Nat. Hist. Trans. 19:107-120.

Schram, F. R., and Schram, J. M., 1979. Some shrimp of the Madera Formation (Pennsylvanian) Manzanita Mountains, New Mexico. J. Paleontol. 53:169-174.

Sepkoski, J. J., Jr., 1979. A kinetic model of Phanerozoic taxonomic diversity: II. Early Phanerozoic families and multiple equilibria. Paleobiol. 5: 222-251.

Sepkoski, J. J., Jr., 1981. A factor analytic description of the Phanerozoic marine fossil record. Paleobiol. 7:36-53.

Sepkoski, J. J., Jr., 1982a. Mass extinctions in the Phanerozoic oceans: A review. In Silver, L. T., and Schultz, P. H. (eds.), Geological implications of impacts of large asteroids and comets on the earth, Geol. Soc. Amer. Spec. Pap. 190.

Sepkoski, J. J., Jr., 1982b. A compendium of fossil marine families. Milwaukee Publ. Mus. Contr. Biol. Geol. 51, 125.

Simpson, G. G., 1953. The major features of evolution. New York: Columbia Univ. Press, 434.

Solem, A., and Yochelson, E. L., 1979. North American Paleozoic land snails with a summary of other Paleozoic nonmarine snails. U.S. Geol. Surv. Prof. Pap. 1072, 42.

Southcott, R. V., and Lange, R. T., 1971. Acarine and other microfossils from the Maslin Eocene, South Australia. S. Aust. Mus. Rec. 16(7):1-21.

Steel, R., 1970. Saurischia. Encyclopedia of paleoherpetology, Pt. 14. Stuttgart: Gustav Fischer Verlag, 87.

Steel, R., 1973. Crocodylia. Encyclopedia of paleoherpetology, Pt. 16. Stuttgart: Gustav Fischer Verlag, 116.

Taylor, D. W., and Sohl, N. F., 1962. An outline of gastropod classification. Malacologia 1:7-32.

Thompson, I., and Jones, D. S., 1980. A possible onychophoran from the middle Pennsylvanian Mazon Creek beds of northern Illinois. J. Paleontol. 54:588-596.

Whalley, P. E. S., 1980. Neuroptera (Insecta) in amber from the Lower Cretaceous of Lebanon. Br. Mus. Nat. Hist. Bull. (Geol.) 33:157-164.

Wighton, D. C., 1982. Middle Paleocene insect fossils from south-central Alberta. Third N. Am. Paleont. Conv. Proc. 2:577-578.

Wilson, M. V. H., 1978. Paleogene insect faunas of western North America. Quaest. Ent. 14:13-34.

Zangerl, R., 1981. Chondrichthyes. I. Paleozoic Elasmobranchii. Handbook of Paleoichthyology. Stuttgart: Gustav Fischer Verlag, 3A, 115.

Chapter 2

TERRESTRIAL VERTEBRATE DIVERSITY: EPISODES AND INSIGHTS

K. PADIAN and W. A. CLEMENS

Department of Paleontology, University of California, Berkeley

Dedicated to the late George Gaylord Simpson

INTRODUCTION

One of the most important contributions to the advancement of paleobiology in the past ten years has been an increased understanding of the character and pace of change of biotic diversity through time. This has been initiated by comprehensive analysis of taxonomic patterns and interpretation of the apparent results with respect to underlying evolutionary histories and processes as well as to possible sources of sampling and other biases (reviewed in Simpson, 1960; Raup, 1976, 1979b; Thomson, 1976; Hallam, 1977; Sepkoski et al., 1981; see also other papers in this volume).

Many questions and paradoxes remain in the study of paleovertebrate diversity. To what extent are some apparent patterns the result of physical processes: post-mortem sorting of bones and teeth, differences in depositional environments, or current availability of outcrops of fossiliferous strata? How do differences in intensity of study bias our perception of the fossil record, and how can nonstochastic patterns be recognized? In short, to what extent can we use the apparent fossil record of changing diversity of taxa to infer evolutionary processes, including adaptation, selection, and species replacement?

The purpose of this paper is to separate that portion of the data that comprises the fossil record of terrestrial vertebrates, and to examine some of the above questions in light of the possibilities and difficulties of the record. A

Phanerozoic Diversity Patterns:
Profiles in Macroevolution

basic discussion of some of these problems may explain why the available data take the shapes seen in tabulations of diversity, and show what kinds of evolutionary questions can be addressed to this record. In some cases, new advances in the field, such as application of Sepkoski's (1981) factor-analytic approach to the terrestrial data, may help to go beyond the biases to a fuller understanding of evolutionary patterns. Recent advances in many areas of vertebrate paleontology were reviewed by Hopson and Radinsky (1980); the uses of phylogenetic analysis as a baseline for other kinds of evolutionary inquiries can be found in Cracraft (1981), Lauder (1981), Fisher (1982), and Padian (1982). This paper centers on patterns, biases, and questions that arise in both general and specific analyses of vertebrate history.

Several useful studies of diversity through time in specific vertebrate groups have formed the basis for part of the present work. Fishes are largely excluded from this survey because they have been extensively considered by Thomson (1977), and to separate marine and nonmarine types here would not give a representative picture of their patterns of diversity. We have incorporated some of the tabulations and interpretations of Carroll (1977) on amphibians and Gingerich (1977) on mammals. Like these authors, we have relied largely on Romer (1966) and Harland et al. (1967) for compilation of systematic diversity, correcting somewhat for systematic revisions made during the past fifteen years. We have also referred to the taxonomic tabulations of Lillegraven (1972) on fossil mammals. Finally, we wish to acknowledge gratefully access to J. J. Sepkoski's previously unpublished tabulations of vertebrate diversity, which were compiled largely from the above sources. The phrase "fossil record" is used here in reference to those fossils in hand and available for study. As Durham (1967) and others have emphasized, these are but a small part of the total assemblage of records of prehistoric organisms preserved in the rocks of the earth's crust.

THE TERRESTRIAL VERTEBRATE RECORD

The terrestrial fossil record is far poorer than the marine record in many ways, although it is still amenable to many kinds of questions. Terrestrial sediments provide a relatively small portion of the available fossil record (Raup, 1976a,b; Niklas et al., 1980). Their representation improves toward the Recent, though biases of sampling, preservation, and taxonomic interest cannot be discounted. The terrestrial record has never been completely assessed with these factors in mind, although some data are available (e.g., Ronov, 1959, 1982) and new approaches have been offered (e.g., Bakker, 1977).

The relative representation of the marine and fossil records has not been extensively studied with respect to amount of preserved area through time.

The size relationship of the shelf area to the inner continental area depends on sea level, the slope of the shelf, the absolute area of the continental mass, and other factors. Available shelf area should increase with subdivision of the continents by epicontinental seas or fragmentation of the continents by drift. Since the Permian, subdivision of the continents has increased, and at times epicontinental seas have also covered much of the continental surface. However, there are only on the order of 10,000 species of fossil terrestrial vertebrates, whereas marine fossils number on the order of 180,000 species (Raup, 1976). The great discrepancy in known diversity of marine and terrestrial species through time seems not to be related mainly to available living area, but to environmental potential for fossilization and preservation. Even the record of "terrestrial" vertebrates depends to some extent on the marine record. The Lower Jurassic limestones of southwest England and the Upper Jurassic limestones of the Solnhofen region of West Germany provide examples. The first records deposition in a transgressive sea, the second in an ancient lagoon; yet both preserve records of "terrestrial" reptiles (pterosaurs, crocodiles, and dinosaurs). In some cases such marine deposits provide most of what we know of the "terrestrial" vertebrates of one age and area (see Buffetaut, et al., 1982 for a report on the Middle Cretaceous of Europe).

The same rules for preservation of organic remains govern marine and terrestrial regimes: quick burial, quiet sedimentation, and no destruction after deposition. The paucity of terrestrial fossils reflects several major biases. In subaerial terrestrial habitats, biotic remains tend to be destroyed by oxidation, decomposition, and reworking of potential fossils and their sedimentary environments. Furthermore, in comparison to the marine realm, the terrestrial environment is primarily erosional, not depositional. On the average, terrestrial sediments are less continuously accumulated than are marine sediments (Sadler, 1981), though some terrestrial environments, such as lakes, floodplains, and stream channels, can produce relatively complete records over a long period of time if the environment remains stable and if the deposited sediments escape subsequent erosion (Behrensmeyer, 1982; Dingus and Sadler, 1982). Deposits formed in aquatic environments, such as the geographically extensive and geologically long-lived Newark Supergroup of the eastern United States, the Eocene deposits of Messel, West Germany, and the Pleistocene La Brea "tar pits" of Los Angeles, provide most of the terrestrial fossil record. But such environments occupy only a small part of the continental surface, are ephemeral in duration, and may be unlikely to preserve any records of certain kinds of organisms and their habitats. Such deposits provide the most detailed insights into fossil history, but are potentially misleading about the history and pace of terrestrial life (Bakker, 1977; see below). Their paleodemographic data cannot be taken at face value, because such Lagerstätten seldom preserve organisms from arid or upland environments.

Every form of life has its own potential for preservation dictated by the structure of its body. "Soft parts" are seldom fossilized, almost by definition, although what is "soft" to an animal (or its depositional environment) may be more durable in a plant, of which leaves, stems, waxy cuticles, and even nuclei may all fossilize. It is rare to find fossil traces of nonskeletal tissues in vertebrates, although compressions or impressions of feathers, fur, wings, gills, and skin are notable, if uncommon, exceptions. Most frequently, isolated elements or partial skeletons are preserved; size, durability, depositional environment, and post-mortem selection by predators and decomposers are among the more important factors determining taphonomic survivorship (Behrensmeyer and Hill, 1980).

Because fossil skeletons are seldom complete and sample sizes rarely large, we have poor control over what defines a vertebrate paleontological species. We do not have very reliable data on intraspecific variation in the form of dinosaur bones, for instance, even though isolated bones have often been the basis of new genera and species. On the other hand, the greater number of parts and morphologic features in vertebrate skeletons provide many more opportunities for determination of "specific" level differences among specimens than is usually the case in invertebrates. Unfortunately, few isolated skeletal elements are diagnostic to the species level. Consequently, as *morphological species* the vertebrates may be relatively oversplit compared to the invertebrates; as *biological species*, however, the current taxonomy may represent or even underrepresent species diversity, because the record is discontinuous enough to cast doubts upon actual breeding potential of many organisms placed in the same species. For these reasons it may be more reliable to use genus- or family-level data for geologic ranges and derived estimates of standing diversity, and this practice is followed here.

Another problem in estimating the tempo and mode of vertebrate fossil diversity is the difficulty in correlating terrestrial and marine faunas. Most time-stratigraphic units of the Standard Geological Time Scale were established on fossiliferous sequences of marine deposits in Europe. Correlations of these units with those found in different areas and formed in different environments decrease in precision with increasing age and stratigraphic incompleteness of the deposits. For most of the Phanerozoic the vertebrate fossil record lacks the quickly evolving and well-distributed groups that enhance the precision of marine biostratigraphy. Usually, a combination of methods, including radiometric and relative geologic dating, marine and terrestrial interfingering, correlations of widely distributed and (preferably) distantly related vertebrate groups, and palynologic analysis, can be employed. These syntheses have made time calibration of the Cenozoic terrestrial fossil record increasingly precise (Savage, 1975, 1977), and some of these methods have also been applied to episodes in the Mesozoic.

Finally, a human factor has to be added to the equation when the invertebrate and vertebrate records are compared. Throughout the history of the field, specialists in vertebrate paleontology have been fewer in number than their invertebrate colleagues (although it may seem that there are quite a lot working on a single sub-phylum). Although stratigraphic studies of fossil vertebrates, the basis for research on patterns of change in diversity, have tended to lag, this tendency has recently been reversed, notably for the Cenozoic, and increasingly so for the Mesozoic. However, it is unlikely that the entire terrestrial record will ever approach the marine record in accuracy of correlation, because of phylogenetic and depositional factors. There are, after all, few vertebrate equivalents of ammonites before the Tertiary radiations of horses, rodents, and other rapidly evolving mammalian groups.

In spite of these limitations, many theoretical advances in paleobiology have been propounded by vertebrate specialists, from Cuvier and Owen to Cope, Osborn and Simpson. Simpson, in the eyes of some, is almost single-handedly responsible for the founding of the modern science of paleobiology (Gould, 1980). Many vertebrate and invertebrate paleontologists, beginning as far back as Brinkmann (1929a,b), produced studies of evolution that were grist for the mill of the Modern Synthesis, but it was Simpson who articulated in the 1940s the theoretical basis of modern macroevolution by showing that any satisfactory theory of evolution has to explain the patterns of the fossil record—not the other way around. Interest in Simpson's theoretical work on vertebrate diversity is long overdue for a renaissance, and in this paper we merely ask again some of the questions he first posed forty years ago.

ANALYSES OF FOSSIL VERTEBRATE DIVERSITY

Like so many other macroevolutionary problems, the modern study of the tempo and mode of vertebrate evolution had its beginnings in Simpson's theoretical writings (1944, 1952). Simpson was preoccupied not by the question of *whether* diversity had changed through time, but *how* it had changed: taxonomic diversity, morphologic diversity, and ecological-adaptive diversity were all considered in turn. On the basis of data summarizing first and last appearances of vertebrate groups at several taxonomic levels, Simpson (1952) rejected the hypothesis that changes in diversity coincided with period and era boundaries (see also Camp, 1952). He showed that such boundaries rarely coincided with important changes in vertebrate history, and that each different vertebrate group seemed to follow its own history, irrespective of the waxing and waning of others.

A common pattern, however, was seen in the tempo of successive diversification of the hierarchical levels of classification (Figure 1). Classes appeared

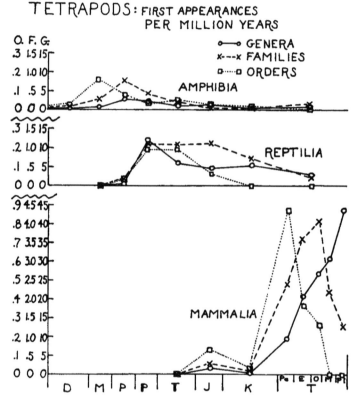

Figure 1. Simpson's (1952) graph of first appearances of tetrapods per million years, showing genera, families, and orders.

in the fossil record some 25 to 30 million years before they achieved maximum ordinal diversity; after a similar interval, the orders achieved maximum generic diversity. Following these peaks, taxic origination rates tended to drop just as quickly as they had risen, a pattern also noted by Thomson (1976; see Figure 2). Simpson (1952) viewed this, as he had in *Tempo and Mode in Evolution* (1944), as ecological opportunism manifesting itself in the adaptive radiation of new major groups. In support of this concept he cited the long fossil histories of birds and mammals that preceded their evolutionary explosions in the Tertiary, relating this to the fact that taxonomic cycles of vertebrates did not usually coincide with geologic periods. He did not believe that the domination of the dinosaurs and other Mesozoic reptiles was responsible for the suppression of diversity in contemporaneous birds and mammals, and emphasized instead the considerable lag between the

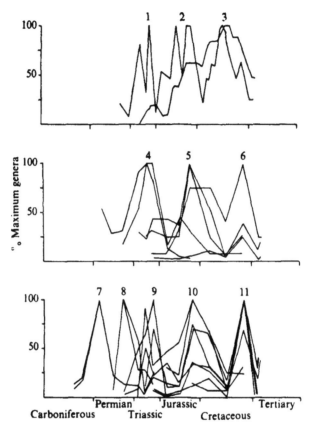

Figure 2. Group diversity curves for the late Paleozoic and Mesozoic. The peak diversities of different groups are identified by number: 1, ammonoids; 2, ammonoids; 3, ammonoids and corals; 4, paleonisciform and semionotiform fishes; 5, amiiform and pholidophoriform fishes; 6, pycnodont and teleost fishes; 7, cotylosaurs and pelycosaurs; 8, cotylosaurs and eosuchians; 9, ichthyosaurs, thecodonts, and saurischians; 10, ptero-saurs, ichthyosaurs, saurischians, and crocodiles; 11, turtles, crocodiles, saurischians, ornithischians, sauropterygians, and squamatans. From Thomson (1976).

disappearance of the dinosaurs and the filling of many of their adaptive zones by mammalian analogs.

A classic example of supposed "explosive" evolution and its correlation with diastrophism is the "explosion" of mammals supposed to initiate the Cenozoic and to be simultaneous with if not caused by the Laramide Re-volution. But, in fact the most basic event, the origin of placentals, occurred sometime well before the end of the Cretaceous. Most of the orders of early

Cenozoic mammals did not appear at the Cretaceous-Paleocene boundary, but straggled in over a span of some 20 million years. The rate of appearance of new genera was low in the Paleocene and it did not reach its climax, its truly "explosive" phase, until the Pliocene, perhaps 60 million years after the end of the Cretaceous. When did the "explosion" occur? Certainly not at the Cretaceous-Paleocene boundary, and claimed relationship to the Laramide Revolution must surely be viewed with suspicion. (Simpson 1952: 365.)

Nor did Simpson take literally the apparent drops in taxonomic diversity at the Permo-Triassic and Cretaceous-Tertiary boundary. Instead, he cautioned readers that the imprecision and incompleteness of the data and his analytical methods would tend to lump, at an inappropriately broad scale, events that might have occurred well apart in time (1952:363). In these conclusions he was again echoed by Camp (1952), who emphasized the probability that faunal associations depended more on geographic ecology than time, and that, therefore, correlation of horizons on the basis of first and last appearances of vertebrates bordered on circular reasoning:

. . .we call the period of great extinction the end of the Cretaceous, we assume that many of the extinctions were world-wide and simultaneous, and therefore when we find dinosaurs, we automatically place them in the pre-Tertiary. Then we are prone to proceed a step farther to postulate some universal diastrophic or catastrophic event and/or change of climate to account for the extinctions. . .We must await the discovery of some more universal means of getting an absolute time scale. Only then shall we discover how accurate our long range correlations may be. (Camp 1952:354-6)

Later attempts to document vertebrate diversity through time profited from larger collections, improved correlations, and publication of taxonomic revisions. In *The Fossil Record* (Harland et al. 1967), a group of vertebrate specialists charted the ranges of all known fossil vertebrate groups; these were tabulated and analyzed by computer (Cutbill and Funnell 1967), and patterns of taxonomic origination, extinction, and standing diversity were given. Contrary to recent efforts on the invertebrate record, these data were not standardized by taxonomic level: genera, families, orders, and classes were given equal shrift. Probably few would argue that the Linnean hierarchy is an equivalent measure of taxonomic, morphologic, or ecologic diversity across all phyletic lines, but it seems equally difficult to gain an accurate measure of shifting diversity from an analysis of data in which all hierarchical levels have been lumped. As a compendium of data on first and last appearances, the treatment of Harland et al. is more precise than another commonly employed source, Romer's *Vertebrate Paleontology* (3rd ed., 1966), because Harland et al. tabulated the stratigraphic occurrences by stage, whereas Romer used only Lower, Middle, and Upper periodic divisions. Apparent bursts or drops in diversity on a coarser scale may turn out to be either more or less drastic when analyzed at the finer level of resolution. Examples of this are discussed below in the sections on the Permo-Triassic and Cretaceous-Tertiary boundaries; see also Bakker (1977:440-441).

The limitations of fossil data, therefore, can result from artifacts of the temporal, geographic, and taxonomic scales used. For reasons explained above, the species level may be unsuitable for analyses of fossil vertebrate diversity; like Simpson (1952), we have undertaken analyses at the level of genus, family, and order, so that a variety of comparisons can be made and taxonomic artifacts avoided as much as possible. These preliminary analyses are intended only to suggest what can be seen now and where further useful work might be done.

Figure 3 is a tabulation of the standing diversity of the orders of vertebrates through the Phanerozoic, based on Romer's (1966) classification, and amended somewhat by subsequent discoveries and taxonomic revisions of the last 15 years. The graph is accompanied by an enumeration of ordinal originations (a) and extinctions (Ω) in each periodic division. The ordinal level is not useful in analyzing any but the coarsest patterns in vertebrate diversity (see also Van Valen, 1973b; Charig, 1973). *Within* the taxa shown here there is little turnover at the ordinal level; new orders do not appear rapidly to replace older orders of the same class that have become extinct. (Exceptions to this pattern occur in the mammals, for reasons discussed below.) Patterns of diversity at the generic level within orders are more meaningful for most questions of evolutionary change (*contra* Simpson, 1952). For example, birds are currently divided into some 33 orders, whereas traditionally the very diverse rodents and the marsupials each comprise only one, as do the depauperate aardvarks and monotremes. This reflects an artifact of the pre-evolutionary Linnean hierarchy. In Table 1, the numbers of fossil genera of the major vertebrate groups are listed along with the first appearance of each class. Birds can be directly compared with mammals, despite the longer history of the latter, because only about 3% of the genera in both classes occur in the Mesozoic. Mammals still outnumber birds 3 to 1 at the generic level during the Cenozoic, although the ordinal totals of both groups number in the mid-thirties (depending on the scheme of classification). Recognized ordinal diversity peaks in the Eocene for mammals, but increases through the Tertiary for birds (Table 2), largely due to what is apparently a lack of extinction in the bird orders. Mammals have had a higher ordinal turnover, two epochs (Eocene and Oligocene) in which significant ordinal extinctions have occurred, and a very low rate of ordinal origination since the Eocene. Given Simpson's note of the lag time between ordinal and generic peaks in diversity

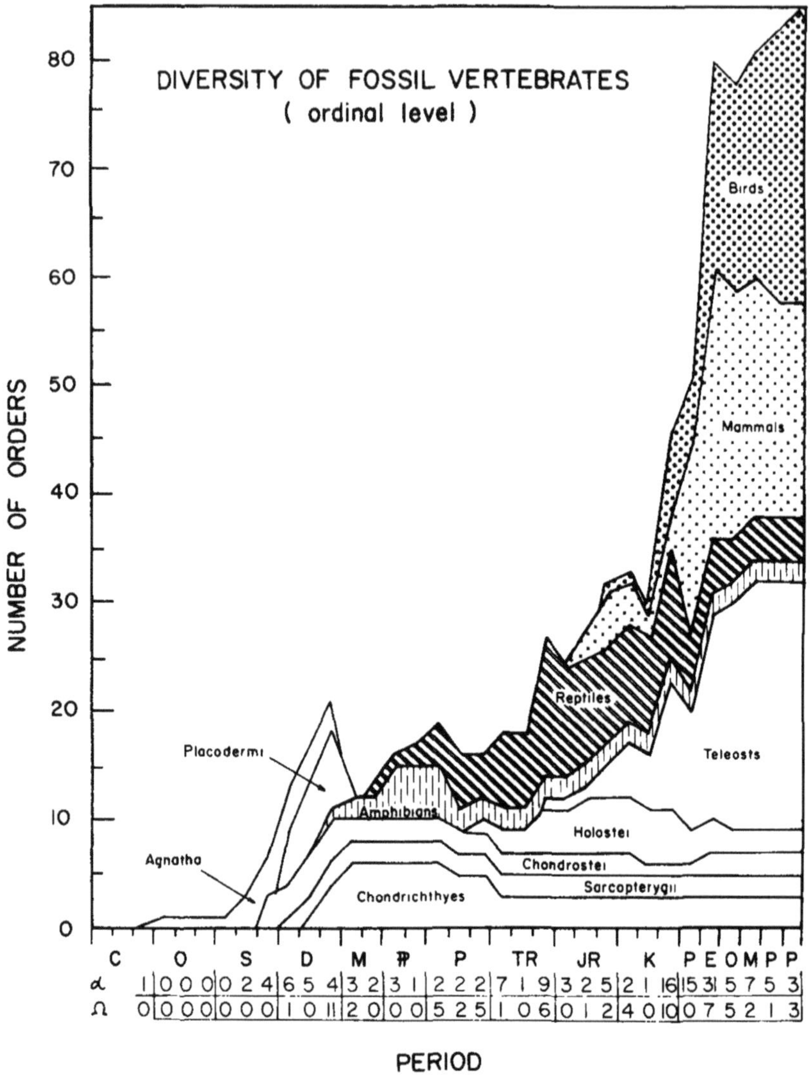

Figure 3. Ordinal diversity of fossil vertebrates through the Phanerozoic, based mainly on data from Romer (1966). Below the graph are figures for new originations (α) and extinctions (Ω) for each interval. Principal periods of turnover at this scale include the Late Devonian, Early Permian, Late Permian, Late Triassic, Early Cretaceous, Late Cretaceous, Eocene, and Oligocene, but these are to be interpreted with caution (see text).

Table 1. First appearances and total generic diversity of the major
groups of vertebrates represented by fossils.

	first appearance	(MYBP)	genera
"Fishes"	Late Cambrian	500	2100
Amphibians	Late Devonian	350	350
Reptiles	Late Pennsylv.	290	1260
Mammals	Late Triassic	200	3240
Birds	Upper Jurassic	150	900

Table 2. Cenozoic ordinal diversity of birds and mammals.

	UK	Pal	Eoc	Olig	Mio	Plio	Pleis
Birds							
Standing crop	7	6	19	19	21	25	27
Originations	5	1	13	1	3	4	2
Extinctions	2	0	1	0	0	0	1
Mammals							
Standing crop	8	18	25	23	22	20	20
Originations	5	13	9	2	2	0	0
Extinctions	2	0	4	5	2	1	2

(from which birds were excluded from consideration) it is more realistic to
consider the avian orders equivalents of mammalian families, branching out
after an initial Cretaceous-Eocene peak of higher-category diversification. If,
within birds, only the Paleognathae and Neognathae were accorded ordinal
status, there would be no Cenozoic originations of avian orders.

Within relatively small clades such as the Tetrapoda, therefore, artifacts of
taxonomic rank may play a great role in creating apparent patterns of diversity. In Figure 4 the pace of change in ordinal diversity in terrestrial vertebrates, excluding birds, is analyzed. The bar graphs show standing diversity of
orders, listed for each period as if each order lasted the entire period. These
bars have been broken down into amphibian, reptilian, and mammalian
components. Two lines on the graph trace other measures of the pace of
ordinal diversity. The broken line tracks the number of new orders that
appear in each period (calibrated on the left side of the graph). In the Triassic, most of these are reptiles; afterward, they are mostly mammals. (Adding
the avian record would scatter another 30 "orders" through the Tertiary.)
The solid line approximates the number of new orders that appear per million
years (calibrated on the right side of the graph). Both lines provide average

DIVERSITY OF ORDERS

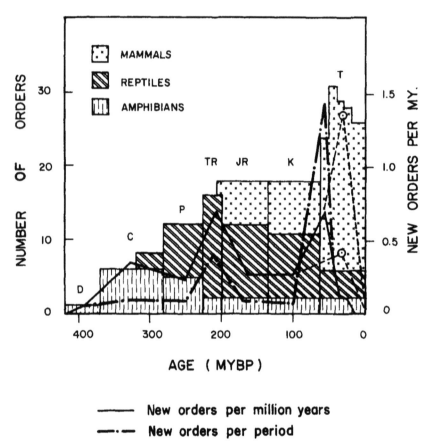

Figure 4. Ordinal diversity of terrestrial vertebrates (excluding birds) through the Phanerozoic, averaged for periods (epochs in the Tertiary). For explanation see text.

values for the entire Tertiary (lower dotted line) and values per epochal division within the Tertiary (upper dotted line). Ordering the data on a scale calibrated in epochs shows the importance of the Paleocene and Eocene in the origination of mammalian orders; a finer-scaled temporal discrimination through geologic time would produce a very different graph for all classes. Charig (1973, Figure 1) approached this problem by graphing ordinal originations of amniotes per million years for epochal divisions of the geologic periods; his graph is reproduced here as Figure 5.

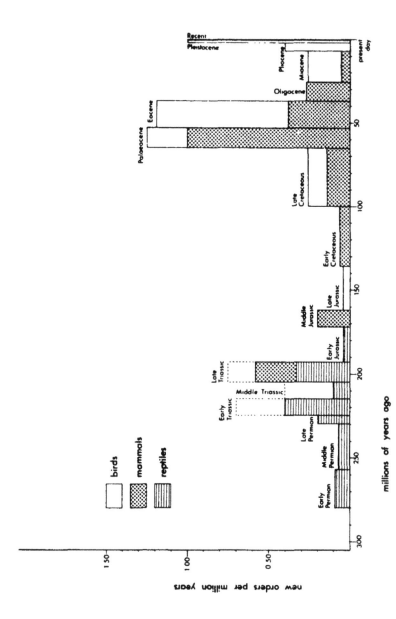

Figure 5. Average ordinal origination rate for amniote orders, given for standard divisions of geologic periods. From Charig (1973).

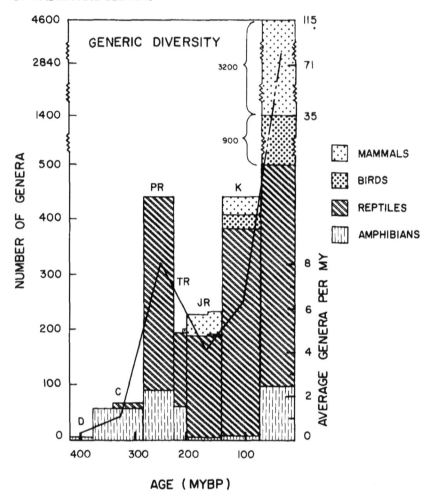

Figure 6. Histogram of generic diversity of terrestrial vertebrates through the Phanerozoic, averaged for periods (scale on left side of graph). The line tracks average generic diversity per million years for each period (scale on right side of graph).

Figure 6 graphs generic diversity per period for all terrestrial vertebrates. Generic diversity is broken down into component classes, with the Tertiary birds and mammals greatly compressed: their estimated total of over 4,000 fossil genera is eight times as large as the estimated total of amphibians and reptiles combined (500), and is nearly ten times as large as the total diversity of any earlier period. The greatest recorded generic diversities before the Tertiary are in the Permian and Cretaceous. The line charting generic diversity per million years indicates that the Triassic and Cretaceous are apparently comparable in diversity for the intervals of time they represent, while the Jurassic and pre-Permian are more poorly represented.

PHYLOGENETIC COMPONENTS OF VERTEBRATE DIVERSITY

Diversity through time can be usefully analyzed by charting the histories of component clades (here, the four tetrapod classes) for each time interval considered. Extensive discussions of the early history of terrestrial vertebrates can be found in Panchen (1980); the early history of mammals is reviewed in Lillegraven et al. (1979), and of birds in Feduccia (1980).

"Amphibian" is a traditional term for anamniote tetrapods, a paraphyletic grade of organization. Amphibian diversity through time has been well documented by Carroll (1977). His graph of family diversity (Figure 7) is very similar to the pattern seen here at the generic level (Figure 6). A bimodal curve is formed by the ancient Late Devonian-Late Triassic genera (7 extinct orders) and more modern Jurassic-Recent genera (the two modern orders of Anura and Urodela; the Caecilia have virtually no fossil record). The ordinal graph (Figure 3) does not reflect this bimodality, but it is real at the family level: the last of the Paleozoic groups disappeared by the end of the Triassic, about the time frogs and salamanders evolved, presumably from dissorophoid ancestors (Bolt, 1977). Their fossil record is initially poor but improves toward the Recent. The subsequent lower level taxonomic diversification of these two basic body plans has proceeded without any new organizational type deemed worthy of ordinal status. The preponderance of Tertiary genera (Figure 6) occurs only since the Miocene (Romer 1966:364).

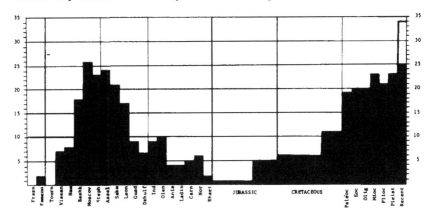

Figure 7. Diversity of the families of amphibians through time, from Carroll (1977).

The reptilian story is more complex. In Figure 8 the "Reptiles," another paraphyletic grade of organization, are factored into component clades. "Anapsida" has been retained as a stem group, including the cotylosaurs (now mostly classified as amphibians: Heaton, 1980), the millerosaurs and protothyridids, the placoderms, procolophonids, mesosaurs, bolosaurs, and captorhinomorphs, but turtles have been segregated from the anapsids as a clearly monophyletic group of uncertain origin. The "Eosuchia," an unde-fined wastebasket for primitive diapsids, has been dissolved and its members assigned to Archosauria or Lepidosauria (both *sensu lato*: J. A. Gauthier, MS.). The appearance of the diapsid *Petrolacosaurus* in the Late Pennsylvan-ian of Kansas (Reisz, 1981) indicates the longevity of the Diapsida, but *Petro-lacosaurus* can be placed in neither the Lepidosauria nor the Archosauria, and there is no further history of either group until the late Permian (e.g., Carroll, 1975a, b, 1978). The Pelycosauria, though clearly paraphyletic as the primi-tive sister group of the Therapsida, is retained.

Two major divisions of "reptiles" make up the bulk of Permian and Meso-zoic diversity. The Permian is dominated by the Synapsida (the Pelycosauria, including the familiar "sail-backed" forms, plus the Therapsida, or "mammal-like" reptiles), but these decline rapidly in the Lower Triassic and are nearly gone by the end of the Triassic; one lineage persists to what may be the Mid-dle Jurassic of China (Sun Ai-Lin, pers. comm.). Most of the Upper Permian and Lower Triassic therapsids are known from the rich deposits of South Africa and the U.S.S.R. with similar, if so far less productive, expanses in Asia and Antarctica. There is no latest Permian (Dzhulfian) in North America, but a good Lower and early Upper Permian is responsible for much of the known pelycosaurian diversity of that time (Olson, 1962). Beginning in the Lower Triassic, archosaurs begin to diversify and predominate, and by the end of that period all the remaining orders of reptiles (except the problematic champsosaurs) originated, including the lizards, sphenodontids, turtles, ich-thyosaurs, sauropterygians (plesiosaurs and their relatives), pterosaurs, both orders of dinosaurs, and the crocodiles. The thecodonts ("stem" archosaurs, another paraphyletic taxon), placodonts, and rhynchosaurs are almost strictly Triassic orders; the others persist to the end (or near the end) of the Creta-ceous. Snakes appear well before the end of the Cretaceous, and they, the lizards, and the crocodiles diversify through the Tertiary and survive to the present. Of *Sphenodon*, which now lives only on a few islands off the coast of New Zealand, no Tertiary fossil record exists, although sphenodontids are known from the Upper Triassic, Jurassic, and Cretaceous. It is impossible to tell whether the group was so geographically restricted through the Tertiary or widespread but only poorly preserved.

Figure 8 reveals other interesting features. The peaks in "reptilian" diversity seem to coincide with four major taxonomic innovations and revolutions:

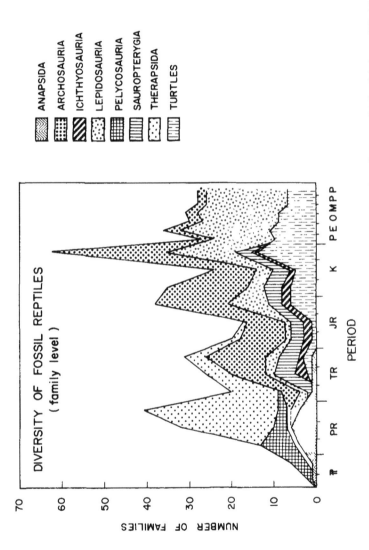

Figure 8. Family diversity of the major clades of reptiles through time, based on data from Romer (1966), slightly modified from later sources. The white space at the base of the Archosauria-Lepidosauria junction, extending back to the *Late Pennsylvanian*, represents Diapsida undifferentiated into these two major subclasses.

(1) Upper Permian: climax of the mammal-like reptiles, and the first appearance of archosaurs and lepidosaurs;
(2) Upper Triassic: climax of the thecodontian radiation and the origination of the other diapsid orders;
(3) Upper Jurassic-Lower Cretaceous: well-known, highly diverse faunas of dinosaurs and other archosaurs in a variety of environments;
(4) Late Cretaceous: the last radiation of dinosaurs and the rise of the modern families of reptiles.

Nevertheless, even these peaks cannot be accepted uncritically. The drops in diversity between the Upper Permian and the Lower Triassic and between the Upper Triassic and Lower Jurassic are evidently real. The first is the crash in therapsid diversity, while the second is primarily the demise of the Triassic thecodont families and most of the remaining mammal-like reptiles. But no obvious biotic factors explain why diversity should continue to drop into the mid-Jurassic, nor why the mid-Cretaceous should witness a plummet from 19 to 10 archosaurian families, only to rise again to 24 in the Late Cretaceous. In these two cases, as with the Middle Triassic, the answer is almost certainly a lack of fossiliferous terrestrial exposures. Buffetaut et al. (1982), reporting on the mid-Cretaceous, state:

> Vertebrate faunas from the mid-Cretaceous are still poorly known in Europe (this is equally true elsewhere, at least for tetrapods, on a worldwide scale). The main reason for this fact is the rarity of outcrops . . . if the fishes, Selachians and Teleostomes, are relatively well represented, sometimes by abundant faunas, it is not the same for tetrapods; among these, some really important groups, which we know should be present, have not been discovered in the Middle Cretaceous of Europe: this is the case notably for the Mammals. In addition, the different stages of the mid-Cretaceous are not equally productive . . . (our translation).

They conclude that the mid-Cretaceous must have been a time of transition for the terrestrial vertebrates, because the Lower and Upper Cretaceous faunas differ so extensively from each other. Compared to these exposures and to those of the Upper Jurassic, the mid-Cretaceous environments that have been sampled are both smaller in extent and diversity of habitat. Consequently, the mid-Cretaceous "crash" should be viewed skeptically. There is also a paucity of extensively explored or accurately dated terrestrial fossil localities from the mid-Jurassic, and similar deficiencies exist for the mid-Triassic and mid-Permian.

Of all terrestrial vertebrates, the archosaurs were those hardest hit by the Late Cretaceous extinctions. Some 28 families known from the Late Cretaceous (Santonian-Maestrichtian) went extinct, leaving only three families of crocodiles in the Paleocene. This does not mean that they all went extinct at once: in the Maestrichtian (Hell Creek) sediments of Montana, only 10 of these families are unequivocally present, representing about 21 well-defined genera (R. A. Long and D. A. Russell, work in progress). Records of the 18 other families present elsewhere during the Maestrichtian have not been confirmed up to the K-T boundary. There is also a slight drop—perhaps from climatic cooling (Hutchison, 1982), or again simply the result of relatively few exposures—in the diversity of lizards, snakes, and turtles sometime in the mid-Paleocene, but they recover somewhat in the Eocene, a rich period for the Squamata especially. Among these groups, family origination was nearly complete by the Eocene, with records of most families extending back to the Upper Cretaceous. (Only the colubroid snakes lack a fossil record before the Miocene.) Like most groups, the squamates follow Simpson's pattern of initially low diversity at first appearance (Late Triassic), a family peak some time later (Late Cretaceous-Eocene), and a generic peak still later (Miocene-Recent). The only families that apparently did not survive into the Tertiary were the two groups of sea-going lizards (Dolichosauridae and Mosasauridae), the Simoliophidae, and the Dinilysiidae, an early family of booid snakes.

The fossil diversity of birds is problematic because their bones are so thin and because they commonly live in habitats that are not likely to be represented in the fossil record (such as forests and open fields). *Archaeopteryx*, as much a dinosaur as a bird, appears alone in the Upper Jurassic, in lagunal deposits. Jensen (1981) has reported a more modern avian synsacrum from the Upper Morrison Formation of Utah, possibly as old as *Archaeopteryx*; but it is not clear that all the Morrison is of Jurassic age, and palynologic data suggest that the type section in Colorado may be Cretaceous (B. Cornet, *in litt.*). Most Cretaceous birds are found in shallow marine or floodplain sediments, and are evidently of the "shorebird" adaptive type (the flightless *Hesperornis* is an extreme example). The first representatives of the living orders of birds are also shorebirds, perhaps questionably placed with the loons, grebes, gannets, cormorants, flamingoes, gulls, and terns.[1] The preservational bias is obvious. The adaptive radiation of birds was rapid: by the time *Hesperornis* appears (Campanian), some birds had already abandoned flight and taken up a diving existence, though retaining the teeth of their dinosaurian ancestors. What sort of land birds might have existed by that time, and what relationships might they have borne to the modern orders of birds,

---

[1] Feduccia (1980) notes an important problem with convergence of birds and theropods in the Cretaceous (actually not convergence, but retention of characters shared by both groups): "The oldest fossils described as owls have been assigned to a distinctive family, the Bradycnemidae, from the Late Cretaceous of England (Harrison and Walker 1975), but these fossils along with two other Cretaceous genera, *Wyleyia* (Harrison and Walker 1973) and *Caenagnathus* (Cracraft 1971) are in reality small dinosaurs."

particularly the ratites? The recent discovery of paleognathous carinate birds from the Early Tertiary of North America (Houde and Olson, 1981) is an indication of much undiscovered diversity in the history of birds.

Mammalian diversity through time has been documented extensively by many authors, notably Simpson (1949, 1952), Butler et al. (1967), Lillegraven (1972), and Gingerich (1977). The last two references include discussions of the data and interpretation of more recent ideas advanced to explain them; these obviate a long review here. Perhaps the most striking thing about mammalian history is that it began at about the same time as that of the dinosaurs (Late Triassic), yet none of the living mammalian orders is unquestionably recorded before the Upper Cretaceous, and most are known only from the Eocene or later. Even in the Upper Cretaceous only two or three living mammalian orders are represented (the didelphoid marsupials, the "prosimian" primates, and possibly a lipotyphlan "insectivore"). Lillegraven et al. (1979) ironically subtitled their book on Mesozoic mammals "The first two-thirds of mammalian history": during this time (about 140 my) a little over 100 mammalian genera are recognized (Clemens et al., 1979), whereas well over 3000 are known from 60 my of the Tertiary (Romer, 1966:379-396). There are at least 10 orders of Mesozoic mammals, depending on classification; the number of families represented is difficult to assess. Refined techniques of collection, notably screenwashing, are improving the Mesozoic record, but it still consists primarily of teeth, with the occasional jaw fragment or end of a limb bone. Very little skull material is known and reasonably complete skeletons have been found at only a few sites before the Late Cretaceous. The historical gaps that plague the study of Mesozoic reptiles apply equally to Mesozoic mammals; however, the fairly low ordinal turnover from the Upper Jurassic through the early Tertiary suggests, on a very gross scale, a good sample of the mammalian types present from what are now Holarctic areas. By contrast, although Jurassic and Cretaceous mammals were probably present in the Southern Hemisphere, essentially nothing is known of their diversity. Most Mesozoic mammals were quite small (the "giant" is a badger-sized Early Cretaceous form), they are as different from each other in dental structure as are the modern orders of mammals, and it is possible to glean from them a good idea of the evolution of the masticatory apparatus as a functional unit (Bown and Kraus, 1979, and references therein). But the phylogenetic interrelationships of these groups, as well as their relationships to Tertiary orders, are poorly understood.

The Cenozoic Era is often called the Age of Mammals, and yet when species diversity alone is considered, the modern descendants of the diapsid branch of the amniotes (the living reptiles plus the birds) are more numerous today than the mammals, which descended from synapsids. Still, the great structural and ecological diversity of mammals has equalled that of the Mesozoic reptiles, whose descendants have in that sense diversified relatively little in the Tertiary. Gingerich (1977, Figure 2) graphed generic diversity of Tertiary mammals (Figure 9), and this graph appears to indicate a successively higher diversity of mammals at each subdivision of the Tertiary epochs (see, however, Harper [1975] on normalization of standing diversity measures). Gingerich noted that, when corrected for absolute duration of these intervals, the curve appeared to be more logarithmic than arithmetic, but he explained that this is almost certainly (once again) due to the increasing probability of finding fossils in more recent strata.

Gingerich also showed a high correlation between the curves describing rates of origination and extinction for several groups of Tertiary mammals (Figure 10). He attributed this correlation to the expectations of equilibrium theory

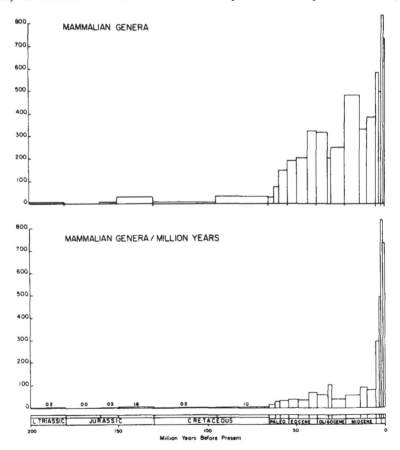

Figure 9. Generic diversity of mammals through time: (above) per period and (below) per million years. From Gingerich (1977).

and to the stability of generic diversity within each group through the Tertiary. It seems equally likely, however, that while net turnover rates have fluctuated considerably for the mammalian orders, sampling is causing the apparent peaks and drops in the curves: this would explain not only why origination and extinction rates track each other so closely, but why the curves are so similar for all the groups assessed. Lillegraven (1972) showed that the total known familial and ordinal diversity of mammals on any given continent has been relatively stable through the Tertiary; this also may indicate constant turnover rates, not necessarily faunal equilibrium. Species-area effects are rejected because there appears to be no clear relationship between continental size and taxonomic diversity. Gingerich pointed out that the times of highest faunal turnover in the Tertiary (Early Eocene, Early Oligocene, Early Miocene, and Early Pliocene) correspond to major periods of climatic change in moisture and temperature. Lillegraven also observed this correspondence, and linked patterns in mammalian ordinal and familial diversity to appearances of angiospermous taxa in the fossil record. When the climate deteriorated rapidly,

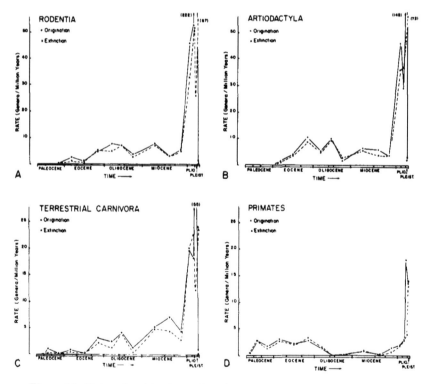

Figure 10. Rates of origination and extinction of genera within four orders of Cenozoic mammals. From Gingerich (1977).

diversity of plants and mammals dropped accordingly, and extinction rates in mammals increased (but not in plants: K. J. Niklas, pers. comm.).

When the patterns of diversity among the classes of terrestrial vertebrates are considered, then, several points must be kept in mind. First, as with the marine record, the apparent patterns of diversity should not be interpreted literally, because fossilization is less likely, and fewer kinds of environments will have a good fossil record. For some long intervals, including most mid-periods, there is almost no record at all. After the Eocene, the terrestrial vertebrate record is heavily biased in favor of faunas of temperate and boreal areas. Thus, statistics on duration and extinctions of groups are particularly vulnerable, because so little is known of the histories of the occupants of tropical areas. Second, because tetrapods consist of only a few classes, trends among classes are not easy to identify, although trends within classes may reflect real patterns of diversity if factors of temporal and preservational bias are taken into account. By comparing patterns within clades at successively lower levels, we may be able to form hypotheses about the character and pace of evolutionary change, or to recognize artifacts of sampling bias (as Gingerich has done for fossil mammals). Evolutionary hypotheses that attempt to account for unusual patterns in the history of life must consider within-group and between-group patterns of diversity measured for a sufficiently long geologic time and calibrated at an appropriately fine scale. Because there are no hard and fast criteria established for such efforts, the modifiers of "long" and "fine" are inevitably vague. Third, whereas the apparent patterns of vertebrate diversity must be considered suspect for the Phanerozoic record as a whole, for *individual episodes* in time the record of fossil vertebrates may be quite good. Some episodes in Phanerozoic time appear to be high in diversity for nearly all groups (e.g., Upper Permian, Upper Triassic, Upper Jurassic, Upper Cretaceous), while other episodes appear to have uniformly low diversity (e.g., Middle Triassic, Middle Jurassic, Middle Cretaceous). Comparison of these peaks and valleys may reveal the extent to which sampling bias is prevalent. Examination of paleoecologic patterns may further reveal the extent to which certain types of environments (e.g., floodplains) have been preferentially preserved at various times, and therefore whether the absence of some ecologically similar taxa is the result of sampling, or even taxonomic interest. Anomalies in the apparent patterns of vertebrate diversity through time mainly point out the well-known intervals: places where further study is likely to be reliable and productive. These intervals may shed light on others that are not so well represented in the fossil record.

At present, vertebrate paleontology can further studies of diversity by concentrating less on *numbers* of species and more on the *phylogenetic components* of these statistics. Cladistic analyses of many vertebrate groups have split paraphyletic taxa into components that have clarified group memberships and stratigraphic ranges. Sepkoski (1981) clearly demonstrated the potential of clade-oriented analysis by "factoring" Phanerozoic marine diversity into component clades that replaced one another through time. This approach could be especially useful with vertebrates because it circumvents sampling bias inherent in empirical counts of species diversity, especially in smaller groups with less complete records.

IMPLICATIONS OF EPISODES FOR LARGER PATTERNS

In the following pages four episodes associated with major extinctions or originations of vertebrate groups are discussed. In some of these, understanding of the underlying component patterns of diversity have recently advanced; in others, the principal questions are left unanswered or are just beginning to be asked. Each episode is characterized by unique patterns that can be approached using data and methods from several disciplines, and each episode reflects problems that seem to be fundamental to a better understanding of vertebrate diversity through time. The four episodes considered are the Permo-Triassic boundary, the Triassic-Jurassic boundary, the Cretaceous-Tertiary boundary, and the Pleistocene.

THE PERMO-TRIASSIC BOUNDARY:
HOW SEVERE FOR LAND VERTEBRATES?

Although the Mesozoic belonged to the diapsid archosaurs, which first appeared at the very end of the Paleozoic, the Permian seems to have been dominated completely by the synapsid pelycosaurs (sail-backed reptiles) and mammal-like reptiles, which were still common and diverse in the early Triassic. In an often overlooked review, Pitrat (1973) explored the components of Permo-Triassic vertebrate diversity and offered several insights into the flush of therapsid taxa that culminated in an abrupt crash at the Permo-Triassic boundary, a time that witnessed the most severe marine extinction in the Phanerozoic (Schopf, 1974; Simberloff, 1974; Raup, 1979a; Raup and Sepkoski, 1982). What does the record of vertebrate change indicate? Were the land and marine extinctions simultaneous, and can they be attributed to a single factor or set of factors?

Pitrat assessed all vertebrate families for which sufficient data were available from the Devonian through the end of the Triassic. Data from each group were compiled from Harland et al. (1967). Pitrat's survey indicated that most groups of marine fishes were severely reduced in diversity during the Late Permian, but that freshwater fishes and amphibians were not so strongly affected and even appear to have been increasing from a slightly earlier diversity crisis. The reptiles, particularly the mammal-like reptiles, experienced a severe crash in the latest Permian (Dzhulfian stage) from which they never recovered; reptilian diversity in the Triassic comprised a very different group of reptiles, notably archosaurs.

Changes in the diversity of these taxa appear to reflect in part conditions of sampling, ecological bias, climatic change, and taxonomic interest. These factors suggest a complex pattern of ecologic and evolutionary replacement. The problems of identifying relative salinity in the environments of fishes have been mentioned, and these data and Pitrat's interpretations were reviewed by Thomson (1977). The cartilaginous fishes, which appear to have been mostly marine, were declining in the Late Permian and were rather more severely affected by the boundary extinction than other fishes, except for the marine holosteans and chondrosteans. Fluctuations in diversity of the freshwater and euryhaline fishes were less severe and occurred some 15 to 20 million years earlier (Figure 11A). A dip in amphibian diversity (Figure 11B) seems to have paralleled that of the latter group of fishes, and is seen by Pitrat as correlated with increasing aridity in preserved environments from Leonardian to Guadalupian time (Olson, 1962; see also Figure 11C). Such a pattern may in part explain the fortunes of the mammal-like reptiles (Figure 11D), which vary in faunal composition both systematically and ecologically, as reflected by sedimentologic change in the Permian Beaufort Series of South Africa (Hotton, 1967). These sediments change gradually from deltaic to fluviatile, but considerable admixtures of both types have resulted in revision of time-stratigraphic zones: when *Lystrosaurus* and *Procolophon* were found in the same deposits, the distinction of their respective "zone" assemblages appeared to be better explained on ecologic grounds than on temporal ones.

A nagging problem in the assessment of Permian therapsid diversity, then, is the extent to which diversity only appears to change because environments are preserved and sampled differentially. Could the apparent taxonomic change be a widespread ecological, and not evolutionary, trend? One possible indication of the evolutionary reality is the similarity of faunal replacement in the Leonardian and Guadalupian Permian beds of both the Soviet Union and the United States, reviewed by Olson (1962). But worldwide ecological trends are also possible, and the problem of how to assess what we do not yet know about these complex ecological and faunal changes has no easy solution.

The problem of how to measure the effects of preservational bias through time remains thorny. Pitrat (1973) showed graphically that the climb in therapsid diversity was restricted to the Leonardian, Guadalupian, and Dzhulfian, with the last stage accounting for the largest flush and crash (Figure 11D). Extinctions track originations so closely throughout this interval that little overall change in stage-to-stage standing crop is evident. Pitrat tabulated diversity by adding new families to existing ones at the beginning of the stage,

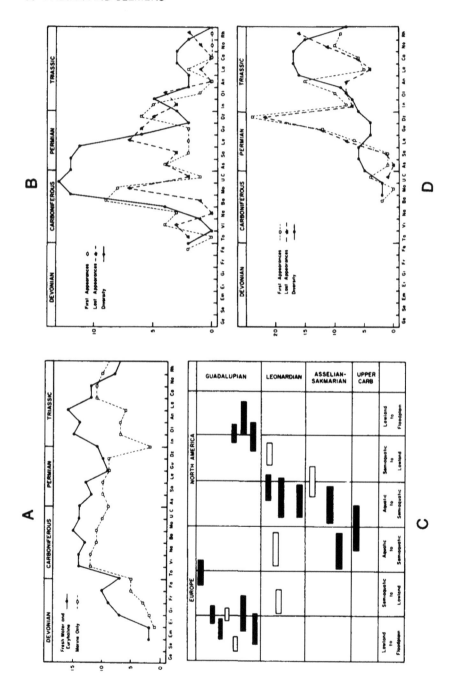

and subtracting last appearances during that stage. The result is a track of diversity at the end of each interval sampled. However, the intervals sampled are not always coincident with the biostratigraphic boundaries. In the Dzhulfian, for example, Pitrat included both the *Daptocephalus* and *Cistecephalus* Zones, and the Dzhulfian in his graph shows twice the diversity of the preceding stage. This is almost surely an artifact of the choice and resolution of temporal scale.

A long-standing tendency to proliferate taxonomic divisions among the mammal-like reptiles has perhaps added to the effect of rapid replacement (or ecological sampling) through the Upper Permian (Figure 8). Some headway on this problem has been made since mid-century, when Haughton and Brink (1954) listed 111 valid species of the Dzhulfian genus *Dicynodon* (20 other species of *Dicynodon* had already been reassigned to other genera and, sometimes, families). Cluver and Hotton (1981) recognized 20 of the 111 species as potentially valid for *Dicynodon*, based on Owen's type skull, and suggested that the existing number of species for three related genera would probably be reduced upon close examination. Sigogneau (1968) reduced 18 families of gorgonopsian therapsids usually recognized to two (with the second, the Ictidorhinidae, admittedly dubious), and synonymized 30 genera with 25 others regarded as valid, leaving 11 "uncertain." If taxonomic revisions like these continue, the Therapsida may become whittled down to manageable proportions in short order. Their meteoric history will be no less real, however, and the final major question to be answered is whether their rapid decline coincided in any particulars with the marine extinctions at the end of the Permian.

Certainly there were waves of faunal replacement on land, as in the ocean, if the terrestrial data is anything but entirely ecologic in bias. However, correlations are difficult to establish. Unfortunately, in South Africa, where the diversity is greatest, assignments of age are based almost entirely on internal associations of vertebrates. The correlations of marine and terrestrial sediments are very complex in the U.S. and U.S.S.R. (Olson, 1962:145-155). Pitrat

---

Figure 11 (facing page). A. Familial diversity of marine and non-marine fishes leading up to and through the Permo-Triassic boundary. B. The same graph for the families of Amphibia. C. Environmental interpretation of faunal change in the Upper Carboniferous and Permian of North America and Europe. Black rectangles indicate relatively complete faunas; white rectangles represent less completely known faunas. D. Familial diversity of reptiles through the Permian and Triassic. All figures from Pitrat (1973); C modified from Olson (1962).

(1973) followed Harland et al. (1967) on the stage divisions of the marine Upper Permian (known only as "zones" on land), but there is no definite evidence to correlate the *Tapinocephalus* Zone with the Guadalupian stage, the *Cistecephalus* and *Daptocephalus* Zones with the Dzhulfian Stage, the *Lystrosaurus* Zone with the Induan Stage, and the *Cynognathus* Zone with the Olenekian Stage. It is therefore not clear that the 18 therapsid groups confined to the *Cistecephalus* and *Daptocephalus* Zones lumped together in the Dzhulfian by Harland et al. all arose and died out at the same time, nor that any of their demises were contemporaneous with the marine extinctions. Given the physical changes involved in major marine regressions at the end of the Permian (Schopf 1974; Robinson 1971; Simberloff 1974), considerable biotic effects on land faunas might be expected; but at present it is difficult to separate the factors contributing to apparent patterns of change in diversity. Pitrat (1973) concluded that the Dzhulfian therapsid boom and bust was short-lived and due to "local factors," and that any connection with events in the marine realm was indirect. Perhaps the question of "local factors" will be testable when the Chinese Late Permian faunal sequence is better known. The possibility of an "indirect" connection with marine events is most intriguing, but for now there appears to be no definite causal relationship between changes in marine and terrestrial diversity.

THE TRIASSIC-JURASSIC BOUNDARY:
THE CASE OF THE MISSING LOWER JURASSIC

When did the Age of Dinosaurs begin? The Mesozoic Era is universally known as the Age of Reptiles, but the dominant archosaur groups (dinosaurs, pterosaurs, crocodiles) did not appear until the Carnian (Upper Triassic) at least, replacing in time a spectrum of other archosaurs found from the base of the Triassic (or even latest Permian) onward. The conventional explanation of the Late Triassic faunal replacement is that the more primitive archosaurian ("thecodont") groups coexisted with the dinosaurs during the Upper Triassic, but that the dinosaurs became the more dominant and abundant members of the fauna toward the end of the Triassic. At this point, the thecodonts went extinct, along with the rhynchosaurs, placodonts, procolophonids, most mammal-like reptiles, and prosauropod dinosaurs, and the labyrinthodont amphibians, as well as many Triassic families of other reptilian orders. Unfortunately there was no way to trace vertebrate history into the period directly following, when the dinosaurs first had the world to themselves, because there was virtually no terrestrial lower Jurassic record. Romer (1968:241-2) lamented this absence: "One of the greatest desiderata (apart from the early Paleozoic) in the vertebrate story would be a full-fledged terrestrial fauna from the Early Jurassic—a time when, it is obvious, important advances were taking place in certain of the reptilian groups and, most especially, notable events were surely occurring in mammalian evolution." But, Romer admitted, the picture looked bleak, even for nonmarine fishes. The terrestrial tetrapod record only became strong in the Upper Jurassic, with the great dinosaur faunas of the Morrison Formation (United States) and the Tendaguru region of South Africa.

The change in this general picture has only recently come about. Two discoveries have prompted reexamination of the criteria for recognizing the Triassic-Jurassic terrestrial boundary and the apparent absence of Lower Jurassic deposits. The first is that many horizons traditionally placed in the Upper Triassic are more properly regarded as Lower Jurassic; the second is that most faunas correctly dated as Late Triassic are not dominated by dinosaurs, but by thecodonts. In addition, the tetrapod extinctions were not simultaneous, but appear to have taken place at different times in different places, some before the Rhaetic, some during, and some after (Olsen and Galton, 1977; see Figure 12). Most apparently took place before the latest Triassic marine extinctions, as Hallam (1981) points out. These discoveries have implications for biostratigraphy and for the understanding of the timing of archosaurian diversity changes in the early Mesozoic.

The temporal reassignment of many "Triassic" beds to the Jurassic was based on reassessment of the faunal composition of these horizons and the diagnostic value of some of their taxa for biostratigraphic correlation (Olsen and Galton, 1977). The European type section is largely marine in the Triassic and almost entirely marine in the Jurassic; it lacks most of the terrestrial vertebrate taxa found elsewhere. Those present are long-ranging or unique to the European section, which is dwarfed in extent and thickness by its temporal equivalents in the U.S. and South Africa. Terrestrial faunal changes are well recognized in footprint faunas of Bunter, Keuper, and Rhaetic age (the terrestrial German Triassic), which complement the largely fragmentary or stratigraphically ambiguous skeletal remains. Using these kinds of data, Olsen and Galton (1977) recognized three footprint zones in the (supposedly Triassic) Newark Supergroup of Eastern North America. The lowest two are equivalent to zones of the middle Middle Keuper and the Rhaetic, respectively. The third zone of the Newark (containing the typical "Connecticut Valley" dinosaur footprint assemblages) has no European equivalent, but pollen evidence from the Newark Supergroup (Cornet, Traverse, and MacDonald, 1973; Cornet and Traverse, 1975) establishes a lower Jurassic date for the third zone and strengthens the correlation of the lower two zones with the European equivalents. Footprint faunas recognizable as equivalent to the third zone of the Newark are found in the Glen Canyon Group of the American Southwest (S. P. Welles, unpublished data) and in the Upper Stormberg of Africa (Ellenberger et al. 1967), and on the basis of the vertebrate faunas of these areas

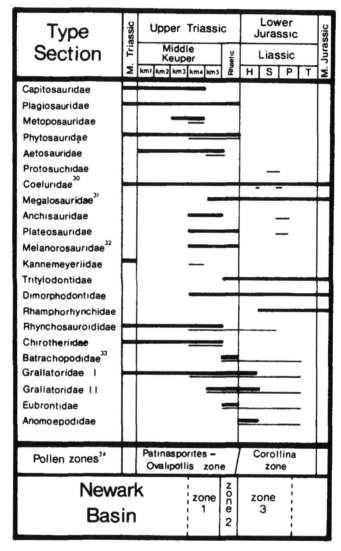

Figure 12. Correlation of vertebrate groups in the Newark Supergroup (light lines, zones at bottom of graph) and in the European type section (bold lines, divisions at top of graph). From Olsen and Galton (1977).

Olsen and Galton suggested that these formations were also Jurassic in age (Figure 13; Olsen and Galton, in press). Their stratigraphic reevaluation prompted a revision of the accepted picture of temporal ranges of many taxa, and changed the concept of Late Triassic vertebrate extinctions (Figure 14). Recently paleontologic opinion in China seems to have shifted to reclassifying the Lower Lufeng series from Triassic to Jurassic (Sigogneau-Russell and Sun, 1981), based on vertebrate and palynologic evidence, and many of the "Rhaeto-Liassic" fissure fills of Great Britain are also now considered Jurassic (Kermack 1975). The worldwide pollen record, where available, appears to be either ambiguous or supportive of these conclusions, but the data are not all in. Evidence from radiometric dating and magnetostratigraphy supports the general age framework proposed by Olsen and Galton (1977; see also references in Olsen, McCune and Thomson, 1982; Puffer et al., 1981; and Olsen, Hubert and Mertz, 1981). A finer correlation of the Newark Supergroup has recently been offered (Olsen, McCune, and Thomson, 1982), based mostly on abundant fish faunas, which were not treated extensively in the original analysis; de Broin et al. (1982) have reported agreement with the correlation of these faunas worldwide on the basis of the earliest appearances of turtles.

The complex logic of correlating deposits on the basis of a set of partially represented and individually insufficient lines of evidence makes it very difficult to sort out which hypotheses are the more robust. Colbert (1981),

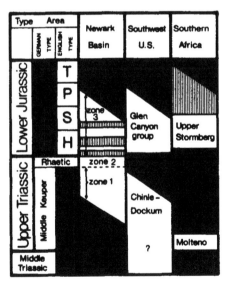

Figure 13. Correlation of Triassic and Jurassic outcrops in the Newark Supergroup, the Southwest U.S., and South Africa. Vertically ruled portions represent radiometrically dated extrusive basalts. From Olsen and Galton (1977).

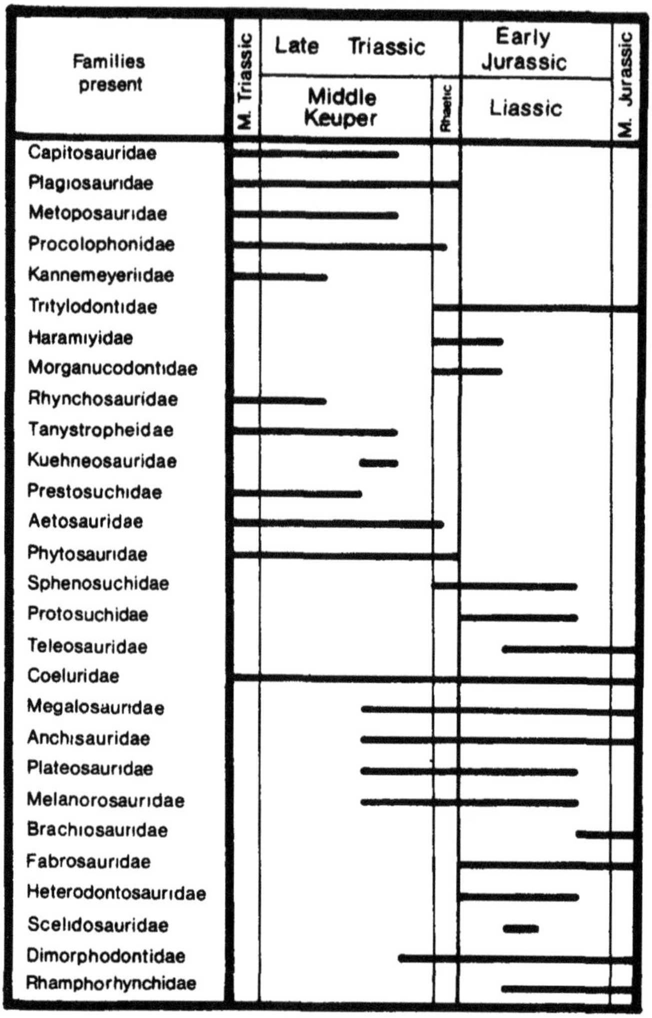

Families present	M. Triassic	Late Triassic		Rhaetic	Early Jurassic	M. Jurassic
		Middle Keuper			Liassic	
Capitosauridae						
Plagiosauridae						
Metoposauridae						
Procolophonidae						
Kannemeyeriidae						
Tritylodontidae						
Haramiyidae						
Morganucodontidae						
Rhynchosauridae						
Tanystropheidae						
Kuehneosauridae						
Prestosuchidae						
Aetosauridae						
Phytosauridae						
Sphenosuchidae						
Protosuchidae						
Teleosauridae						
Coeluridae						
Megalosauridae						
Anchisauridae						
Plateosauridae						
Melanorosauridae						
Brachiosauridae						
Fabrosauridae						
Heterodontosauridae						
Scelidosauridae						
Dimorphodontidae						
Rhamphorhynchidae						

Figure 14. Revised ranges of tetrapods, based on correlations in Figures 12 and 13 From Olsen and Galton (1977).

reviewing the history of thought on the age of the Glen Canyon Group, acknowledges the noncommittal judgments of most workers, but argues strongly for a Triassic age (*contra* Olsen and Galton). His reasoning is that ornithischian dinosaurs, tritylodontid therapsids, and protosuchid crocodiles from the Kayenta most closely resemble similar taxa from the Upper Red Beds (Elliot Fm.) and Cave Sandstone (Clarens Fm.) of the South African Stormberg Series and from the Lower Lufeng of China, formations that traditionally have been considered Triassic. The similarity of these faunas is not in doubt, but the reasoning behind assigning them to the Late Triassic was the focus of Olsen and Galton's paper, so the question reverts to the validity of the correlation based on the logic of their evaluation.

A different line of evidence contributing to the understanding of the Early Mesozoic comes from reexamination of much of the Upper Triassic vertebrate remains. The principal bone-bearing horizons in the United States are the Chinle Formation of Arizona, New Mexico, Colorado, and Utah, and the penecontemporaneous and lithologically similar Dockum Formation of New Mexico and Texas. Collections from these horizons were made mostly in the first half of the 20th century; the material, and its stratigraphic associations, have remained largely unstudied. Tentative identification of much of the material as "dinosaurian" has given the general impression that the entire Upper Triassic contained a good mixture of dinosaurs and thecodonts, with the dinosaurs completely replacing the thecodonts by the end of the Triassic. However, recent examination of much of the material in these collections shows that their referral to the small early carnivorous dinosaur *Coelophysis* (or to "dinosaur *indet.*") is unjustified: most of this material can be referred diagnostically to phytosaurs, aetosaurs, rauisuchids, or poposaurs (all thecodonts). R. A. Long (pers. comm. and work in progress) estimates that, of the 10,000 or more bones collected by Charles M. Camp from the "*Placerias* Quarry" of the Blue Hills (Chinle Formation, Northern Arizona), only about fifteen are positively identifiable as dinosaurian, whereas the great majority are diagnostic of various thecodont taxa. Only the *Coelophysis* "graveyard" of the Ghost Ranch (Chinle Formation of New Mexico) preserves a dinosaur-dominated thanatocoenosis, and its position within the Triassic, while late, has not been fully established. Apparently, dinosaurs were present in Carnian times, but they did not dominate most faunas until the Late Triassic, perhaps in Rhaetic times; these conclusions are supported by Olsen and Galton's documentation of changes in footprint faunal associations across the Triassic-Jurassic boundary in the Newark Supergroup.

The question is now of the consistency and universality of these two hypotheses of biostratigraphic correlation and evolutionary replacement. Some Triassic formations, such as the Los Colorados of South America, have faunas that do not fit the new Triassic-Jurassic dichotomy very closely. Do such faunas indicate evolutionary "transitions" of faunas, ecological differences, or biogeographic endemism? These questions are more difficult to assess in light of the emphasis on footprints for establishing correlations. Like any other fossil datum, footprints are diagnostic and useful for biostratigraphy only to a certain level; what counts is the repeatability of the associations. A larger problem is determining the chronostratigraphic range of taxonomic groups that are poorly understood phylogenetically: "prosauropods," phytosaurs, aetosaurs, and several poorly-known "thecodont" groups, to name only a few. Diagnostic characterizations of lower taxa in these groups would be of great assistance in correlating first appearances of taxa among widely-spread geographic areas. Olsen, McCune, and Thomson (1982:13-15) used the faunal associations of certain morphotypes of the *Semionotus* group of holostean fishes to establish correlation of the more than ten fossil lake basins of the Newark Supergroup; yet they admit, in view of the taxonomic confusion of these fishes, that the use of formal names is impossible, and so their five zones are recognized on the basis of "informal" taxonomic and biostrati-graphic associations of semionotids and other fishes. The larger correlations of horizons figuring in the Triassic-Jurassic boundary question can be tested by phylogenetic analyses of taxa, even long-ranging ones, from which certain relatively short-lived but widespread members can be identified to a limited taxonomic level. At present, the principal questions are which assumptions are dependent upon which associations, the circumstances under which each subhypothesis would collapse, and the effect this might have on the strati-graphic model as a whole.

THE CRETACEOUS-TERTIARY BOUNDARY: MECHANISMS IN SEARCH OF A PATTERN?

In terrestrial deposits, the end of the Cretaceous Period is usually equated with the last records of several groups of reptiles often informally grouped and dubbed "dinosaurs." But stratigraphers have yet to concur on a basic def-inition of the Cretaceous-Tertiary (K-T) boundary. It is generally agreed that this boundary should be defined in terms of changes (usually extinctions) in the marine biota, but whether in microorganisms (various planktic foramin-ifera and calcareous nannoplankton), ammonites, rudistid bivalves, or some other biotic component has not been resolved. In terrestrial deposits at least three approaches are currently used to define the K-T boundary. The last, stratigraphically highest record of typical Late Cretaceous reptilian faunas, including dinosaurs, is sometimes employed. However, because of the im-probability of preservation of even one individual in a population, the last local record of these reptiles probably does not coincide with the precise time of extinction of the group. Recognizing this taphonomic problem, many stratigraphers working in the Western Interior of North America have employed a formula advocated by the paleobotanist Roland Brown (1952). He suggested that the K-T boundary be drawn at the base of the first coal, no matter how thin, above the last record of dinosaurs. This tacitly assumes that the inception of deposition of these coals and lignites throughout the Western Interior was a contemporaneous event. Other paleobotanists use the last occurrence of various kinds of plants to mark the boundary. On a temporal scale with sufficient resolution to discriminate *biologically* significant periods of time—years, decades, centuries—do these three procedures for identification of the K-T boundary result in recognition of the same instant in earth history?

Magnetostratigraphic and radiometric correlations indicate that the extinctions of terrestrial and marine organisms were at least broadly contemporaneous, occurring within a period of several hundreds of thousands of years or, possibly, a million or so years. They do not provide greater precision in correlation. No section of interdigitating marine and terrestrial strata has been found in which it can be shown that the local extinction of dinosaurs and other members of the terrestrial biota, or the inception of deposition of lignites, was precisely contemporaneous with changes in the marine macro- or microbiota taken to be definitive of the Cretaceous-Tertiary boundary. In the northern Western Interior the last local records of dinosaurs and the lowest overlying lignites are found in strata of reversed magnetic polarity (Archibald et al., 1982), now thought to represent the interval 29R. More detailed stratigraphic studies by L. Dingus (1983) demonstrate the uncertainties in attempts to employ inception of lignite deposition as a temporal marker on the biologically significant time scale needed to test hypotheses of causal factors of extinctions of members of the terrestrial biota. Though Roland Brown's formula for recognition of the K-T boundary—the base of the first coal above the last record of dinosaurs—may still be useful in coarse-scaled stratigraphic studies, it must be rejected for use in fine-scaled correlations. Thus, the precise contemporaneity on a biologically significant time scale of these extinctions in the terrestrial and marine biotas has yet to be established. The assertion, based in part on their presumed contemporaneity, that these extinctions were the products of the same causal factors remains an assertion.

To what extent has the severity of the Cretaceous-Tertiary extinctions of terrestrial organisms been artificially magnified by imprecise calibration and interpretation of the fossil record? Were these extraordinary events requiring special explanation? Although a great deal has been written about the extinction of dinosaurs and other groups of terrestrial vertebrates, the fossil record of these events is remarkably meager. For many years, terrestrial faunas of earliest Paleocene age were known only from a few sites in the San Juan Basin

(New Mexico) and the Bighorn Basin (Wyoming). On other continents there was—and in most cases continues to be—a wide stratigraphic gap separating faunas of Late Cretaceous age from the next younger samples of the terrestrial biota. Only in a few areas of western North America is there a reasonably continuous sequence of fossiliferous terrestrial deposits of latest Cretaceous and early Paleocene age. Thus the paleontological record currently can be studied from a global perspective only on a coarse scale of temporal and taxonomic change. Patterns of origins and extinctions or other changes in diversity can be analyzed at the familial or sub-ordinal levels. Research at a more refined scale must be limited to geographically restricted study areas, which are known today primarily within western North America. Detailed stratigraphic studies of geographically limited areas, such as the exposures of the Hell Creek and Tullock Formations in eastern Montana (Archibald, 1982), circumvent some of the problems of imprecision of correlation between fossil localities in different parts of the world. On the other hand, it is too early to determine if all of the local patterns of extinction and survival here are typical of those in other areas.

Patterns of appearance and extinction of terrestrial vertebrates in this area were environmentally controlled (Clemens, Archibald, and Hickey, 1981; Archibald and Clemens, 1982). In floodplain environments there appears to have been no modification in the composition of the fauna until the abrupt extinction of many lineages. On the other hand, in stream valley environments during deposition of the upper part of the Hell Creek Formation, lineages of dinosaurs apparently disappeared gradually from the record as new lineages of placental and multituberculate mammals appeared (note Van Valen and Sloan, 1977). This local record appears to illustrate a complex of changes in biotic composition that seem to have taken place over a geologically short but biologically significant period of time.

As a result of recent studies in the Western Interior of North America and refinements in the stratigraphic records from other areas, understanding of the terminal Cretaceous extinctions of terrestrial vertebrates is undergoing considerable change. At the *ordinal* level (Figure 3), a *global* summary of current records shows not a net decrease in diversity of vertebrates, but an increase. Terminal Cretaceous losses in diversity of archosaurian reptiles and, apparently, in the poorly-known teleosts were compensated by an increase in mammalian and avian diversity during the Paleocene. In Figure 4 comparisons are limited to the three most abundantly sampled and well studied classes of terrestrial vertebrates. The ordinal data suggest that from the Cretaceous to the Paleocene the decrease in reptilian diversity was concomitant with an increase in mammalian diversity. But the time scale of the graph is coarse and the temporal lag between some extinctions and first appearances documented in the fossil record is masked. Attention must also be called to the relatively high numbers of both *first* and last records of orders during this interval (Figure 3). Why are extinctions during this interval of high turnover stressed to the exclusion of originations? The appearance of so many new types is no less a biological phenomenon than the disappearance of many archosaurs.

These charts of global ordinal diversity reflect only three measures of biotic change—records of extinction, first appearance, and total diversity. But note the time scale being considered: summations of Cretaceous (Figure 4) or Late Cretaceous (Figure 3) diversity are being compared with summed Paleocene diversity. If the temporal scale was finer, a short-term decrease in mammalian diversity at the K-T boundary would probably be perceived, because the first records of many orders of mammals occur in the middle and late Paleocene. Artifacts of sampling must also be considered. Until recently, early Paleocene faunas were known from but a few samples obtained in western North America and biased in favor of larger vertebrates. In contrast, those of middle and late Paleocene age are much more numerous, geographically widespread, and include abundant representations of smaller vertebrates. Therefore, additional knowledge of the smaller vertebrates, particularly the mammals, in early Paleocene faunas might significantly increase ordinal diversity recorded at the beginning of the Tertiary.

What is the current resolution of Latest Cretaceous vertebrate extinctions? North American records at the familial and probably generic levels show that most groups of freshwater vertebrates (fishes, amphibians, turtles, crocodilians, and champsosaurs) survived unscathed until the mid-Paleocene. Some of these animals were quite large. They form a "constant" foundation maintaining relatively unchanged turnover rates throughout the K-T transition. Among nonaquatic groups, the patterns of extinction and survival are more complex. In North America, as in other parts of the world, dinosaurs and other large, terrestrial vertebrates (e.g., tortoise-like turtles) became extinct.[1] In contrast,

---

[1] Among the Crocodilia: Paralligatoridae, Goniophilidae, Notosuchidae, Baurusuchidae, and Stomatosuchidae; among the Pterosauria: "Pterodactylidae" (including the giant *Titanopteryx* and *Quetzalcoatlus*); among the Saurischian dinosaurs: Elmisauridae, Ornithomimidae, Oviraptoridae, Caenagnathidae, Dryptosauridae, Tyrannosauridae, Segnosauridae, Ovimimidae, Dromaeosauridae, Stenonychosauridae, Brachiosauridae, Chipitosauridae, Diplodocidae, and Titanosauridae; among the Ornithischian dinosaurs: Hypsilophodontidae, Hadrosauridae, Iguanodontidae, Pachycephalosauridae, Nodosauridae, Ankylosauridae, Protoceratopsidae, and Ceratopsidae. *N.B.*: this is a *global* summation. Many of these "families" are monotypic or contain only two or three latest Cretaceous forms; they are often based on poor material, and the validity of the family designation is dubious. In many cases they are known from only a limited geographic area, and a number of them are currently known just from faunas of Maestrichtian age: a *caveat lector*.

most late Cretaceous families of lizards and even some genera are represented in the Paleocene. Within the Mammalia in North America almost all genera (over 90%) of Late Cretaceous marsupials disappeared, but the severity of extinction among multituberculates (about 25%) and, particularly, placentals was much less (below 10%). The rodent-like multituberculates appear to have replaced their losses quickly, reaching an acme of generic diversity in the middle Paleocene. The diversity of placentals was increasing before the extinction of dinosaurs, with the appearance of proteutherian and possibly lipotyphlan "insectivores," condylarths, and primates. Few, if any, Late Cretaceous genera lack Paleocene relatives.

Thus, the K-T transition is characterized by complex patterns of extinction and survival. In general, species with the largest adult body size disappeared. No dinosaurs survived into the Paleocene (by definition), but not all dinosaurs were large, so size alone will not describe the pattern. Large specimens have tended to be preferentially preserved and collected, but size at reproductive age is unknown, as are the habits of immature dinosaurs. Among species of somewhat smaller adult body size, crocodilians for example, extinction is not closely correlated with size. And, whether the dinosaurs were "warm-blooded" or "cold-blooded," neither alternative will explain their extinction, since other warm- and cold-blooded groups survived without much change. Also, except for apparently greater survival of aquatic vertebrates, the patterns of extinction cannot be directly correlated with characteristics of presumed ecologic niches. In fact, no combination of biotic factors evident from taxonomic surveys alone is sufficient to explain the observed pattern. As for floral changes, paleobotanists (e.g., Hickey, 1981; Jarzen and Dilcher in Russell and Rice, 1982:48, 51-53, 119-120, 122-124; Niklas et al., this volume) have repeatedly pointed out that the cladography of plants across the K-T boundary is heterogeneous and complex. Most workers recognize from intensive studies in western North America that the change from typical Late Cretaceous to typical Paleocene pollen floras was not abrupt and instantaneously complete. Hickey (1981), considering global patterns of floral change, emphasized that, depending upon the phytogeographic province, the K-T boundary can be characterized by everything from severe (ca. 70%) regional extinctions, to no apparent change in composition, to an increase in diversity. In western North America, within the *Aquilapollenites* pollen province, he found that the K-T boundary as defined on floral evidence consistently lies several meters above the last local records of dinosaurs. This repeated pattern suggests (as recognized by Russell, 1982) that the changes in terrestrial faunas and floras were not precisely contemporaneous. How can these two biological patterns be reconciled temporally and causally?

Studies of the record of change in the physical environment have spawned a number of hypotheses of large-scale abiotic changes to "account for" the extinction of dinosaurs (see Archibald and Clemens, 1982, or Jepsen, 1963).

These are usually characterized by the assumptions that the extinctions were globally contemporaneous and the results of a biologically "instantaneous" (a matter of a few months or years) catastrophic change in the physical environment. These assumptions have been repeated so often that some workers have come to treat them as facts. It must be stressed that, while this possibility is not yet ruled out, methods of age determination and long-distance correlation of events that occurred some 65 million years ago have resolutions not more refined than several hundred thousand years, even in the best situations. There is a significant biological difference between hypotheses that call for global extinctions during a period of a few months or years and those suggesting that extinctions of groups began in limited areas and may have continued for 10^5-10^6 years before the last affected members disappeared.

One method of solving both the problem of establishing a virtually instantaneous K-T boundary and the problem of identifying the causal factors of the extinction of Late Cretaceous taxa, including the dinosaurs, was proposed by Alvarez et al. (1980). Anomalously high concentrations of the rare element iridium were found at or very near the traditionally recognized K-T boundary in many marine and terrestrial sections in different parts of the world, often in conjunction with or near abrupt lithologic changes or a major change in faunal composition. The iridium anomaly and its worldwide distribution were explained as the result of the impact of a giant asteroid colliding with the earth. Alvarez et al. (1980) extrapolated the physical effects of such a collision and hypothesized the spread of fragmentary material throughout the global atmosphere. Calculations suggest that the resulting dust cloud might have blocked out sunlight for a period of a few months or years, temporarily lowered ambient temperatures, and directly or indirectly caused the extinction of many groups that did not survive into the Paleocene. The globally synchronous iridium marker, they contend, can be correlated with the simultaneous extinction of marine and terrestrial taxa, and the effects of this dust cloud explain at least much of the sudden change in biological diversity across the K-T boundary.

This "asteroid extinction" hypothesis is really a combination of two inferences: that an asteroid struck the earth some 65 million years ago and had these physical effects and that, in turn, these were the causal factors of the extinctions utilized to mark the K-T boundary in marine and terrestrial deposits. Paleontological evidence cannot bear on the interpretation that the anomalously high concentrations of iridium seen in some geologic sections at or near the currently delineated K-T boundary are the signature of a giant meteorite. As of this writing, several lines of geological evidence appear to converge on that conclusion, and we accept it as a working hypothesis for the present. However, clearly some facets of this hypothesis require further investigation (Padian et al., 1984).

First, operational definitions of "anomalies" in noble metal concentrations must be established, and procedures for measuring them standardized. Their geochemical associations with other such elements, indicative of an extra-terrestrial source, must be distinguishable from terrestrial sources. Second, the postdepositional behavior of iridium in sediments is only poorly known. Are some types of deposits more likely to contain iridium than others? Will its presence be especially abundant in organic-rich sediments, as some geo-chemists suggest? Third, there is a need for detailed control data on distribu-tion of anomalies in geologic sections with and without obvious evidence of sedimentologic or biologic change. In addition, all hypotheses of extinction must face three problems: time resolution, taxonomic resolution, and bio-logic causality. These will be discussed below.

Historical circumstances often conspire against resolution of paleobiological patterns. As Officer and Drake (1983) noted, it is unfortunate that the pro-posed asteroid chose the end of the Cretaceous Period to collide with the earth, because the earth's surface and its biota were then undergoing some very noisy changes. It is difficult to disentangle proposed extra-terrestrial ef-fects from those caused by many well-documented terrestrial changes in the latest Cretaceous, including extensive marine transgression-regression cycles, climatic deterioration, and long-term changes in biotic composition that be-gan well before the end of the Cretaceous. To date, searches for iridium anomalies have focused on horizons at or near actual or apparent extinction events, where a taxonomic change has already been witnessed. But iridium has not been found associated with other major terrestrial or marine extinc-tions, and records of iridium in stratigraphic sections without apparently abrupt extinctions have not been obtained. It is too early to tell whether the asteroid was causal or coincidental to faunal change, and predictions of the magnitude of biotic effects have not been tested in other cases. Astrophysical calculations suggest that from three to ten extraterrestrial bodies larger than 10 km diameter would have been expected to fall within the past 10^8 years. If the biologic effects of such events were as large as predicted, why are there not more mass extinctions in the Tertiary? Ganapathy (1982) noted an iridium-rich marine horizon about 30 cm below a microtektite concentration in one deep-sea core in the Caribbean, associated with the disappearance of a few radiolarian marker species. This layer was dated at about 34 mybp, near one recognized placement of the marine Eocene-Oligocene boundary. Alvarez et al. (1982) suggested that these extinctions were asteroid-caused, and might be correlated with late Eocene terrestrial mammalian extinctions. But terres-trial records suggest a closer correlation of the latter with the marine Pria-bonian-Rupelian stage Boundary, at about 37 mybp (Savage and Russell, 1983), and for this reason both the magnitude of the marine extinctions and the correlation of terrestrial and marine extinctions are as yet unsubstantiated

(Clemens and Padian, MS.). So far, no other biotic effects of extra-terrestrial events within the last 10^8 years have been recognized or proposed. Consequently, substantial revisions may be required in the calculated frequency of large-body impacts, in the predictions of their physical consequences, in the assessments of their biotic effects, or in all three.

As mentioned above, catastrophic models such as the "asteroid hypothesis" entail both physical and biological components. With regard to the second, any biological ramifications predicted by a catastrophic model must correspond to observed biotic events, not prescribe them. A convincing causal mechanism must account for the differential survival of taxonomic groups in different areas on plausible biologic grounds. There is an important pitfall to avoid here: the approach frequently has been to predict the physical effects of a catastrophic model, then list the organisms that went extinct, and account for the demise of the second in terms of the first (e.g., Alvarez et al., 1980; Hsü et al., 1982). Organisms that survived the "holocaust," whether the model predicted that they should have or not, are then given alibis—reasons why they "were able" to avoid certain death. However, these alibis are not extended to those groups that were equally likely to have availed themselves of the same excuses, but instead went extinct. A comprehensive extinction scenario cannot simply point to those extinctions that seem to accord with its predictions, and ignore those that do not.

It is difficult to see how any organisms could have survived some of the purely physical predictions of the extreme versions of the impact model. Observed biological patterns, however, do not accord with even the milder scenarios. The extinctions were selective; any causal model, if it can be invoked, must accord with these biological patterns. Ideally, the organisms that survived and those that did not should be separated into two groups. Within each group, some unifying biological pattern should be sought. The differences between these patterns should then be shown to be lethal under conditions that an asteroid impact, or any proposed catastrophic model, could explicitly provide. Only in this way can a hypothesis of causality be established.

In practice, the situation is far more complex. Raup and Sepkoski (1982) examined extinctions of marine organisms through geologic time and noted a two-factored pattern consisting of a stochastically decreasing "background level" of extinction punctuated by intervals of varying duration when extinction rates were so anomalously high that the descriptive term "mass extinction" becomes appropriate. If the background levels of extinction around the K-T boundary were measured, the selective effects of a catstrophic model could be examined for potential validity. One way to recognize background noise might be to calculate mean extinction rates over their entire geologic duration for each taxonomic group involved in the extinctions, and to see which groups show anomalously high extinction rates during that interval.

These might be recognized as severely affected, whereas groups that showed a constant or lower rate of extinction would be recognized as relatively unaffected (within stochastic limits) by the postulated catastrophe. Until this is done, paleontologists will not be able to provide an operational definition of a "mass extinction" (Padian et al., 1984), and the efforts of physical scientists to examine catastrophic events in earth history will be untestable in biologic terms.

While such detailed studies have not yet been done, it is clear that the archosaurs, which show a decline from some 28 latest Cretaceous families to three in the Paleocene, were foremost among those most affected at the end of the Cretaceous. However, the *real* biotic loss may not have been as unusual or as extreme as it first appears. Some (but not all) of these Late Cretaceous families had long histories; yet few had more than one or two genera present in the Maestrichtian, and current estimates of dinosaur diversity in Hell Creek (Latest Maestrichtian) beds hover at around twenty genera (R. A. Long and D. A. Russell, in progress). Twenty families going extinct sounds more impressive than twenty genera, especially if the families and genera are highly depauperate. The question of taxonomic scale, therefore, plays some part in perception of evolutionary tempo.

Let us take this analysis a step further. The disappearance of some twenty genera (="kinds") of Maestrichtian dinosaurs is regarded as an event worthy of special explanation. Yet, in the entire history of dinosaurs, known from several hundred genera and many diverse and well-studied faunas, almost no well-established genera (ignoring wastebasket taxa like *"Megalosaurus"*) are known from more than one formation, or have geologic ranges that extend beyond a single stage. If dinosaur genera are characteristically short-lived, then at this scale of analysis it is equally reasonable to conclude that dinosaurs experienced "mass extinctions" all through their history. For example, one of the richest dinosaur faunas comes from the Upper Jurassic Morrison Formation of Colorado, Wyoming, Utah, and adjacent states. Dodson et al. (1980) recognized 18 valid genera (and one other undescribed) from environmentally heterogeneous beds deposited over a million square kilometers. These 18 genera belong to 11 different families, yet only one genus (*Brachiosaurus*) appears to be definitely known from elsewhere (the broadly contemporaneous Tendaguru beds of South Africa), and no genera apparently survive after Morrison time. However, no one has ever proposed a "mass extinction" of Morrison dinosaurs. No Morrison genera are definitely ancestral to those of later formations, yet dinosaurs continued to diversify through the Cretaceous. High turnover rates, on the order of a few million years or less, appear to have been the rule, not the exception, for dinosaurian genera throughout their history. As such, at the end of the Mesozoic the significant biological question is not one of increased extinction rate, but of a drop in origination rate—the failure to replace old taxa with new ones. Certainly the dinosaurs went extinct, completely and finally, sometime around the end of the Cretaceous Period; but the latest Cretaceous extinction itself may turn out to be nothing extraordinary in terms of rate or magnitude of extinction. This problem cannot be solved until the tempo of dinosaur diversity through time is adequately assessed.

THE PLEISTOCENE EXTINCTIONS: A QUESTION OF BALANCE

Of the several terrestrial megafaunal mass extinctions, the most recent are the later Pleistocene extinctions. Within a period of a few thousand years in various parts of the world, some 200 genera of mammals, most of them larger than 50 kg (adult body weight), disappeared; many other genera lost species without becoming completely extinct. Because these events fall largely within the range of the Carbon-14 method of age determination, a relatively high-resolution chronology of these extinctions is being established. It is now clear that extinctions were not exactly synchronous on all continents and major islands. However, there seems to be a common pattern of survival and extinction in these events.

As usual in terrestrial extinctions, "lower" vertebrates were not much affected, except where they were among the larger members of their biotas. Among the birds, usually the largest were those hardest hit, including the giant flightless Dinornithiform birds and their allies from New Zealand, and a few large species elsewhere. In the mammals, well-scattered extinctions in all orders were supplemented by some especially severe ones, particularly amid species of large adult body size. The edentates (including ground and tree sloths and armadillos) and proboscideans (mammoths, mastodons, and elephants) were decimated, and their orders have today only a shadow of their former diversities. In other orders, certain families suffered severely. Among the perissodactyls, five genera of Equidae perished, leaving only the modern *Equus* (the horses, zebras, and donkeys). Of the artiodactyls, the camels and antilocaprids (pronghorns) followed the horse pattern, while deer and bovids were significantly, but not as drastically, reduced. Lemuroid and indrid primates of Madagascar were affected, especially the larger, possibly diurnal species, a pattern that cannot be explained away by the usually poor fossil record of most primates. No continents appear to have escaped some degree of decimation in the later Pleistocene extinctions. Even Australia's terrestrial fauna was reduced by loss of some larger marsupials such as the rhinoceros-like diprotodonts. In some instances, smaller lineages of marsupials underwent further reduction in body size (Marshall and Corruccini, 1978).

Two camps, whose lines were drawn in the early 1960s have framed the question of extinction pattern in bipolar terms. One camp invokes climatic or other environmental change as "the cause," while the other singles out human predation. The state of the art was summarized in a stimulating collection of papers some fifteen years ago (Martin and Wright, 1967). No consensus was then or has since been reached, but the complexity of the pattern is universally acknowledged. Even some of the strongest advocates of human predation recognize that the differential survival of some "prey" groups in the presence of man cannot be explained by the "overkill" model. Conversely, those who put faith in the sufficiency of environmental deterioration to account for most megafaunal extinctions cannot deny the influence and efficacy of prehistoric man as a hunter. Separating the effects of the two influences is not easy.

As Savage and Russell (1983) point out for the Cenozoic as a whole, faunal lists are inadequate to "explain" extinction patterns. The relationship of ecology and behavior to environmental change affects individual species in different ways. Guilday (1967) documented the tendency for larger mammalian species to be more affected by environmental stresses in recent biotas, and suggested this as a general model for Pleistocene patterns. In Africa, the combined effects of Saharan desiccation, human predation, and encroachment of available grassland by domesticated animals have affected megafaunal diversity in some areas; yet despite the ancient presence of man (and his ancestors) in Africa, this continent continues to maintain a Pleistocene level of diversity, compared to other continents. Africa was not as severely affected by glaciation as were the northern continents. South American mammals were still suffering the effects of the faunal interchanges brought about approximately 3 mybp by the connection with North America; but abruptly, during the transition from the late Pleistocene to Recent about 15,000 years ago, some 22% of the known late Pleistocene mammalian families and 37% of the genera became extinct (Marshall, 1981).

As with the K-T boundary extinctions, fine-scaled analysis of environmentally and geographically heterogeneous cases may reveal some larger overall patterns and fail to substantiate others. The global chronology of the extinctions has been cited as evidence in favor of the "overkill" model, for it appears (Martin, 1967) that the megafaunal extinctions on many continents and islands closely correlate with patterns of human dispersal and cultural change. However, these events must be viewed in the context of evolution of late Cenozoic faunas.

In North America, for example, since the Oligocene the climate has become cool and the grasslands of the High Plains may be only the culmination of the reduction of forested areas (Gregory, 1971). During the Pleistocene, in addition to the oscillations of the glaciers, areas of desert in intermontane basins, whose borders were heightened by tectonic movements, waxed and waned. These complex changes in the physical environment appear to be reflected in the evolution of the North American mammalian fauna. Webb (1969) called attention to the high rate of mammalian extinctions in these areas during the Pliocene—late Hemphillian extinctions overtook some 55 mammalian genera whereas the late Pleistocene toll was only 25 genera. Of these genera it appears that approximately two-thirds died out about 8,000 to 15,000 years ago during a period of climatic change after the maximum Wisconsin glaciations. Two-thirds of these had been present since at least the Irvingtonian, and nearly a third since the Blancan. Of the animals of less than 50 kg body weight that became extinct, their disappearance occurred more regularly throughout the Blancan and Irvingtonian: only four genera are recognized in the Rancholabrean, and only one of these is present during the Wisconsin glaciation (Martin, 1967, Table 1). The patterns in large and small mammals are significantly different.

The history of late Cenozoic faunal change in North America might mirror the changes in other continents, but much information must be gathered in other areas so that local and global patterns can be clearly separated and then contrasted and compared. Parenthetically, it should be noted that research on Pleistocene extinctions is calling for resolution of correlations and age determinations on scales calibrated in increments of thousands of years or less, intervals that are of too fine a scale to even be attempted in studies of major extinctions in the much more distant past.

However, extinctions are but one factor in the causes of change in diversity. As Gingerich (1977) pointed out, what is astonishing is not so much the late Pleistocene extinctions, but the early Pleistocene originations (see Figure 10). The reasons for the taxonomic explosion in that brief interval have yet to receive the attention accorded the extinctions. However, any interpretations of the Pleistocene pattern should be set in the context of evolutionary rates for the entire Tertiary.

The two "modern" ungulate orders, the Perissodactyla and Artiodactyla, have provided the opportunity to study ecologically similar groups with similarly long histories in the Tertiary. Both orders appeared in the early Eocene, but today the perissodactyls are composed only of the horses, tapirs, and rhinos, while the artiodactyls include all the other hoofed animals (pigs, cows, and giraffes, to name only a few). Cifelli (1981) charted generic and familial diversity of these two orders through the Cenozoic, and concluded that their histories could not be explained by progressive competitive replacement of perissodactyls by artiodactyls: the two groups had independent histories and unique rates of origination and extinction. The difference was that, after the Eocene, the perissodactyls did not maintain their generic origination rate, but the artiodactyls did (Figure 15). When viewed at the familial level (Figure 16), the artiodactyls seem to have performed like the perissodactyls, except in greater numbers. Both groups show an apparent plummet in generic

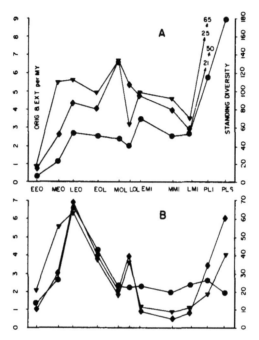

Figure 15. Generic diversity (dots) and rates of origination (triangles) and extinctions (diamonds) of artiodactyls (A) and perissodactyls (B) through the Eocene (EO), Oligocene (OL), Miocene (MI), Pliocene (PLI), and Pleistocene (PLS). E, early; M, middle; L, late. From Cifelli (1981).

origination rate at the end of the Oligocene and an explosion in originations (though not in diversity) through the Pliocene and Pleistocene; apart from this, their patterns and rates are not similar. As Gingerich (1977) and Lillegraven (1972) also noted, Cifelli showed that extinction rates within faunas closely tracked origination rates; like the earlier authors, he concluded from this pattern that some sort of biogeographic faunal equilibrium may have operated. We find these conclusions difficult to qualify, because similar patterns appear among unrelated and ecologically dissimilar taxa (Gingerich 1977; see Figure 10). These patterns appear to uphold the view of Savage and Russell (1983) that patterns of diversity from the Puercan (early Paleocene) through the Chadronian (early Oligocene) are simply the result of sample sizes, not of fluctuating evolutionary or immigrational rates. They also show that very few species are present in more than one North American land-mammal age, though generic longevity varies.

Apparently, "equilibrium" effects cannot be separated from those due to

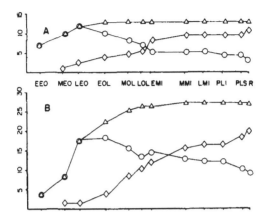

Figure 16. Familial diversity (circles), originations (triangles), and extinctions (diamonds) of perissodactyls (A) and artiodactyls (B). Originations and extinctions are cumulative. R, Recent; other abbreviations as in Figure 15. From Cifelli (1981).

sampling biases if the longevity of taxa is sufficiently short and regular. It seems more than coincidental that various mammalian groups show similar curves of diversity and such close tracking of origination rates by extinction rates, irrespective of age, area, climate, ecology, or clade size. These considerations, when applied to the Pleistocene, suggest that the high extinction rates of the Late Pleistocene may not be unexpected. Because fossil taxa appear to have a characteristic range of duration (Van Valen, 1973a), it should be possible to see to what extent extinction rates in the Pleistocene deviate from expected probabilities for each of their component taxonomic groups. If the genera going extinct at the end of the Pleistocene had relatively "normal" taxonomic durations, this would shift drastically the focus of evolutionary study. Thomson (1976) suggested that the periods in earth history during which rapid diversification of unrelated groups took place deserve more attention, and this accords with Gingerich's perception of the Pleistocene problem. It is unlikely that the relatively short Pleistocene Epoch will be fully understood until both sides of its faunal question (origination and extinction) are equally considered in the full context of late Cenozoic evolution.

CONCLUSIONS

1. The apparent patterns of vertebrate diversity gleaned directly from the fossil record show many episodes of high diversity in a variety of clades, many episodes of low diversity in nearly all clades, and some episodes

wherein a few clades are much more abundant and diverse than others. Only the first pattern can be regarded as a reasonable approximation of actual diversity. The other patterns are often the result of preservational or collection bias, although the apparently rapid evolutionary explosions or mass extinctions of some clades may be quite real. In the Tertiary, where the terrestrial record is best preserved, extinction rates appear to track origination rates quite closely. This pattern appears to hold for the Phanerozoic in general, and suggests that it is largely due to sampling bias and spotty preservation.

2. Samples of fossil vertebrates are generally small. For most taxa we have no firm idea of morphologic variability within species. Taxonomic discrimination is such that species-level data are not as realistic as generic-level data for comparing patterns over large time scales. It is useful to compare patterns of diversity within groups at several taxonomic levels (genus, family, order) because these patterns can reveal information about the timing of diversification, the temporal distribution of clades analyzed at various levels, and the appearance of morphological novelties underlying the arbitrary recognition of the higher Linnean categories.

3. For many reasons, the fossil record for terrestrial organisms is not as good as it is for marine organisms. Still, for certain intervals of geologic time, the terrestrial record is adequate to ask questions of significance about patterns of vertebrate evolution. The questions must be designed to fit the data base. Determination of the timing, both relative and specific, of changes in marine and terrestrial biotas in various parts of the world is of great importance in trying to assess possible causal interrelationships. These studies must begin with an assessment of the degrees of resolution and precision of the various methods of biostratigraphic, radiometric, or magnetostratigraphic methods utilized to establish the timing of events and correlation of fossiliferous strata.

4. Because Pangaea began to fragment soon after the emergence of terrestrial vertebrates, increasing provinciality with occasional renewed land contacts has been a frequent pattern in biogeographic history. Tabulations of vertebrate diversity should take into account this provinciality in order to avoid statistical but nonbiological patterns united only by temporal synchroneity. For example, it is less interesting to note that marsupials and monotremes survive to the present day than to note that they appear to have been essentially limited to the "island continents" of South America and Australia since the end of the Cretaceous. Traditional charts of diversity do not show this, and so an important qualification in biogeographic history is lost (Lillegraven, 1972).

5. Several episodes of major change in vertebrate history can be analyzed for their changes in patterns of diversity, both real and apparent. New

conclusions have already surfaced in some, and much more work is needed on most. The therapsid "flush-crash" in the Late Permian is partly a taxonomic artifact and perhaps to some extent the effect of ecologic sampling; yet more needs to be known about terrestrial climatic change and the nearly concomitant changes in the marine realm before allowing acceptance or rejection of a connection between the two. At the Triassic-Jurassic boundary, an extensive terrestrial Lower Jurassic record has been discovered in sediments previously assigned to the Upper Triassic; still unsolved is the problem of the growing role of dinosaurs in the Upper Triassic and the decline of the thecodonts they replaced. The Cretaceous-Tertiary boundary continues to draw attention, most recently because of the suggestion that an asteroid may have landed at about the same time that extensive marine and terrestrial extinctions took place. However, it must be stressed that predictions about biologic change derived from physical events must make sense in biologic terms. The explanations of physical sciences are not sufficient in themselves to explain biologic patterns and processes, either in theory or in observed practice. Finally, as Gingerich (1977) has pointed out, the spectacular Late Pleistocene extinction of the mammalian megafauna was preceded by an even more spectacular early Pleistocene radiation; questions of cause and pattern for both events are complex, and the largely unstudied originations provide a challenge to evolutionary theory that has often been overshadowed by interest in large-scale extinctions.

Analysis of vertebrate cladographic patterns, in conjunction with hypotheses of phylogenetic and biogeographic evolution, is a most important area of research for the understanding of the history of terrestrial vertebrate diversity. Recent advances in methods and concepts have encouraged new studies of the history of terrestrial vertebrates. Thanks to these advances we anticipate in the near future more new questions, and more than a few new answers.

ACKNOWLEDGMENTS

Reviews of this kind, superficial as they must inevitably be, cannot be written without the advice and cooperation of many people. We thank K. J. Niklas, J. M. Parrish, D. M. Raup, and J. J. Sepkoski for reviewing the manuscript in various stages, and C. A. Brunner, J. M. Clark, L. W. Dingus, R. A. Long, A. R. McCune, P. E. Olsen, W. D. Riedel, T. Rowe, D. E. Savage, and J. W. Valentine for assistance, information, and critical comments. Our gratitude to these people does not, of course, necessarily imply their agreement with our conclusions and views, and any errors or misinterpretations are our responsibility. For permission to reproduce their illustrations we would like to thank the authors and their publishers, including Elsevier Publishing Co. (Figures 7, 9, 10 and 11), Academic Press (Figure 5), MacMillan Journals (Figure 2), and the editors of *Evolution* (Figures 15 and 16). Figures 12, 13, and 14 are Copyright 1977 by the American Association for the Advancement of Science. Figures 3, 4, 6, 7 and 8 were drafted by Lily Shen. This report was inspired by a NASA workshop on the evolution of complex life, and funded in part by the donors of the Petroleum Research Fund of the American Chemical Society (Grant 13577-G2 to K. P. and 12487-AC2 to W. A. C.) and a grant from the National Science Foundation (DEB 81-19217 to W A. C.).

REFERENCES

Alvarez, L. W., Alvarez, W., Asaro, F., and Michel, H., 1980. Extraterrestrial cause for the Cretaceous-Tertiary extinction. Science 208:1095-1108.

Alvarez, W., Asaro, F., Michel, H. V., Alvarez, L. W., 1982. Iridium anomaly approximately synchronous with terminal Eocene extinction. Science 216:886-888.

Archibald, J. D., 1982. A study of Mammalia and geology across the Cretaceous-Tertiary boundary in Garfield County, Montana. Univ. Calif. Publ., Geol. Sci. (122), 286.

Archibald, J. D., Butler, R. F., Lindsay, E. H., Clemens, W. A., and Dingus, L., 1982. Upper Cretaceous-Paleocene biostratigraphy and magnetostratigraphy, Hell Creek and Tullock Formations, northeastern Montana. Geol. 10:153-159.

Archibald, J. D., and Clemens, W. A., 1982. Late Cretaceous extinctions. Am. Scientist 70 (4):377-385.

Bakker, R. T., 1977. Tetrapod mass extinctions—a model of the regulation of speciation rates and immigration by cycles of topographic diversity. In Hallam, A. (ed.), Patterns of evolution, as illustrated by the fossil record, Amsterdam: Elsevier, 439-468.

Behrensmeyer, A. K., 1982. Time resolution in fluvial vertebrate assemblages. Paleobiol. 8:211-227.

Behrensmeyer, A. K., and Hill, A. P., eds., 1980. Fossils in the making: Vertebrate taphonomy and paleoecology. Chicago: Univ. Chicago Press, 338.

Bolt, J. R., 1977. Dissorophoid relationships and ontogeny, and the origin of the Lissamphibia. Paleontol. J. 51:235-249.

Bown, T. M., and Kraus, M. J., 1979. Origin of the tribosphenic molar and metatherian and eutherian dental formulae. In Lillegraven, J. A., Kielan-Jaworowska, Z., and Clemens, W. A. (eds.), Mesozoic mammals: The first two-thirds of mammalian history, Berkeley: Univ. Calif. Press, 172-181.

Brinkmann, R., 1929a. Statistisch-biostratigraphische Untersuchungen an mitteljurassischen Ammoniten über Artbegriff und Stammesentwicklung. Abh., Gesellschaft der Wissenschaften zu Gottingen, Math-Phys. Klasse, N.F. 13(3):1-249.

Brinkmann, R., 1929b. Monographie der Gattung Kosmoceras. Abh., Gesellschaft der Wissenschaften zu Göttingen, Math-Phys. Klasse, N.F. 13(4):1-123.

Brown, R. W., 1952. Tertiary strata in eastern Montana and western North and South Dakota. Billings Geol. Soc. Guidebook 3:89-92.

Buffetaut, E., Cappetta, J., Gayet, M., Martin, M., Moody, R. T. J., Rage, J. C., Taquet, P., and Wellnhofer, P., 1982. Les vertébrés de la partie moyenne du Crétacé en Europe. Cretaceous Res. 2:275-281.

Butler, P. M. et al., 1967. Mammalia. In Harland, W. B. et al., The Fossil record, London: Geol. Soc. 763-787.

Camp, C. L., 1952. Geological boundaries in relation to faunal changes and diastrophism. Paleont. J. 26:353-358.

Carroll, R. L., 1975a. The early differentiation of diapsid reptiles. Centre National de Recherche Scientifique, Colloque International 218:443-449.

Carroll, R. L., 1975b. Permo-Triassic "lizards" from the Karroo. Palaeontologia Africana 18:71-87.

Carroll, R. L., 1977. Patterns of amphibian evolution: An extended example of the incompleteness of the fossil record. In Hallam, A. (ed.), Patterns of evolution as illustrated by the fossil record, Amsterdam: Elsevier, 405-438.

Carroll, R. L., 1978. Permo-Triassic "lizards" from the Karoo system. II. A gliding reptile from the Upper Permian of Madagascar. Palaeontologia Africana 21:143-159.

Charig, A. J., 1973. Kurtén's theory of ordinal variety and the number of the continents. In Tarling, D. H., and Runcorn, S. K., Implications of continental drift to the earth sciences, New York: Academic Press, 231-245.

Cifelli, R. L., 1981. Patterns of evolution among the Artiodactyla and Perissodactyla (Mammalia). Evol. 35:433-440.

Clemens, W. A., Archibald, J. D., and Hickey, L. J., 1981. Out with a whimper not a bang. Paleobiol. 7:293-298.

Clemens, W. A., Lillegraven, J. A., Lindsay, E. H., and Simpson, G. G., 1979. Where, when, and what—a survey of known Mesozoic mammal distribution. In Lillegraven, J. A., Kielan-Jaworowska, Z., and Clemens, W. A. (eds.), Mesozoic mammals: The first two-thirds of mammalian history, Berkeley: Univ. Calif. Press, 7-58.

Cluver, M. A., and Hotton, N., III, 1981. The genera *Dicynodon* and *Diictodon* and their bearing on the classification of the Dicynodontia (Reptilia, Therapsida). So. Afri. Mus. Ann. 83(6):99-146.

Colbert, E. H., 1981. A primitive ornithischian dinosaur from the Kayenta Formation of Arizona. Mus. No. Ariz. Bull. 53:1-61.

Cornet, B., and Traverse, A., 1975. Palynological contribution to the chronology and stratigraphy of the Hartford Basin in Connecticut and Massachusetts. Geoscience and Man 11:1-33.

Cornet, B., Traverse, A., and McDonald, N. H., 1973. Fossil spores, pollen, and fishes from Connecticut indicate Early Jurassic age for part of the Newark Group. Science 182:1243-1246.

Cracraft, J., 1981. Pattern and process in paleobiology: The role of cladistic analysis in systematic paleontology. Paleobiol. 7(4):456-468.

Cutbill, J. L., and Funnell, B. M., 1967. Computer analysis of the fossil record. In Harland, W. B., et al., The fossil record: A symposium with documentation, London: Geol. Soc., 791-822.

DeBroin, F., Ingavat, R., Janvier, P., and Sattayarak, N., 1982. Triassic turtle remains from northeastern Thailand. Vert. Paleontol. J. 2:41-46.

Dingus, L., 1983. A stratigraphic review and analysis for selected marine and terrestrial sections spanning the Cretaceous-Tertiary boundary. Ph.D. dissertation, Univ. California, Berkeley.

Dingus, L., and Sadler, P. M., 1982. The effects of stratigraphic completeness on estimates of evolutionary rates. System. Zool. 31:400-412.

Dodson, P., Behrensmeyer, A. K., Bakker, R. T., and McIntosh, J. S., 1980. Taphonomy and paleoecology of the dinosaur beds of the Jurassic Morrison Formation. Paleobiol. 6:208-232.

Durham, J. W., 1967. Presidential address: The incompleteness of our knowledge of the fossil record. J. Paleontol. 41:559-565.

Ellenberger, F., Ellenberger, P., and Ginsburg, L., 1967. The appearance and evolution of dinosaurs in the Trias and Lias: A comparison between South African Upper Karroo and Western Europe based on vertebrate footprints. In Gondwana stratigraphy, Paris: UNESCO, IUGS Symposium, 1967, 333-354.

Feduccia, A., 1980. The age of birds. Cambridge: Harvard Univ. Press, 196.

Fisher, D. C., 1982. Phylogenetic and macroevolutionary patterns within the Xiphosurida. Third No. Amer. Paleontol. Conv. Proc. 1:175-180.

Ganapathy, R., 1982. Evidence for a major meteorite impact on the earth 34 million years ago: Implication for Eocene extinctions. Science 216:885-886.

Gingerich, P. D., 1977. Patterns of evolution in the mammalian fossil record. In Hallam, A. (ed.), Patterns of evolution as illustrated by the fossil record, Amsterdam: Elsevier, 469-500.

Gould, S. J., 1980. G. G. Simpson, paleontology, and the modern synthesis. In Mayr, E., and Provine, W. B. (eds.), The evolutionary synthesis, Cambridge: Harvard Univ. Press, 153-172.

Gregory, J. T., 1971. Speculations on the significance of fossil vertebrates for the antiquity of the Great Plains of North America. Abh., Hessisches Landesamt fur Bodenforschung 60:64-72.

Guilday, J. E., 1967. Differential extinction during Late Pleistocene and Recent times. In Martin, P. E., and Wright, H. E. (eds.), Pleistocene extinctions: The search for a cause, New Haven: Yale Univ. Press, 121-140.

Hallam, A., 1977. Patterns of evolution as illustrated by the fossil record. Amsterdam: Elsevier, 591.

Hallam, A., 1981. The end-Triassic bivalve extinction event. Palaeogeog.,

Palaeoclimatol., Palaeoecol. 35:1-44.

Harland, W. B., et al., 1967. The fossil record: A symposium with documentation. London: Geol. Soc.

Harper, C. W., Jr., 1975. Standing diversity of fossil groups in successive intervals of geologic time: A new measure. J. Paleontol. 49:752-757.

Haughton, S. H., and Brink, A. S., 1954. A bibliographic list of Reptilia from the Karroo beds of Africa. Palaeontologia Africana 2:1-187.

Heaton, M. J., 1980. The Cotylosauria: A reconsideration of a group of archaic tetrapods. In Panchen, A. L. (ed.), The terrestrial environment and the origin of land vertebrates, New York: Academic Press, Systematics Assoc. Spec. 15:497-551.

Hickey, L. J., 1981. Land plant evidence compatible with gradual, not catastrophic, change at the end of the Cretaceous. Nature 292:529-531.

Hopson, J. A., and Radinsky, L. B., 1980. Vertebrate paleontology: New approaches and new insights. Paleobiology 6:250-270.

Hotton, N., III, 1967. Stratigraphy and sedimentation in the Beaufort Series (Permian-Triassic), South Africa. In Essays in paleontology and stratigraphy, Lawrence: Univ. Kansas, Geol. Dept. Spec. Publ. (2):390-428.

Houde, P., and Olson, S. L., 1981. Paleognathous carinate birds from the early Tertiary of North America. Science 214:1236-1237.

Hsu, K. J., et al., 1982. Mass mortality and its environmental and evolutionary consequences. Science 216:249-256.

Hutchison, J. H., 1982. Turtle, crocodilian, and champsosaur diversity changes in the Cenozoic of the North-Central region of western United States. Palaeogeog., Palaeoclimat., Palaeoecol. 37:149-164.

Jensen, J. A., 1981. Another look at *Archaeopteryx* as the "oldest" bird. Encyclia 58:109-128.

Jepsen, G. L., 1963. Terrible lizards revisited. Princeton Alumni Weekly 64:6-10, 17-19.

Kermack, K. A. 1975. The complex of early mammals. Cent. Nat. de Recherche Scientifique, Colloque International 218:563-672.

Lauder, G. V., 1981. Form and function: Structural analysis in evolutionary morphology. Paleobiology 7:430-442.

Lillegraven, J. A., 1972. Ordinal and familial diversity of Cenozoic mammals. Taxon 21:261-274.

Lillegraven, J. A., Kielan-Jaworowska, Z., and Clemens, W. A., eds., 1979. Mesozoic mammals: The first two-thirds of mammalian history. Berkeley: Univ. Calif. Press, 311.

Marshall, L. G., 1981. The great American interchange—an invasion induced crisis for South American mammals. In Nitecki, M. H. (ed.), Biotic crises in ecological and evolutionary time, New York: Academic Press, 133-229.

Marshall, L. G., and Corruccini, R., 1978. Variability, evolutionary rates, and allometry in dwarfing lineages. Paleobiology 4:101-119.

Martin, P. S., 1967. Prehistoric overkill. In Martin, P. S., and Wright, H. E., Jr. (eds.), Pleistocene extinctions: The search for a cause, New Haven: Yale Univ. Press, 75-120.

Martin, P. S., and Wright, H. E., Jr., eds., 1967. Pleistocene extinctions: The search for a cause. New Haven: Yale Univ. Press, 453.

Niklas, K. J., Tiffney, B. H., and Knoll, A. H., 1980. Apparent changes in the diversity of fossil plants. Evol. Biol. 12:1-89.

Niklas, K. J., Tiffney, B. H., and Knoll, A. H., 1985. Patterns in vascular land plant diversification: An analysis at the species level. In Valentine, J. W. (ed.), Phanerozoic diversity patterns: Profiles in macroevolution, Princeton, N.J.: Princeton Univ. Press and Amer. Assoc. Adv. Sci. (this volume).

Officer, C. B., and Drake, C. L., 1983. The Cretaceous-Tertiary transition. Science 219:1383-1390.

Olsen, P. E., and Galton, P. M., 1977. Triassic-Jurassic tetrapod extinctions: are they real? Science 197:983-986.

Olsen, P. E., and Galton, P. M., in press. A review of the reptile and amphibian assemblages from the Stormberg Group of southern Africa, with special emphasis on the footprints and the age of the Stormberg. Palaeontologia Africana.

Olsen, P. E., Hubert, J. F., and Mertz, K. A., 1981. Eolian dune field of Late Triassic age, Fundy Basin, Nova Scotia: Discussion and reply. Geology 9: 557-558.

Olsen, P. E., McCune, A. R., and Thomson, K. S., 1982. Correlation of the Early Mesozoic Newark Supergroup by vertebrates, principally fishes. Amer. J. Sci. 282:1-44.

Olson, E. C., 1962. Late Permian terrestrial vertebrates, U.S.A. and U.S.S.R. Amer. Philos. Soc. (N.S.) Trans. 52:1-224.

Padian, K., 1982. Macroevolution and the origin of major adaptations: Vertebrate flight as a paradigm for the analysis of patterns. Third No. Amer. Paleontol. Conv. Proc. 2:387-392.

Padian, K. (rapporteur), et al., 1984. The possible influences of sudden events on biological radiations and extinctions. In Holland, H. D., and Trendall, A. F. (eds.), Patterns of change in earth evolution (Dahlem Konferenzen), Berlin, Heidelberg, New York, Tokyo: Springer-Verlag, 77-102.

Panchen, A. L., ed., 1980. The terrestrial environment and the origin of land vertebrates. New York: Academic Press, Systematics Assoc. Spec. 15.

Pitrat, C. W., 1973. Vertebrates and the Permo-Triassic extinctions. Palaeogeog., Palaeoclimat., Palaeoecol. 14:249-264.

Puffer, J. H., Hurtubise, D. O., Geiger, F. J., and Lechler, P., 1981. Chemical composition and stratigraphic correlation of Mesozoic basalt units of the Newark Basin, New Jersey, and the Hartford Basin, Connecticut: Summary. Geol. Soc. Amer. Bull. Pt. 1, 92:155-159.

Raup, D. M., 1976. Species diversity in the Phanerozoic: A tabulation. Paleobiology 2:279-288.

Raup, D. M., 1979a. Size of the Permo-Triassic bottleneck and its evolutionary implications. Science 206:217-218.

Raup, D. M., 1979b. Biases in the fossil record of species and genera. Carnegie Mus. Nat. Hist. Bull. 13:85-91.

Raup, D. M., and Sepkoski, J. J., Jr., 1982. Mass extinctions in the marine fossil record. Science 215:1501-1503.

Reisz, R. R., 1981. A diapsid reptile from the Pennsylvanian of Kansas. Lawrence, Kan.: Univ. Kansas Mus. Nat. Hist. Spec. Publ. 7:1-74.

Robinson, P. L., 1971. A problem of faunal replacement on Permo-Triassic continents. Palaeontology 14:131-153.

Romer, A. S., 1966. Vertebrate paleontology (3rd ed.). Chicago: Univ. Chicago Press.

Romer A. S., 1968. Notes and comments on vertebrate paleontology. Chicago: Univ. Chicago Press.

Ronov, A. B., 1959. On the post-Precambrian geochemical history of the atmosphere and hydrosphere. Geochemistry 5:493-506.

Ronov, A. B., 1982. The earth's sedimentary shell (quantitative patterns of its structure, compositions, and evolution). Internat. Geol. Rev. 24:1313-1388.

Russell, D. A., 1982. The mass extinctions of the Late Mesozoic. Sci. Amer. 246(1):58-65.

Russell, D. A., and Rice, G., eds., 1982. K-TEC II: Cretaceous-Tertiary extinctions and possible terrestrial and extraterrestrial causes. Syllogeus 39: 1-151.

Sadler, P. M., 1981. Sediment accumulation rates and the completeness of stratigraphic sections. Jour. Geol. 89:569-584.

Savage, D. E., 1975. Cenozoic–the Primate episode. Contr. Primatol. 5:2-27.

Savage, D. E., 1977. Aspects of vertebrate paleontological stratigraphy and geochronology. In Kauffman, E. G., and Hazel, J. E. (eds.), Concepts and methods of biostratigraphy, Stroudsburg, P.A.: Dowden, Hutchison, and Ross, 427-442.

Savage, D. E., and Russell, D. E., 1983. Mammalian paleofaunas of the world. Reading, M.A.: Addison-Wesley Publ.

Schopf, T. J. M., 1974. Permo-Triassic extinctions: Relation to sea-floor spreading. Jour. Geol. 82:129-143.

Sepkoski, J. J., Jr., 1981. A factor-analytic description of the Phanerozoic marine fossil record. Paleobiology 7:36-53.

Sepkoski, J. J., Jr., Bambach, R. K., Raup, D. M., and Valentine, J. W., 1981. Phanerozoic marine diversity and the fossil record. Nature 293:435-437.

Sigogneau, D., 1968. On the classification of the Gorgonopsia. Palaeontol. Africana 11:33-46.

Sigogneau-Russell, D., and Ai-Lin, S., 1981. A brief review of Chinese synapsids. Géobios 14:275-279.

Simberloff, D. S., 1974. Permo-Triassic extinctions: Effects of area on biotic equilibrium. Jour. Geol. 82:267-274.

Simpson, G. G., 1944. Tempo and Mode in Evolution. New York: Columbia Univ. Press, 237.

Simpson, G. G., 1952. Periodicity in vertebrate evolution. Jour. Paleontol. 26:359-370.

Simpson, G. G., 1960. The history of life. In Tax, S. (ed.), Evolution after Darwin, Chicago: Univ. Chicago Press, 117-180.

Thomson, K. S., 1976. Explanation of large scale extinctions of lower vertebrates. Nature 261:578-580.

Thomson, K. S., 1977. The pattern of diversification among fishes. In Hallam,
A. (ed.), Patterns of evolution as illustrated by the fossil record, Amsterdam: Elsevier, 377-404.

Van Valen, L., 1973a. A new evolutionary law. Evol. Theory 1:1-30.

Van Valen, L., 1973b. Are categories in different phyla comparable? Taxon
22:333-373.

Van Valen, L., and Sloan, R. E., 1977. Ecology and the extinction of the
dinosaurs. Evol. Theory 2:37-64.

Webb, S. D., 1969. Extinction-origination equilibria in late Cenozoic land
mammals of North America. Evolution 23:688-702.

Chapter 3

PATTERNS IN VASCULAR LAND PLANT DIVERSIFICATION: AN ANALYSIS AT THE SPECIES LEVEL

KARL J. NIKLAS

Division of Biological Sciences, Section of Plant Biology, Cornell University

BRUCE H. TIFFNEY

Peabody Museum of Natural History and Department of Biology, Yale University

ANDREW H. KNOLL

Department of Organismic and Evolutionary Biology, Harvard University

INTRODUCTION

This paper attempts to define the tempo and pattern of vascular land plant diversification for the last 410 million years, and to place the relevant paleobotanical data into apposition with similar studies based on changes in the diversity and faunal composition of marine invertebrates (Bambach, 1977; Raup, 1972, 1976; Sepkoski, 1979, 1981; Sepkoski et al., 1981; Valentine, 1968, 1977). The objectives of this study are to determine (1) if there exists a unique or reiterative pattern in the exploitation by plants of a new habitat—the land surface, (2) if the pattern demonstrated by tracheophytes has parallels with that of the exploitation of the marine environment by a totally different grade of organism—animals, and (3) to suggest proximate causes for observed parallels or differences in these two patterns.

By a factor analytic description of the Phanerozoic marine fossil record, Sepkoski suggested that the adaptive radiation of heterotrophs in the sea could be dissected into the successive appearance of three major "evolutionary faunas": a trilobite-dominated Cambrian fauna, an articulate brachiopod-dominated later Paleozoic fauna, and a mollusc-dominated Mesozoic-Cenozoic or "modern" fauna (Sepkoski, 1981). Sepkoski noted that, with the appearance and expansion of each new fauna, the previous assemblage appears to decline. He attributed this to a competitive ecologic displacement. In its broadest context, the observed Phanerozoic marine fauna pattern has parallels with the appearance and subsequent decline of major Phanerozoic vascular plant groups (Knoll et al., 1979; Niklas et al., 1980; Tiffney, 1981).

Conventional paleobotanical treatments have long recognized that the first tracheophytes, dominated by a free-sporing mode of reproduction, were augmented in Carboniferous-Permian times by a gymnosperm component that dominated through the early Cretaceous, and that by Cretaceous times a new component, the angiosperms, appeared and gradually achieved dominance of the world floras by the late Tertiary. As early as 1954 these three phases in plant evolution were referred to as the Paleophytic, Mesophytic, and Cenophytic ages (Gothan and Weyland, 1954). Quantification of these intuitively determined trends indicated an expansionary and decay phase to each major tracheophyte flora except for the last occurring angiosperm grade (Niklas et al., 1980). Subsequent analyses of the Phanerozoic patterns of the sexual and asexual components to reproduction further suggested that the grade of reproduction may be a significant factor in the observed temporal trends in land plant diversification (Tiffney and Niklas, in press). Collectively, analysis of tracheophyte evolution indicates that what appears to be a complex pattern can be degraded into smaller components.

SOURCES OF DATA AND METHODS OF ANALYSIS

Analyses of land plant diversification patterns are based on the data initially reported by Niklas (1978), Knoll et al. (1979), Niklas et al. (1980), and Tiffney (1981). Since the publication of these reports, the data base for terrestrial plant fossils has been expanded to encompass over 18,000 paleospecies, of which over 80% are vascular taxa. The data for each species of plant is derived directly from a primary literature citation; however secondary sources have been used to canvas the available literature, e.g., *Traité de Paléobotanique*. Under most circumstances, attempts were made to correct for synonomies resulting from organ genera. In many cases, however, there were reasons to suppose that the morphologic basis for the recognition of an organ species was in fact the recognition of a legitimate paleospecies. Wherever possible form genera were ignored for purposes of computing species numbers (cf. Niklas et al., 1980).

One serious limitation of the data results from the fact that the fossil localities representing the majority of the data are from North America, Europe, and Russia. Thus, during the Cretaceous-Tertiary many "diversification" trends may reflect biogeographic biases, e.g. the pteridophyte graph in Figure 3 shows a gradual decline through the K-T in spite of our knowledge that ferns rapidly diversified in tropical and neotropical latitudes. Analyses of the data for time dependent and time independent biases are discussed by Niklas et al. (1980).

In addition to megafossil data, the following references were used to gather information on the species- or genus-origination rates of spores or pollen: Chaloner (1967), McGregor (1977, 1979), Muller (1970); Richardson (1974).

RATES OF ORIGINATION

A number of workers in paleontology have argued that species have evolutionary properties that are analogous to individuals and as such are amenable to analyses similar to those used in classical population ecology: (1) Eldredge and Gould (1972), Stanley (1975), and Gould and Eldredge (1977) have argued that species are discrete and fairly static entities with relatively constant or moderately changing morphologies during their durations; (2) Van Valen (1973) has used life-table techniques to analyze taxonomic extinction; (3) Stanley (1979) has borrowed from demography the standard exponential equation for population growth to compute rates of radiation for different taxa; and (4) Sepkoski (1978) has assumed that a simple exponential model of diversification is applicable to a variety of radiations in the fossil record.

For the present report, the rate of diversification, R, can be taken as the difference between r_s and r_e, such that

$$r_s = \frac{1}{D} \cdot \frac{S}{\Delta t} \qquad \text{Equation 1}$$

$$r_e = \frac{1}{D} \cdot \frac{E}{\Delta t} \qquad \text{Equation 2}$$

$$R = r_s - r_e \qquad \text{Equation 3}$$

where S is the number of originations, E is the number of extinctions, and D is the diversity per interval of time, t (cf. Sepkoski, 1978:225). In an earlier treatment of tracheophyte diversification, a more complex equation for diversification, based on stochastic probability functions was used (cf. Niklas et al., 1980:17-18). Although this equation was appropriate, given the context of resolving random versus non-random components of the data set, the complexity of this function was deemed inappropriate for the data presented here. The values for species-origination rates, R, given in Table 1 are based, therefore, on Equations 1-3. These R-values can be compared directly to those published for mammals and bivalves (Stanley, 1979:106-107) and to

Table 1. Species duration (D) and origination rates (R) for various
tracheophyte groups.

	D(Myr)	R(Myr $^{-1}$)
EARLY DEVONIAN PLANT GROUPS		
Rhyniophytina	14.1(5.8)*	0.04*
Trimerophytina	6.7*	0.10*
Progymnospermophyta		
Archaeopteridales	14.8	0.16
Aneurophytales	14.7	0.16
Pteridophyta		
Rhacophytopsida	8.2	0.17
Coenopteridopsida	8.0	0.18
Sphenophyta	10.1	0.08
Zosterophyllophytina	11.8*	0.07*
Lycophytina		
Protolepidodendrales	16.7	0.08
Lepidodendrales	18.8	0.08
CARBONIFEROUS-PERMIAN PLANT GROUPS		
Lycophyta		
Protolepidodendrales	18.3	0.05
Lepidodendrales	18.8	0.05
Lycopodiales	28.3	0.04
Selaginellales	10.2	0.06
Pteridophyta		
Coenopteridopsida	8.0	0.11
Marattiales	5.9	0.14
Filiales		
Osmundaceae	7.3	0.15
Tedeleaceae	7.8	0.13
Sphenophyta		
Sphenophyllales	8.2	0.14
Equisetales	10.7	0.04
Pteridospermophyta		
Lyginopteridales	4.8	0.12
Medullosales	4.0	0.08
Callistophytales	3.5	0.15
Calamopityales	4.5	0.10
Cordaitopsida		
Cordaitales	7.2	0.13
Cyacodophyta	10.9	0.12
Cycadeoidophyta	10.0	0.13
Ginkophyta	5.0	0.17
Coniferopsida		
Voltziales	4.0	0.20

Table 1 (continued).

TRIASSIC-JURASSIC PLANT GROUPS

Pteridophyta		
Filicales		
Osmundaceae	8.2	0.11
Schizaeaceae	5.2	0.12
Gleicheniaceae	5.0	0.12
Cyatheaceae	5.0	0.12
Polypodiaceae	7.0	0.13
Pteridospermophyta		
Caytoniales	5.6	0.11
Corystospermales	6.2	0.09
Peltaspermales	6.3	0.09
Glossopteridales	6.0	0.10
Cycadophyta	15.0	0.10
Cycadeoidophyta	15.3	0.09
Ginkophyta	7.2	0.14
Coniferopsida		
Voltziales	6.0	0.13
Coniferales	3.8	0.16
Taxales	4.2	0.14

CRETACEOUS-TERTIARY PLANT GROUPS

	pollen	seed-fruit	pollen	seed-fruit
Pteridophytes				
Filicales				
Osmundaceae	9.0		0.08	
Schizaeaceae	6.0		0.08	
Gleicheniaceae	6.0		0.09	
Cyatheaceae	6.0		0.09	
Polypodiaceae	8.0		0.10	
Cycadophyta	15.0		0.08	
Ginkgophyta	10.0		0.08	
Coniferopsida				
Coniferales	5.3		0.10	
Taxales	5.0		0.11	
Anthrophyta				
Monocots	7.3	6.3	0.33	0.37
Araliaceae**	6.3	5.2	0.26	0.29
Cyperaceae	6.5	5.0	0.26	0.25
Hydrocharitaceae	8.9	7.4	0.05	0.09
Potomogetonaceae	9.0	6.7	0.05	0.08
Dicots	3.8	3.8	0.26	0.47
Compositae**	–	1.7	–	0.52
Umbelliferae	–	1.8	–	0.61
Aceraceae	4.3	4.1	0.09	0.14
Fagaceae	4.3	5.3	0.04	0.15

*   Based on genera rather than species.
** Representative families.

predictions based on a kinetic model of taxonomic diversity (cf. Sepkoski, 1978). Various sources were used to determine species durations. Usually taxonomic ranges were computed directly from primary citations. For Devonian taxa, we used the data supplied by Banks (1980) and Chaloner and Sherrin (1979).

THE DEVONIAN RADIATION OF VASCULAR PLANTS

The diversification of the early vascular land plants during the Late Silurian and Devonian shows a relatively simple temporal pattern of variation (Figure 1). The oldest vascular plant group, the rhyniophytes, has been reported from several horizons through the Pridolian of Wales but earlier, Middle Ludlovian, material may yet yield vascular tissues (cf. Edwards et al., 1979). The Rhyniophytina disappeared as a recognizable group of plants in the Givetian (Figure 1). By the Gedinnian-Siegenian, three other major plant groups appeared: the Trimerophytina, Zosterophyllophytina, and Lycophytina (Banks, 1968; Gensel, 1977). All of these suprageneric vascular plant taxa, save for the lycopods, disappeared by the Fammenian. The lycopods diversified rapidly during the Givetian and became a major component of the subsequent Carboniferous swamp floras. During the Givetian, other plant groups appeared which later became increasingly diverse during the Carboniferous—pteridophytes (ferns), sphenopsids (articulates), and progymnosperms. The Fammenian marks the appearance of the first seed plants, a grade of organization that would dominate the world flora from the Permian through to the Tertiary.

This spurt of adaptation and diversification to life on land in the Devonian quickly brought plants of different genetic lineages into close contact, initiating competition for resources including light, soil nutrients and soil moisture. The last resource is particularly important, as all of these early land plant groups were limited to areas with sufficient moisture to permit sexual reproduction by swimming sperm, and because of the (at least initially) poorly-developed vascular systems of the sporophytes. The earliest land plants were presumably very small and erect, or they were prostrate. With the evolution of vascular tissue, many of their successors developed a rhizomatous habit (Tiffney and Niklas, in press). In its simplest form, an early land plant community was composed of a single stratum of low rhizomatous plants, each occupying its own, frequently extensive, piece of ground. However, the development of vascular supporting tissues, together with competition for light and the selective value of increased efficiency of spore dispersal, led rapidly to the appearance of taller and larger plants. The mode of origin of taller, non-rhizomatous plants is conjectural (Tiffney and Niklas, in press), but they clearly

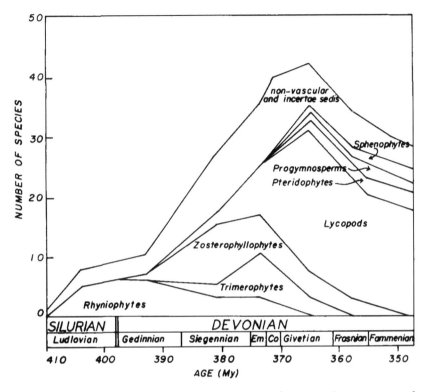

Figure 1. Changes in the diversity (at the species level) of suprageneric taxa represented in the Silurian-Devonian. The rhyniophytes, trimerophytes, and zosterophyllophytes collectively define Factor I, while lycopods and other plant groups mark the appearance of taxa ascribable to Factor II. The upper region of the graph, labelled "non-vascular and *incertae sedis*" include thallophytes, bryophytes, and vascular taxa presently unassigned to any of the recognized tracheophyte groupings.

diversified such that by the Late Devonian, the beginnings of a multi-storied plant community were established, providing a great increase in the total number of plants which could occupy a given unit of space. As a direct result of this increasing adaptation to land, competition became refined, and while the selective pressures of plant vs. environment remain important to the present day, those resulting from competition between plants become dominant. The ascension of biotic competition is central in the evolution and canalization of new morphologic and reproductive characters, and thus of new lineages. We suspect that the jump in diversity from the Devonian into the Carboniferous is a direct reflection of the diversification of habit types and the rise of biotic competition, although clearly other factors may have been of significance.

It is possible to present a crude estimate of the evolutionary tempo of these lineages, based upon values for rates of species-origination and longevity. The mean duration (D) and mean origination rates (R) for some of the early Devonian plant groups are given in Table 1. The rhyniophytes have mean generic and species durations of 15 and 14.1 Myr, respectively. The mean generic duration is significantly reduced to 5.8 Myr, if *Cooksonia* (D=34 Myr) and *Taeniocrada* (D=45 Myr), which may be form genera, are removed from the sample set. The mean generic and species-origination rate of the rhyniophytes are computed to be 0.04 and 0.06 Myr^{-1}, respectively. The maximum diversity of this group is seen in the Emsian (371 Myr) with a subsequent decline in percent contribution to the Devonian flora, owing to the appearance of other suprageneric groups (Table 1). Computation of R- and D-values for the trimerophytes, zosterophyllophytes, early lycopods, and subsequent Upper Devonian plant groups indicates a general trend of increasing R-values which is interpreted as evidence for an accelerated and rapid diversification of the vascular flora.

The apparently low R-value (0.04 Myr^{-1}) for the rhyniophytes may reflect an inability to identify and therefore include more complex, directly derived taxa within this anestral line, which on the basis of their morphology would be systematically assigned to other plant groups such as the trimerophytes. Such a phenomenon would be the suprageneric equivalent to pseudoextinction. If the rhyniophyte-trimerophyte taxa are grouped together, then an R-value comparable to the zosterophyllophyte-lycopod data set is computed (0.08 Myr^{-1}).

Analyses indicate that the shortest lived species and genera within a lineage occur in the earliest phases of its diversification. Typically, over 50% of the species occurring in the first 10 to 25 Myr of the Devonian lineages examined have durations of 1 to 8 Myr. As the subsequent diversification of a lineage is followed, the longevity of its species increases, until the last surviving species typically have durations of 16 to 23 Myr. A similar pattern is seen when genus durations are examined. With the exception of the rhyniophytes and progymnosperms, the durations for the first occurring genera are from 6 to 8 Myr, while subsequent genera have longevities ranging from 16 to 26 Myr.

RADIATION OF DERIVED PLANT GROUPS

By the Givetian (370 Myr), the zosterophyllophyte-derived lycopods are joined by three other derived plant groups: (1) the Pteridophyta, represented by the Rhacophytopsida and Coenopteridopsida (primitive fern-like plants), (2) the Sphenophyta or the articulates, from which the arborescent *Calamites* and the rhizomatous *Equisetum* are presumably derived, and (3) a group of free-sporing plants with gymnosperm anatomy, the Progymnospermophyta

(Figure 1). Both the early Pteridophyta and progymnosperms are initially characterized by relatively high species-origination rates (0.17 to 0.18 and 0.16 Myr^{-1}, respectively; cf. Table 1). In terms of species numbers, these derived plant groups, however, contributed less than 25% of the Upper Devonian standing diversity. However, together with the lycopods, these lineages mark the appearance of a floristic association that proliferated and dominated the Carboniferous and Permian landscapes.

SPORE VERSUS MEGAFOSSIL DURATION AND ORIGINATION RATES

The R- and D-values for Devonian spore data provide additional insights into the tempo of early land plant diversification. Based on the data from Chaloner (1967), the mean duration for Devonian spore genera is 25.9 ± 15.1 Myr (n = 73). This value, however, varies depending upon the source of the data: consultation of McGregor (1977; n = 79) provides values of 11.4 ± 10.6 Myr, and of Richardson (1974; n = 44), values of 16.3 ± 14.0 Myr. Various biases enter into these computations. For example, Chaloner gives the duration of a spore genus on the basis of first and last occurrence, even if no intermediate occurrence is known. Similarly, McGregor (1977, 1979) specifically lists spore taxa judged to be of stratigraphic value, biasing the data toward species with short durations. Even so, the D-values computed for spore genera and species appear to be slightly higher than D-values derived from comparable megafossil data. The species-origination rate for Devonian spores is 0.05 Myr^{-1}, which stands in contrast to the cumulative data from Devonian vascular plant species based on megafossils (R = 0.09 Myr^{-1}). On the surface, the D- and R-values for spore and megafossil species suggest that the rate of miospore evolution was slower than the rate of morphologic and anatomic changes in vascular sporophytes. However, very little is known about the taxonomic affinities of the spore taxa reported here, since studies of *in situ* spores are limited (cf. Gensel, 1980). Further, the simplicity and number of available characters in spores, as contrasted to those available for megafossils, may result in the recognition of far fewer miospore "species" than megafossil "species"—an analogous condition to that of clam and mammal species determinations (cf. Schopf et al., 1975).

THE CARBONIFEROUS-PERMIAN RADIATION

A number of major vascular plant groups, which trace their ancestry back to the rhyniophyte-trimerophyte and zosterophyllophyte lineages, appeared by the end of the Devonian and rapidly radiated during the Carboniferous (Figure 2). After the Devonian-Carboniferous transition the terrestrial flora increased in species numbers by a factor of 5. The lycopods, already well represented by Upper Devonian times, proliferated and reached their maximum diversity in the Late Carboniferous-Lower Permian coal swamps. Similarly, the ancestral plexus of the ferns (Rhacophytopsida and Coenopteridopsida), the earliest articulates (Sphenophyta), and seed plants (Pteridospermophyta) show an increase in the Upper Devonian and Lower Carboniferous. These latter groups, representing less than 20% of the Upper Devonian vascular flora, rose in species numbers to account for over 50% of the known Westphalian and Stephanian taxa. The Upper Devonian-Lower Carboniferous also marks the appearance and probable demise of the progymnosperms, a group of free-sporing plants with gymnosperm-like anatomy. Based on their relatively high rates of origination, this group may have undergone a pattern of pronounced morphologic and anatomic diversification, similar to the rhyniophytes, culminating in their pseudoextinction (Table 1). Such an hypothesis is consistent with the speculation of Beck (1981), who would derive the cordaites, lebachiacean conifers, and the pteridophytic Noeggerathiopsida from this stock of plants.

The Carboniferous is the time when most of the major plant groups represented in the Devonian developed specialized arborescent and rhizomatous lines (Tiffney and Niklas, in press). Within the Lycophyta, the protolepidodendralean-derived Lepidodendrales achieved heights in excess of 30 meters, while the Lycopodales and Selaginales continued on with an apparently archaic rhizomatous habit. Similarly, the Sphenophyta produced arborescent (Calamitales) and rhizomatous (Equisetales) taxa. Amongst the seed plants, various pteridosperms developed secondary growth, resulting in shrubby or liana-like plants, while the cordaites and Voltziales attained a tree habit. Even in the pteridophytes, modifications of growth patterns, and the frequency and position of adventitious roots resulted in the attainment of tree-like plants (*Psaronius*), although the majority of ferns remained rhizomatous.

The Carboniferous marks the rapid diversification of seed plants. By the Carboniferous-Permian, the Calamopityales is joined by four other orders of seed-ferns (Lyginopteridales, Medullosales, Callistophytales, and Glossopteridales), as well as the Cordaitopsida and the voltziacean conifers (Table 1; Figure 2). Owing to the retention of the megagametophyte within sporophytic tissues and a release from the necessity of growing near free-standing water, the appearance of the seed-habit may have provided the basis for the diversification of an "upland flora." This flora may be poorly represented in the fossil record, and the species richness of the Carboniferous-Permian given in Figure 2 may represent a conservative estimate, based primarily on plants growing in or near swampy environments.

If the seed was as much of an advance as it is assumed to be, why did gymnosperms, which evolved in the Late Devonian, form such a small portion of

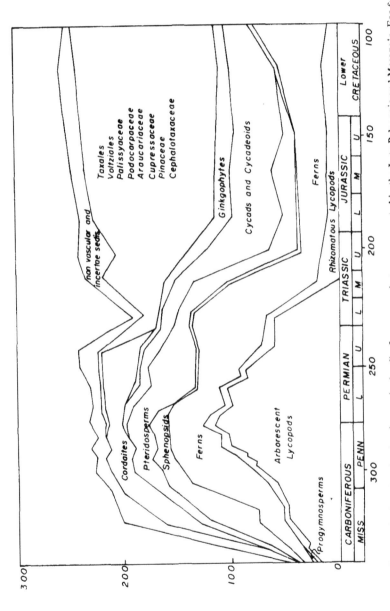

Figure 2. Changes in the diversity (at the species level) of suprageneric taxa represented in the Late Paleozoic and Mesozoic. For further details, see text.

Carboniferous diversity? Our answer to this is speculative: (1) Although the seed allowed the exploitation of upland or drier environments, the success of the seed plants in such habitats is predicated on vegetative characters that permit the survival of sporophytes. The earliest seed plants (pteridosperms) are characterized by pteridophyte-like vegetative morphology, and as such may have been limited to mesic and hydric habitats. Until non-reproductive and reproductive features evolved to act in concert, the upland gymnosperms may have been restricted in their exploitation of mesic or xeric environments, while lowland mesic taxa formed a significant but low diversity group in the early Mesozoic. (2) The reproductive mode of gymnosperms requires extensive stands of individuals for efficient reproduction. One reason for the low diversity of species numbers in the Carboniferous to Jurassic time period may have been that "packing" of gymnosperm taxa, essential for sexual reproduction, reduced alpha diversity. (3) The apparently low species diversity of gymnosperms in the late Paleozoic may be an artifact of preservational biases favoring lowland (swamp) communities.

The dynamics of floristic change associated with the Carboniferous-Permian radiation can be gauged in part by the mean durations of taxa and their rates of species-origination (Table 1). Although the lycopods represented the majority of Carboniferous and Lower Permian species (30% in the later Mississippian; 50% in the late Pennsylvanian), they show in general the lowest rates of species-origination (0.05 Myr^{-1} for rhizomatous and arborescent orders) and longest durations (18.6 Myr and 22 Myr, respectively), suggesting that they had peaked in their diversification. Ferns show higher R-values and shorter D-values than lycopods; however, the seed plants have the highest species-origination rates and shortest durations (Table 1). Within the seed-ferns (=pteridosperms), the Callistophytales had the shortest mean durations of species (3.5 Myr) and the highest origination rate (R=0.15 Myr^{-1}). The Voltziales, an upper Permian group of seed plants, had the highest R-value among any gymnosperm lineage represented in this time frame (R=0.20 Myr^{-1}).

Trend analyses reveal a relatively simple pattern of changing D- and R-values through the tenure of specific lineages. The earliest represented species within the arborescent lycopods (Lepidodendrales) and sphenophytes (Sphenophyllales) characteristically have short durations (~ 8 and ~ 6 Myr, respectively) and higher species-origination rates (0.11 and 0.16 Myr^{-1}, respectively) than do subsequently appearing taxa. The greatest acceleration in the origination of arborescent lycopods occurs during the Pennsylvanian, while the arborescent sphenophytes show their greatest species-origination rates in the Upper Mississippian (Figure 2). The ferns, seed-ferns, and cordaites show a similar pattern through the Carboniferous. With the appearance of some upper Carboniferous and Permian groups (Osmundaceae and Tedeleaceae) the ferns show periods of accelerated speciation and a shortening in their species durations. Although the cordaites and seed ferns increase in floristic importance during the middle Permian, no statistically significant alteration in their species-origination rates is observable. Presumably, the decline in the absolute number of lycopod and fern species may account for the observed differences in the Permian and Carboniferous floristic compositions.

The Permian marks a major change in the world floras. The aborescent lycopods decline in species number (a decrease in R-values and an increase in extinction rates) and the arborescent sphenopsids disappear (Figure 2). These two ancient plant groups continued to be represented, however, by rhizomatous taxa. The Permian also marks the appearance of a number of new seed plant lineages, which include the cycads (Cycadophyta), cycadeoides (Cycadeoidophyta), Vojnovskyales (data not presented) and Ginkgophyta (e.g., *Trichopitys* and *Sphenobaiera*). These new plant groups reiterate the pattern of having initially high species-origination rates (R) and short species durations.

TRIASSIC-JURASSIC RADIATIONS

The early Mesozoic is characterized by a major shift in the composition of floras associated with the adaptive radiations of various seed plant lineages. Over 60% of the plant species known from Middle Triassic strata are gymnosperms, and by the Middle Jurassic this number rose to 80% (Figure 2). Unlike the ascension of the preceding pteridophytic flora during Devonian-Carboniferous times, the Mesozoic change in species composition was not accompanied by an appreciable increase in absolute species numbers—the total species diversity of tracheophytes increased by only 9% from the Permian to the Jurassic. Thus the rise of the gymnosperms is associated with a decline in pteridophyte numbers. Although the Permo-Triassic marks a significant and relatively abrupt change in total land plant numbers (a decline of 20%), the beginnings of the decrease in pteridophyte species can be traced back well into the Lower Permian with a concomitant rise in seed-plant species (Figure 2).

The Triassic marks the extinction of the arborescent lycopods and the cordaites, and a precipitous decline in some fern families. Various pteridosperm lineages and the ginkgophytes increase in species diversity, but the main thrust of adaptive radiation occurs within the Coniferophytes (Voltziales and Coniferales), and the cycads and cycadeoids (Figure 2). These seed plants account for 60% of the world's flora by the Triassic-Jurassic boundary, and are characterized by the possession of significant amounts of secondary wood and the attainment of tree or shrub growth habits.

A *basso profundo* to the gymnosperm theme of the Mesozoic is the continuation from Carboniferous times of various pteridophyte lineages (Osmundaceae, and rhizomatous lycopods and sphenophytes). Although the pteridophytes maintain a relatively constant species number throughout the Upper Triassic-Jurassic (18%), they exhibit significant differences in turnover and interlineage species-origination rates. For example, by the Upper Triassic-Jurassic the ferns are largely represented by new families or Orders (Gleicheniaceae, Schizaeaceae, Cyatheaceae, Polypodiaceae, and Ophioglossales) whereas more ancient fern groups are either extinct (Coenopteridopsida, Botryopteridaceae, Anachoropteridaceae, and Tedeleaceae) or declining in species diversity (Marattiales). Thus the relatively "static" appearance of the "fern," lycopsid, and sphenopsid portions of the graph for the Jurassic shown in Figure 2 is misleading and represents only the cumulative species numbers for an internally, dynamically changing group of plants.

Although most of the early Mesozoic lineages have origination rates ranging from 0.09 (pteridosperm Orders and the cycadeoids) to 0.16 Myr^{-1} (Coniferales), significant changes from preceding origination rates occur. For example, the origination rates of the Mesozoic cycads and cycadeoids (0.10 and 0.09 Myr^{-1}, respectively) have significantly slowed down in comparison to their Permian counterparts. The increase in the total species numbers observed for the Mesozoic is attributable to an increase in their mean duration of species (from $\sim$ 10 Myr in the Permian to 15 Myr in the Jurassic). Similarly, among the Triassic-Jurassic pteridosperms and ginkgophytes, there is a significant decline in species-origination rates and an increase in mean duration compared to Carboniferous-Permian taxa. A similar trend of "rate deceleration" and "increased longevity" is seen for some fern families (Table 1). Once again, the Mesozoic appears to show a reiteration of a now familiar theme—within specific lineages, the first taxa to appear are short lived and speciate rapidly, whereas subsequent groups are longer lived and have slower species-origination rates. Even within the Coniferopsida, some of the more ancient groups, such as the Voltziales, appear to slow down. New Orders within the Coniferopsida, such as the Coniferales and Taxales, show an initially rapid rate of speciation (0.16 and 0.14 Myr^{-1}, respectively) and contain species with the shortest durations ($\sim$ 4 Myr).

The Lower Cretaceous is characterized by a pronounced decline in the number of cycad species and the extinction of the contemporaneous cycadeoids, a change which is numerically "absorbed" by an increase in the number of fern species (Figure 2). Although the Coniferopsida appear to be numerically stable, there are significant changes in the dynamics at the family level—the Palissyaceae go to extinction (Upper Jurassic) and the Pinaceae, Taxodiaceae, and Taxales demonstrate higher species-origination rates and shorter durations.

Currently, the oldest angiosperm megafossils are known from Early Cretaceous strata of Barremian age. Their apperance marks the beginning of a major alteration in the terrestrial flora, which led to the dominance of flowering plants and a perceptible decline in the species numbers of gymnosperms and pteridophytes (Figure 3).

During the Mesozoic and early Cenozoic various gymnosperm lineages declined in species numbers. Most of the seed ferns became extinct by the end of the Triassic, with the Caytoniales surviving until the Early Cretaceous. The cycadeoids were represented by significant species numbers in the early Cretaceous but by the end of the Cretaceous they too became extinct. Similarly, the cycads begin to show a pronounced decline in species number. The deterioration in the diversity of these gymnosperm lineages is noticeable well before the Barremian appearance of angiosperms, and is correlated with a concomitant increase in other gymnosperm taxa, such as the Taxales and Coniferales (Figure 2). Thus, significant alterations in the relative abundance of gymnosperm lineages occurring in the Early Cretaceous are not a priori ascribable to the adaptive radiation of the angiosperms.

Underlying the fluctuations in the relative species abundance of gymnosperm lineages in the Cretaceous and Tertiary is a relatively stable consortium of pteridophytes. The sphenopsids, lycopsids and ferns continue at a low and fairly constant level of diversity through the early Tertiary, whereupon they begin a gentle decline (Figure 2), perhaps in response to the spread of herbaceous angiosperms. As before, considerable turnover exists among the various fern lineages. The Marsileales and Salviniales, two aquatic orders of ferns, appear for the first time in Tertiary and Cretaceous times, respectively, and rapidly diversify, plateauing during the Oligocene. The Polypodiaceae appear to gain dominance among the major fern families during the Oligo-Miocene; however, this trend may be more an artifact of our over-representation of non-tropical fossil localities.

CRETACEOUS RADIATION OF THE ANGIOSPERMS

Between the Lower Cretaceous and Paleocene, the total standing diversity of land plants increased by 45%, of which over 40% were angiosperm taxa. By the later Neogene, the flowering plants constituted almost 80% of all known vascular plant fossils (Figure 3).

The increase in angiosperm species number was initially a gradual process, and the shape of the species number versus time plot is similar to those observed in the initial radiations of other major plant groups. What distinguishes the angiosperms from other plant groups, however, is the rapid rate at which their species number increased (an order of magnitude higher than preceding plant groups). The rise in angiosperm diversity had two major effects on the

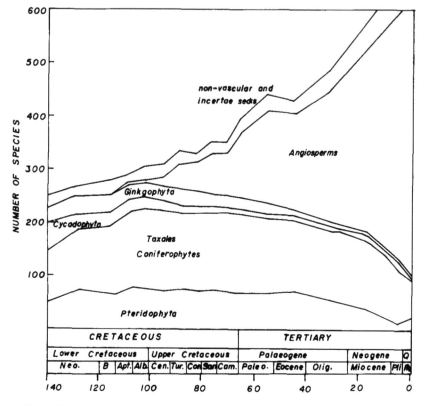

Figure 3. Changes in the diversity (at the species level) of suprageneric taxa represented in the Late Mesozoic and Cenozoic. Owing to the sources of data (predominantly from North America, Europe and Asia), these graphs more properly reflect changes in plant diversification for the Northern Hemisphere. For further details, see text.

species composition of the world floras: (1) a rapid increase in species numbers, the bulk of which is attributable directly to the proliferation of the angiosperms, and (2) a decline in the numbers of other plant lineage species. This observed decline is in part an effect of dilution resulting from the vast numbers of angiosperm fossils found in some Tertiary deposits, as well as a *bona fide* reflection of competitive exclusion. Unfortunately, a precise quantification of the contributions made by both these phenomena is not possible; however the effects of both are noticeable in Figures 2 and 3.

The changes in angiosperm species numbers from the Early Cretaceous to the recent may be analyzed at various taxonomic levels: (1) comparisons among the rate of species-origination and species durations of the angiosperms as a whole and other vascular plant groups, (2) comparisons among the various orders of monocots and dicots, and finally, (3) comparisons among the various families of angiosperms (Table 1). Similarly, rates of species-origination and species diversity may be computed for pollen, leaf, and fruit/seed taxa, providing independent computations based on three organs.

The angiosperms as a whole demonstrate the highest speciation rates of all the plant groups examined. Based on seed and fruit data, average R-values range from $0.37 \, Myr^{-1}$ for monocots to $0.46 \, Myr^{-1}$ for dicots—values that are on the order of 3 to 4 times that of the most rapidly speciating gymnosperms (Taxales, $R=0.11 \, Myr^{-1}$; Coniferales, $R=0.10 \, Myr^{-1}$). The mean species durations for all the angiosperm taxa examined are on the order of 5 Myr—a value comparable to other seed plant groups (Voltziales and pteridosperm lineages, $D=4$ to 5 Myr). However, the average duration can be as low as 1.7 Myr in some families of herbaceous angiosperms. Collectively, the high mean speciation rates and short durations set the angiosperms apart from all other vascular plant groups.

Significant differences exist among the mean speciation rates and species durations within the various orders of angiosperms. Although the mean speciation rate for a dicot order is $0.26 \pm 0.32 \, Myr^{-1}$, some orders yield very low values: Trochodendrales ($0.11 \, Myr^{-1}$), Myricales ($0.11 \, Myr^{-1}$), and Hammamelidales ($0.08 \, Myr^{-1}$). Orders including the highest speciation rates are the Capparales ($1.04 \, Myr^{-1}$), Geraniales ($2.19 \, Myr^{-1}$, and Plumbaginales ($2.44 \, Myr^{-1}$). Significantly, these relatively low and high speciation rates are associated with long and short species durations, respectively.

Figure 4 shows the relationship between the mean species duration and mean species origination rates for representative families of angiosperms. Those families showing the highest speciation rates (Umbelliferae, Compositae, and Labiatae, $R=0.4$ to $0.6 \, Myr^{-1}$) typically show short durations (~2 Myr), while families with low speciation rates (Aceraceae, Fagaceae, Juglandaceae, Moraceae, Hydrocharitaceae, and Actinidiaceae, $R \cong 0.2 \, Myr^{-1}$) show a broad range in mean species durations (4 to 9.4 Myr).

The mean species duration and origination rates given for families in Figure 4 are based exclusively on seed and fruit data. However, comparable calculations can be derived from the fossil record of pollen ascribable to each of the families shown. Linear regression analysis of the pollen and seed/fruit mean speciation rates data yield a coefficient of correlation of 0.85. The average duration for pollen for a family equals 34.9 Myr, while that for an order equals 43.5 Myr.

Some broad ecologic, chemosystematic, and anatomical trends can be seen within the angiosperm families shown in Figure 4. Those families with the highest speciation rates (0.4 to $0.6 \, Myr^{-1}$) are typically dominated by weedy, north temperate species characterized by the production of specific rather than generalized allelotoxins (e.g. Compositae, Umbelliferae; Labiatae).

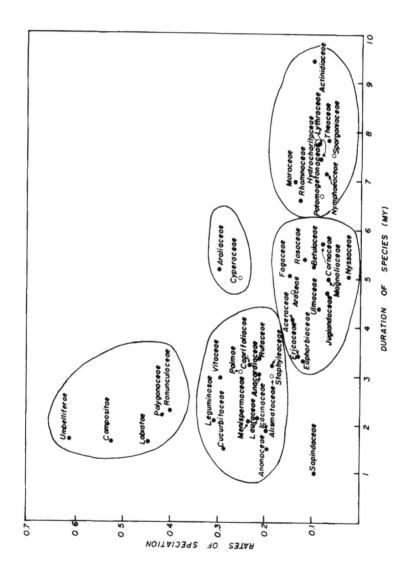

Figure 4. Plot of mean species-origination rate versus mean species duration for representative families of angiosperms (● - dicotyledonous families; ○ - monocotyledonous families). The solid lines drawn around groups of families are subjective. For further details, see text.

Among the families showing the lowest mean speciation rates ($\sim$0.1 Myr^{-1}) and longest durations (4 to 9.4 Myr) are the hydrophytes (Hydrocharitaceae, Potomogetonaceae, and Sparganiaceae), which are characterized by taxa typically showing extensive vegetative propagation and limited habitat. Families with intermediate R- and D-values (R=0.2 to 0.3 Myr^{-1} and D=2 to 4 Myr) are largely tropical woody trees or lianas (e.g. Anacardiaceae, Annonaceae, Vitaceae, and Curcurbitaceae). The significance of various anatomic, reproductive and biogeographic features as they relate to the apparent speciation rates for particular family groups is a topic too broad to cover within the rubric of the present report. However, a cursory examination of the biology of the various angiosperm families studied suggests that the pollination mechanism, growth habit, and chemical defense system have played major roles in defining the relative *rate* influenced by innate characteristics (karyotypic variation) *and* rate of environmental change.

MODES OF PLANT DIVERSIFICATION

Knowing the pattern of tracheophyte diversification tells us very little about the processes or modes involved. Two areas that have bearing on the "mode" of plant diversification are (1) ecologic theory as it relates to diversification and species equilibrium, and (2) the genetic mechanisms for speciation and the factors that relate to the apparent rates of speciation calculated for the various lineages or groups of plants. The questions that can be asked of ecologic theory are "Do the trends seen in the fossil record of land plant diversification conform to those observed for living plants?", and, if they do, "What are the extrinsic factors that influence species diversification?" In essence, what we wish to address is the issue of compatability between the trends seen in the fossil record and those either predicted or observed in living plant communities and biotas. Questions that may be addressed to the literature pertaining to genetics are "What are the intrinsic factors that dictate potential rates of speciation?" and "Are there differences in the life history strategies of various plant groups that can accelerate or decelerate speciation?"

EQUILIBRATION OF SPECIES NUMBERS

The concept of an indefinitely increasing species diversity has been criticized as counterintuitive (Strong, 1974; Rosenzweig, 1975), while that of species equilibration has been explored owing to the possibility for modeling, predicting, and interpreting diversity. Debate on the generality of species equilibration has focused on five levels of geographic scale and time span: (1) succession and climax, (2) component communities, (3) islands, (4) biosphere diversity, and (5) major taxa in geologic time. In his review of each of these five levels of species diversity, Whittaker concluded that "For a given taxon, diversity increase slows, given sufficient evolutionary time; but the evolution of new groups capable of new elaboration of ecotopes permits diversity increase" (1977:53). Similarly, Whittaker concluded that "Evidence and theory are against the general occurrence of species equilibria on the evolutionary scale" (1977:52). Thus neobotanical ecologic data and theory are consistent with the observation that the total number of vascular land plant species increases progressively through the Phanerozoic.

The mechanism permitting the non-occurrence of species equilibrium can be understood by recourse to niche-theory. The principle of competitive exclusion argues that no two species can occupy the same niche in a stable community. However, new species may be added to even a stable community by one of two ways: (1) by evolving into positions of resource use that are extreme or marginal for the species already present, or (2) by evolving toward an intermediary resource use to those of other species. Respectively, these modes result in (1) the "extension" of a pre-existing gradient in the niche hyperspace, and (2) the "intercalation" of a new niche among existing ones. Either or both of these modes may result in an increase in species diversity; however, the quantitative contribution of each to the diversification of specific evolutionary lineages or floras must remain speculative. It is safe to assume that with an evolutionary innovation that permitted the exploitation of a new resource or habitat, as for example the colonization of land by the first land plants (and later the potential to expand into more mesic conditions with the advent of the seed), the primary mode of diversification was initially one of niche "extension." With sufficient time, however, the "intercalation" mode would become dominant resulting in a stable but diversifying community structure. Some evidence for this can be gleaned directly from the fossil record. The average species number per flora appears to increase with time through the Devonian (Niklas et al., 1980, Table 1). Similarly, the number of individual taxa found on exposed surfaces of rock strata increases through geologic time. Morphologic complexity also increases and provides evidence that communities gradually acquired the capacity for greater vertical stratification and attendant niche-partitioning. Thus the increase in total tracheophyte species number may be interpreted ecologically as the "fragmentation" of niche hyperspace into progressively smaller niche-volumes. This scenario is testable, since it predicts that, within limits, we should see a progressive trend in the history of plants from "generalists" to "specialists."

The slowing down in the apparent diversification within a lineage or evolutionary flora may be the result of: (1) niche hyperspace saturation, i.e., with sufficient time, the limits of the morphologic and reproductive versatility of the group may be reached, or (2) the appearance of a new group of plants, characterized by a reproductive, morphologic, or physiologic innovation, would result in a displacement of the preceding flora. (This would be the macroevolutionary equivalent to competitive exclusion at the species level.) The empirically determined decline in the species numbers of some evolutionary floras, with the advent and adaptive radiation of newer floras, is compatible with the second of these two explanations.

Competition among plants is of a type in which there is no standardization of resource use and is therefore quite different from that commonly found in animals (Nicholson, 1958; Whittaker, 1977). A sexually mature animal has a specific quota of food and space necessary for survival which, when met, usually results in non-competitive behavior. By contrast, a given plant, owing to its indeterminate growth, utilizes as much water, inorganic nutrients, and space (relevant to photosynthesis) as can be obtained during the course of its growth. Demands on resources, even for an individual, increase during the life span of a plant. Effectively, all species of plants compete with one another in a direct fashion. Indeed, "competition" may even exist between parts of the same plant. Thus plants differ from animals in ecologically fundamental ways even at the basic level of competitive exclusion. Plant adaptations to environmental fluctuations result in conspicuous differentiation in the time and manner of reproductive and vegetative growth, and can produce a finer subdivision of the intracommunity pattern of microhabitats by plants than that seen in animals.

The extent to which the niche hyperspace was fragmented by an evolutionary flora and to which that fragmentation could be caused to go to instability owing to evolutionary innovations (reproductive strategies, allelochemics, etc.) provide the critical "modes" necessary to interpreting the total diversity of tracheophytes. Clearly, however, these "modes" are more philosophical than mechanical. Although providing proximate explanations, the ultimate causality relevant to speciation is within the context of genetics.

RATES OF SPECIATION AND GENETIC CHANGE

The correlation between the rate of cladogenesis and the rate of genetic change has been explored by a number of authors who have reached equivocal but often polarized conclusions (Ayala, 1975; Levin and Wilson, 1976; Prager et al., 1976). Attempts to resolve this issue have centered around determinations of structural protein and karyotypic variation, and are based on calculations of phyletic distance measures. Based on the state of the literature, Wilson et al. (1974a-b) suggested that there are two types of "molecular evolution": one type involving structural genes, which goes on at an apparently constant rate, and a second for regulatory genes, which is primarily responsible for reproductive incompatibilities and morphologic evolution.

In a study of two fish groups (*Lepomis*), Ayala (1975) concluded that there is no evidence of greater genetic differentiation in highly speciose fish than in those that are species depauperate. Ayala suggested that there is no obvious correlation with the apparent rate of speciation and genetic change, but cautioned that this conclusion is based on a limited data set relevant to a particular group. Prager et al. (1976) concluded that karyotypic evolution in gymnosperms has been remarkably slow (100 times lower than most mammals; 10 times lower than other mammals and mollusks, cf. Wilson et al., 1975), yet the rate of change in amino acid sequences of seed proteins is comparable to that calculated for flowering plants (e.g., the rate of change between two lineages of gymnosperms is about 1% 7.5 Myr^{-1}, while the average for that of flowering plants is between 1% 5-10 Myr^{-1}). Levin and Wilson (1976) have presented comparable analyses of net increase in diversity of karyotypes in seed plants, and conclude that a high correlation exists between karyotypic evolution and rates of speciation. Thus, support for the hypothesis of Wilson et al. (1974a-b) is seen in the conclusion that the rate of structural protein evolution is roughly the same for gymnosperms and for angiosperms, yet karyotypic rates of evolution appear to be low in gymnosperms and high in angiosperms. The positive correlation between the rate of karyotypic change and the rate of speciation within seed plant groups has not been extended to other plant groups. Thus the currently restricted data precludes a generalization and demands a note of caution. However, based on studies of DNA sequences in the fern *Osmunda*, Stein et al. (1979) speculated that the rate of addition of repetitive sequences is probably slower in ferns than in flowering plants. These authors suggest that the slower rates of evolution in ferns is referable in part to generation time (Stein et al., 1979:229).

The various mechanisms or modes of karyotypic modification are reviewed elsewhere and are not elaborated here (cf. Stebbins, 1971).

RATES OF SPECIATION AND POLYMORPHISM

The question of genetic polymorphism as it relates to the ability or expression of a taxon to speciate has prompted some authors to seek correlations between these two phenomena. Ayala and Kiger (1980) and Ayala (1982) indicate that natural populations of most organisms possess large stores of genetic variation. The average heterozygosity in vertebrate groups is between 4 and 8 percent, between 6 and 15 percent for a given invertebrate group, and between 3 and 8 percent for plants. The average polymorphism for self-pollinating and autocrossing plants is 0.231 and 0.344, respectively, and is high compared to mammals (0.206) and birds (0.145). Based on enzyme polymorphism, Hamrick et al. (1979) present data relevant to a polymorphic index (PI) and twelve life history variables. Data for six of the twelve variables are given in Table 2. Gymnosperms show roughly twice as many polymorphic loci than either monocots or dicots. There appears to be a significant correlation between high polymorphism on the one hand, and wind pollination and long-lived perennial growth habits on the other hand (Table 2, variables 5-6). Although on the basis of calculated speciation rates for gymnosperms and angiosperms (Table 1), it appears that there is a negative correlation between enzyme polymorphism and relative rates of speciation (Table 2, variable 1), the limited data are not sufficient to warrant any conclusion.

The difficulty in interpreting these data in an evolutionary context lies in our lack of knowledge concerning the adaptive significance of iso- and allo-enzymes. Allosterically regulated enzymes, critical to the coordination of metabolism, can operate efficiently only over a limited range of conditions. Individuals with multiple (=polymorphic) forms are potentially capable of responding appropriately to the binding of effector molecules over a broader range. Thus enzyme polymorphism is potentially adaptive in the sense that it endows an organism with a greater metabolic latitude. This potential for enzymatic "plasticity" is consistent with the high polymorphic indices of gymnosperms and long-lived perennials in general (Table 2)–plants that must survive in a variable microenvironment for many years before reaching reproductive maturity. Although speculative, polymorphism in long-lived taxa may be adaptive to the individual, but may so "buffer" a plant lineage so as to ameliorate selection pressures thereby reducing apparent speciation.

An alternative explanation for the high polymorphism seen in gymnosperms may be simply that genetic variation tends to accumulate through time (Valentine, 1978). Therefore, older plant groups should be genetically more polymorphic than younger groups, such as angiosperms. (Interestingly, such a conjecture would allow the speculation that monocots are older as a group than are dicots, cf. Table 2, variable 1.) To test this hypothesis, comparable analyses to those of Hamrick et al. (1979) for lycopods, horsetails, and various ancient fern taxa are needed. Regrettably, these analyses are currently not available.

BREEDING STRUCTURE, EFFECTIVE POPULATION SIZE,
AND GENERATION RATES

In their analyses of seed plants, Levin and Wilson (1976) concluded that the rates of evolution at both the karyotypic and organismal level are related to the breeding structure of species and to environmental predictability. In taxa with small, semi-isolated populations occupying transient habitats or ecotones experiencing sharp fluctuations, the probability of fixing chromosomal or morphologic novelties is high and speciation rates are rapid. By contrast, taxa that maintain large continuous populations where abiotic and biotic pressures

Table 2. Levels of variability between categories of six life history and ecologic variables (from Hamrick, Linhart and Mitton, 1979).

	Number of species	Mean number of loci	Polymorphic loci (P) (%)		Number of alleles per locus (A)		Polymorphic index [PI][b]	
			x̄[a]	S.D.	x̄	S.D.	x̄	S.D.
1. Taxonomic Status							**	
Gymnospermae	11	9.2	67.01	7.99	2.12	0.20	0.270	0.041
Dicotyledoneae	74	11.4	31.28	3.31	1.46	0.06	0.113	0.014
Monocotyledoneae	28	11.6	39.70	6.02	2.11	0.19	0.165	0.026
2. Geographic Range							*	
Endemic	17	15.1	23.52	5.06	1.43	0.11	0.086	0.019
Narrow	22	11.4	36.73	6.01	1.60	0.14	0.158	0.030
Regional	39	8.3	55.96	5.13	1.85	0.10	0.185	0.025
Widespread	35	12.5	30.36	5.03	1.58	0.15	0.120	0.021
3. Generation Length							***	
Annual	42	11.2	39.47	4.32	1.72	0.11	0.132	0.017
Biennial	13	17.2	15.78	5.12	1.26	0.09	0.060	0.020
Short-lived Perennial	31	12.0	28.09	5.06	1.46	0.09	0.123	0.023
Long-lived Perennial	27	7.6	65.77	5.08	2.07	0.13	0.267	0.027
4. Pollination Mechanism							NS	
Selfed	33	14.2	18.99	3.51	1.31	0.07	0.058	0.016
Animal	55	9.5	38.83	3.94	1.55	0.07	0.130	0.015
Wind	23	10.7	57.45	6.29	2.27	0.17	0.264	0.028
5. Mode of Reproduction							NS	
Asexual	1	8.0	50.00	0.00	1.91	0.00	0.139	0.000
Sexual	95	11.7	35.64	3.03	1.63	0.07	0.135	0.012
Both	17	8.9	41.71	8.12	1.67	0.14	0.185	0.034
6. Habitat Type							NS	
Xeric	4	8.8	15.39	8.20	1.11	0.09	0.048	0.040
Sub-Mesic	19	10.5	43.68	4.86	1.66	0.08	0.140	0.020
Mesic	82	11.4	36.01	3.61	1.65	0.07	0.146	0.016
Hydric	8	13.0	27.71	10.33	1.59	0.22	0.145	0.050

[a] Weighted means and standard deviations are given for each measure of variability.

[b] Differences in PI between categories tested by ANOVA. Significance: *(P<0.05), **(P>0.01), and ***(PL<0.001); NS=not significant.

are stable through time are likely to be evolutionarily conservative. Levin and Wilson (1976) conclude that the mean rates of increase in chromosomal diversity and species numbers in seed plants are highly correlated (r=0.64; P < 0.01). This correlation is consistent with the Levin and Wilson hypothesis, and may be the result or the proximate cause of relative speciation rates: (1) both processes may be influenced by the same or similar factors, or (2) karyotypic changes set the stage for speciation either ". . . by reducing gene exchange or by altering the regulatory system and thereby providing the species with new phenotypic alternatives that may be selectively favored" (Levin and Wilson,1976:2089). Thus herbs have very high net rates of chromosomal evolution (≈ 0.7) and high net rates of species evolution (≈ 1.0). Similarly, angiosperm shrubs and trees yield net rates of chromosomal (Rc) and species evolution (Rs) comparable to those calculated by Levin and Wilson for cycads and conifers (Rc $\approx$ 0 to 0.01; Rs $\approx$ 0.05 to 0.02).

The conclusions of Levin and Wilson are compatible with our data, particularly when families of angiosperms are compared (Table 1, Figure 4). Angiosperm families dominated by long-lived perennial species (Fagaceae, Aceraceae, Betulaceae, Magnoliaceae, Juglandaceae, and Ulmaceae) have low origination rates (R=0.1$-$0.2 Myr^{-1}), while herbaceous angiosperm families (Umbelliferae, Compositae, Labiatae, Polygonaceae, and Ranunculaceae) have relatively high origination rates (R=0.4$-$0.6 Myr^{-1}). Similarly, cycads, ginkgophytes and conifers have extremely low rates of origination in comparison to flowering plants (Table 1, Figure 5).

At least within seed plants, generation time could play a significant role in effecting rates of chromosomal and species evolution. Since each generation represents an opportunity for selection or random genetic drift, seed plants with the shortest generation times, such as herbs or annuals, may speciate more rapidly than longer-lived taxa, such as shrubby or arborescent angiosperms. If the generation times for angiosperm herbs, shrubs, and trees are taken as one, five and ten years, respectively, then herbs would be expected to have had an opportunity to speciate an order of magnitude faster than trees. Similarly, if the generation time for cycads and conifers is taken to be roughly 20-25 years, then these gymnosperm taxa would speciate less rapidly, at least in theory, than angiosperms. The correlation between generation time and apparent origination rates, however, does not appear to extend beyond seed plants. The average fern requires 3 to 6 years to achieve sexual maturity. A similar time appears to be required to reach maturity for present day sphenopsids and lycopods. Although these generation times are comparable for those seen in agiosperm herbs, the average origination rates of pteridophytes is considerably lower than any calculated for flowering plant families.

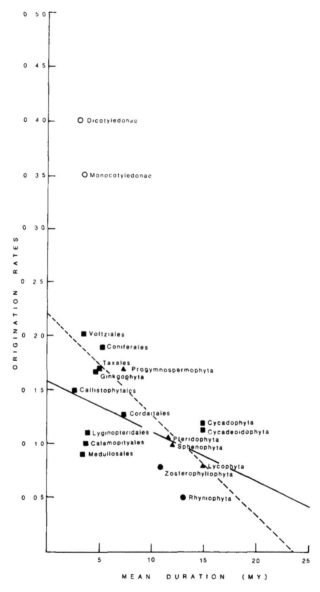

Figure 5. Plot of mean species-origination rates versus mean species durations for various suprageneric plant groups (open dots represent data based on genera rather than species). For further details, see text.

REGULATORY GENES AND EVOLUTION

Recent speculations concerning the roles of regulatory genes in the evolution of major taxa have not been fully explored (Wilson, 1975; Wilson et al., 1974a-b; Maxson and Wilson, 1979; Stebbins, 1982), but offer exciting potential to explain evolutionary patterns in land plants. Two examples where changes in plant regulatory mechanisms may stimulate major evolutionary changes are illustrative. Basile (1979) has shown that phenovariation in liverworts can be induced by extremely low level modifications of the amino acid hydroxyproline. By altering proline-hydroxyproline metabolism, liverworts manifest extreme changes in morphogenesis resulting in leafy and thalloid forms. Regulatory gene action at such a fundamental metabolic level may have played a significant role in the morphogenetic alteration of aquatic plants attending their exploitation of the terrestrial habitat (cf. Niklas, 1976). More recently, analyses of the developmental changes in a species of flowering plants, *Streptocarpus*, has shown that exposure of germinating seeds to growth substances can alter both vegetative and reproductive structure (Rosenblum, 1981). Clearly, the influence of plant growth regulators have profound effects on meristematic and hence morphologic development. For example, Stebbins (1974) has reviewed the role of intercalary meristems in the transition of unfused petals to a bell-shaped or tubular corolla. The production of fused parts, through the process of syngenesis, may alter pollination mechanisms, resulting in sexual isolation and subsequent speciation. Although inferential, the evolution of the integumented seed and the adaptive radiation of seed plant lineages may have stemmed from relatively minor and developmentally quick alterations of meristematic activity (cf. Niklas, 1981a-b). As Stebbins points out: "Rates of evolutionary change usually differ by factors of two or ten . . . when different organs, tissues, cells or molecules are compared . . . When evolutionary rates are similar, either the characters are controlled by pleiotropic effects of the same genes or they contribute to adaptive syndromes that are subject to the same or similar selection pressures." (Stebbins, 1982:13-14).

SUMMARY

Analysis, at the species level, indicates that the temporal pattern of terrestrial plant diversification can be decomposed into four distinct floral components: (1) a Silurian-Early Devonian proliferation of early vascular plants which were morphologically characterized by a simple and presumably primitive construction (Figure 1), (2) a subsequent Late Devonian-Carboniferous radiation of derived plant lineages some of which achieved arborescence and heterospory (Figure 2), (3) the appearance of seed plants in the latest Devonian and their adaptive radiation culminating in a gymnosperm dominated Meso-zoic flora (Figure 2), and (4) the appearance and rise of the flowering plants in the Early Cretaceous and Tertiary (Figure 3). Three of these four phases of tracheophyte evolution are associated with significant increases in the total species numbers of land plants. The first occurred within a 60 Myr period be-tween the appearance of the first vascular plants and the decline of these ear-ly lineages, where the diversity of land plants increased four-fold. The second involved the adaptive radiation of Upper Devonian pteridophytes which re-sulted in another four-fold increase in species numbers by Permian times. The last and third major increase in tracheophyte diversity, resulting in a three-fold increase in species numbers, occurred by the Neogene and is referable to the proliferation of the angiosperms.

Based on the rate of appearance of new species and their durations for the various plant lineages examined, two distinctive patterns in tracheophyte evo-lution become evident: (1) with a few exceptions, the initial occurrence of a new lineage is characterized by species with relatively short durations and by high species-origination rates, followed by a progressive decrease in speciation rates and an increase in species duration (Table 1), and (2) among successively occurring suprageneric groups, there appears to be an increase in mean spe-cies-origination rates and a decrease in mean species duration (Figure 5). Al-though the initial species-origination rates of every plant group is high, the earliest vascular land plants (rhyniophytes and zosterophyllophytes) have the longest species durations (12 to 14 my) and the lowest species-origination rates (0.04 to 0.07 my^{-1}) of any plant lineage. Subsequently evolving pteri-dophyte and seed plant groups show increased mean species-origination rates and decreased mean durations. The angiosperms, standing in marked contrast to the earliest tracheophytes, show the highest speciation rates and the lowest species durations. Even if the "short species durations" of the an-giosperms are considered to be an artifact of observing their initial evolution-ary phase of radiation, their significantly high species-origination rates stand in marked constrast to all other lineages. A linear regression of mean species-duration versus mean species-origination rates for the plant groups examined yields a coefficient of correlation equal to 0.89. If the angiosperm data points are removed, r=0.91.

Examination of the possible mechanisms or "modes" of plant diversifica-tion indicates that the data are consistent with ecologic theory and observa-tions of species diversity at the intercontinental level (Whittaker, 1977). The overall pattern of increasing species numbers through geologic time indicates that tracheophyte species equilibrium has not been achieved, and may be in-terpreted as an expression of the ability of plants to increasingly "fragment" the available niche hyperspace and to evolve into positions of resource use that are extreme or marginal relative to antecedent species. Available data as to "species richness" and paleocommunity structure indicate that, through geologic time, species packing and community stratification has increased. The decline observed in an evolutionary flora is usually concomitant with the appearance of a new flora. Subsequent adaptive radiation of the new flora is associated with the diminution in species diversity of antecedent taxa. This is consistent with the speculation that a macroevolutionary analog to competitive displacement is operative in the succession of evolutionary floras.

Crude estimates of species-origination rates clearly indicate that the mean rates of evolution for various plant lineages are unequal over long periods of time. An examination of the relevant genetic literature leads us to conclude that there may exist a high correlation between the apparent rates of speciation and karyotypic variation. At least for seed plants this correlation is at the level of 0.64 (cf. Levin and Wilson, 1976; Prager et al., 1979). Although the rate of chromosomal variation within a seed-plant lineage correlates with the apparent rate of speciation, karyotypic alteration is not interpretable as ultimate causation, since it is possibly a cytologic symptom of genetic processes attending speciation. Correlations among breeding strategies, karyotypic variation, and mean speciation rates, however, may provide a basis for concluding that taxa with restricted gene flow, low dispersibility, and small effective population size speciate more rapidly than those taxa with high gene flow probabilities, high propagule dispersibility, and large effective population size. Clearly, the influence of environmental stability and heterogeneity play significant roles in determining the probability of speciation. Environmental changes and breeding structure differences are interpreted as proximate, if not ultimate, factors determining evolutionary rates.

ACKNOWLEDGMENTS

The authors wish to acknowledge the support of NSF grants DEB-81-18416 and DEB-81-18749 to KJN, and DEB-79-05082 to BHT. All line drawings were prepared by Miss Robin M. Gowen.

REFERENCES

Ayala, F. J., 1975. Genetic differentiation during the speciation process. In Dobzhansky, T., Hecht, M. K., and Steere, W. C. (eds.), Evolutionary biology, New York: Plenum Press, 8:1-78.

Ayala, F. J., 1982. The genetic structure of genes. In Milkman, R. (ed.), Perspectives on evolution, Sunderland, Mass.: Sinauer Assoc., 60-82.

Ayala, F. J., and Kiger, J. A., 1980. Modern genetics. Menlo Park, Calif.:

Benjamin/Cummings.

Bambach, R. K., 1977. Species richness in marine benthic habitats through the Phanerozoic. Paleobiology 3:152-167.

Banks, H. P., 1968. The early history of land plants. In Drake, E. T. (ed.), Evolution and environment: A symposium of the 100th anniversary of the foundation of Peabody Museum of Natural History at Yale University, New Haven: Yale Univ. Press, 73-107.

Banks, H. P., 1980. Floral assemblages in the Siluro-Devonian. In Dilcher, D. L., and Taylor, T. N. (eds.), Biostratigraphy of fossil plants, Stroudsburg, Penna.: Dowden, Hutchinson, and Ross, 1-24.

Basile, D. V., 1979. Hydroxyproline-induced changes in form, apical development and cell wall protein in the liverwort *Plagiochila arctica*. Amer. J. Bot. 66:776-783.

Beck, C. B., 1981. *Archaeopteris* and its role in vascular plant evolution. In Niklas, K. J. (ed.), Paleobotany, paleoecology, and evolution, New York: Praeger Publishers, 1:193-230.

Chaloner, W. G., 1967. Spores and land plant evolution. Paleobot. Palynol. Rev. 1:83-94.

Chaloner, W. G., and Sheerin, A., 1979. Devonian macrofloras. In House, M. R., Scrutton, C. T., and Bassett, M. G. (eds.), The Devonian system, Paleontol. Spec. Pap. (23):145-161.

Edwards, D., Bassett, M. G., and Rogerson, E. C. W., 1979. The earliest vascular land plants: Continuing the search for proof. Lethaia 12:313-324.

Eldredge, N. and Gould, S. J., 1972. Punctuated equilibria: An alternative to phyletic gradualism. In Schopf, T. J. M. (ed.), Models in paleobiology, San Francisco: Freeman, Cooper, 82-115.

Gensel, P. G., 1977. Morphologic and taxonomic relationships of the Psilotaceae relative to evolutionary lines in early land vascular plants. Brittonia 29:14-29.

Gensel, P. G., 1980. Devonian in situ spores: A survey and discussion. Palaeobot. Palynol. Rev. 30:101-132.

Gothan W., and Weyland, H., 1954. Lehrbuch der Palaobotanik. Berlin: Akademie Verlag.

Gould, S. J., and Eldredge, N., 1977. Punctuated equilibria: The tempo and mode of evolution reconsidered. Paleobiology 3:115-151.

Hamrick, J. L., Linhart, Y. B., and Mitton, J. B., 1979. Relationships between life history characteristics and electrophoretically detectable genetic variation in plants. Ann. Rev. Eco. Sys. 10:173-200.

Hughes, N. F., 1976. Palaeobiology of angiosperm origins. Cambridge: Cambridge Univ. Press.

Knoll, A. H., Niklas, K. J., and Tiffney, B. H., 1979. Phanerozoic land plant diversity in North America. Science 206:1400-1402.

Levin, D. A., and Wilson, A. C., 1976. Rates of evolution in seed plants: Net increase in diversity of chromosome numbers and species numbers through time. Nat. Acad. Sci. Proc., USA 73:2086-2090.

Maxson, L. R., and Wilson, A. C., 1979. Rates of molecular and chromosomal evolution in salamanders. Evolution 33:734-740.

McGregor, D. C., 1977. Lower and Middle Devonian spores of eastern Gaspe, Canada. II. Biostratigraphy. Palaeontographica 163B:111-142.

McGregor, D. C., 1979. Spores in Devonian stratigraphical correlation. In House, M. R., Scrutton, C. T., and Bassett, M. G. (eds.), The Devonian system. Spec. Pap. Palaeontology (23):163-184.

Meeuse, A. D. J., 1975. Aspects of the evolution of the monocotyledons. Acta Bot. Neerl. 24:421-436.

Muller, J., 1970. Palynological evidence on early differentiation of angiosperms. Biol. Rev. 45:417-450.

Nicholson, A. J., 1958. The self-adjustment of populations to change. Cold Spring Harbor Symp. Quant. Biol. 22:153-173.

Niklas, K. J., 1976. The role of morphological biochemical reciprocity in early land plant evolution. Ann. Bot. (London) 40:1239-1254.

Niklas, K. J., 1978. Coupled evolutionary rates and the fossil record. Brittonia 30:373-394.

Niklas, K. J., 1981a. Simulated wind pollination and airflow around ovules of some early seed plants. Science 211:275-277.

Niklas, K. J., 1981b. Airflow patterns around some early seed plant ovules and cupules: Implications concerning efficiency in wind pollination. Amer. J. Bot. 68:635-650.

Niklas, K. J., Tiffney, B. H., and Knoll, A. H., 1980. Apparent changes in the diversity of fossil plants. In Hecht, M. K., Steere, W. C., and Wallace, B. (eds.), Evolutionary biology, New York: Plenum Publishing, 12:1-89.

Prager, E. M., Fowler, D. P., and Wilson, A. C., 1976. Rates of evolution in conifers (Pinaceae). Evolution 30:637-649.

Raup, D. M., 1972. Taxonomic diversity during the Phanerozoic. Science 177:1065-1071.

Raup, D. M., 1976. Species diversity in the Phanerozoic: A tabulation. Paleobiology 2:279-288.

Richardson, J. B., 1974. The stratigraphic utilization of some Silurian and Devonian miospore species in the northern hemisphere: An attempt at synthesis. Intern. Symp. Belgian Micropaleont. Limits (Namur), Publ. (9): 1-13.

Rosenblum, I. M., 1981. An approach toward understanding some of the morphogenetic bases of the phylogeny of *Streptocarpus* (Gesneriaceae). Ph.D. dissertation, City University of New York, Lehman College, New York.

Rosenzweig, M. L., 1975. On continental steady states of species diversity. In Cody, M. L., and Diamond, J. M. (eds.), Ecology and evolution of communities, Cambridge: Harvard Univ. Press, 121-140.

Schopf, T. J. M., Raup, D. M., Gould, S. J., and Simberloff, D. S., 1975. Genomic versus morphologic rates of evolution: Influence of morphologic complexity. Paleobiology 1:63-70.

Sepkoski, J. J., Jr., 1978. A kinetic model of Phanerozoic taxonomic diversity. I. Analysis of marine orders. Paleobiology 4:223-251.

Sepkoski, J. J., Jr., 1979. A kinetic model of Phanerozoic taxonomic diversity. II. Early Phanerozoic families and multiple equilibria. Paleobiology 5: 222-251.

Sepkoski, J. J., Jr., 1981. A factor analytic description of the Phanerozoic marine fossil record. Paleobiology 7:36-53.

Sepkoski, J. J., Jr., Bambach, R. K., Raup, D. M., and Valentine, J. W., 1981. Phanerozoic marine diversity and the fossil record. Nature 293:435-437.

Stanley, S. M., 1975. A theory of evolution above the species level. Nat. Acad. Sci. USA Proc. 72:646-650.

Stanley, S. M., 1979. Macroevolution. San Francisco: W. H. Freeman.

Stebbins, G. L., 1971. Chromosomal evolution in higher plants. London: Edward Arnold (publ.) Ltd.

Stebbins, G. L., 1974. Flowering plants: Evolution above the species level. Cambridge, Mass.: Harvard Univ. Press.

Stebbins, G. L., 1982. Modal themes: A new framework for evolutionary syntheses. In Milkman, R. (ed.), Perspectives on evolution, Sunderland, Mass.: Sinauer Assoc., 1-14.

Stein, D. B., Thompson, W. F., and Bedford, H. S., 1979. Studies on DNA sequences in the Osmundaceae. Mol. Evol. J. 13:215-232.

Strong, D. R., Jr., 1974. Nonasymptotic species richness models and the insects of British trees. Nat. Acad. Sci. USA Proc. 71:2766-2769.

Tiffney, B. H., 1981. Diversity and major events in the evolution of land plants. In Niklas, K. J. (ed.), Paleobotany, paleoecology, and evolution, New York: Praeger Publishers, 2:193-230.

Tiffney, B. H., and Niklas, K. J., in press. The history of clonal growth in land plants—a paleobotanical perspective. In Buss, L., Cook, R., and Jackson, J. B. C. (eds.), Biology of clonal organisms.

Valentine, J. W., 1968. The evolution of ecological units above the population level. J. Paleontol. 42:253-267.

Valentine, J. W., 1977. Genetic strategies of adaptation. In Ayala, F. J. (ed.), Molecular evolution, Sunderland, Mass.: Sinaur Associates, 78-94.

Van Valen, L., 1973. Are categories in different phyla comparable? Taxon 22:333-373.

Whittaker, R. H., 1977. Evolution of species diversity in land communities. In Hecht, M. K., Steere, W. C., and Wallace, B. (eds.), Evolutionary biology, New York: Plenum Press, 10:1-68.

Wilson, A. C., 1975. Evolutionary importance of gene regulation. Stadler Symp. Univ. Missouri 7:117-133.

Wilson, A. C., Maxson, L. R., and Sarich, V. M., 1974a. Two types of molecular evolution: Evidence from studies of interspecific hybridization. Nat. Acad. Sci. USA Proc. 71:2843-2847.

Wilson, A. C., Sarich, V. M., and Maxson, L. R., 1974b. The importance of gene rearrangement in evolution: Evidence from studies on rates of chromosomal, protein, and anatomical evolution. Nat. Acad. Sci. USA Proc. 71:3028-3030.

Chapter 4

REAL AND APPARENT TRENDS IN
SPECIES RICHNESS THROUGH TIME

PHILIP W. SIGNOR, III

Department of Geology, University of California, Davis

INTRODUCTION

Perhaps the greatest pattern in the history of life is the variation of global species richness through time. Paradoxically, the history of species richness has also been one of the most difficult patterns to resolve. Over the past decade a number of paleontologists, using a variety of ingenious arguments, have attempted to estimate trends in species richness through time but no generally accepted model has emerged.

Resolution of actual trends in species richness through the Phanerozoic using counts of described species has been hampered by sampling biases of several types (Simpson, 1960; Raup, 1972, 1976b, 1977; Sheehan, 1977; Koch, 1978; Signor, 1978, 1982). The likelihood of a species being retained in the rock record apparently varies with time (Gregory, 1955; Simpson, 1960; Raup, 1972, 1976b) and the attention devoted by systematists to faunas of different ages is unequal (Cooper and Williams, 1952; Williams, 1957; Raup, 1972, 1976b, 1977; Sheehan, 1977). Other biases, while acting on local or regional scales, are equally important. Species richness at any given locality tends to vary as a function of sampling (Durham, 1967; Stanton and Evans, 1972; Koch, 1978). The habitats occupied by potential fossils and the representation of those habitats in the stratigraphic record will also affect the probability that a potential fossil is retained in the fossil record and available for sampling (Simpson, 1960; Raup, 1972). Trends in apparent species richness (numbers of described species: Figure 1) reflect these biases, and not necessarily trends in actual species richness (Raup, 1972, 1976b; Signor, 1978). Given these problems, paleontologists have sought other means to infer the history of species richness in the Phanerozoic.

FOUR MODELS OF SPECIES RICHNESS IN THE PHANEROZOIC

Four models of species richness in the Phanerozoic have been proposed over the past fourteen years. Each is based on evidence other than apparent global species richness. Hence, the validity of each model is dependent upon the correspondence between actual species richness and the evidence upon which the various models are based. In most cases, arguments in support of these models are based upon counts of higher taxonomic units or patterns of subtaxa richness within higher taxonomic units. Each of the following models refers to global diversity in the limited sense of numbers of fossilizable marine invertebrates.

Valentine (1970) proposed the Empirical model of species richness (Figure 2a), which was based on temporal variations in the ratio of genera to families. He compiled data on the ratio of genera to families of readily preservable marine invertebrates and found a sharp increase in post-Mesozoic genus/family ratios. Valentine reasoned that, if a similar increase in the species/genus ratio

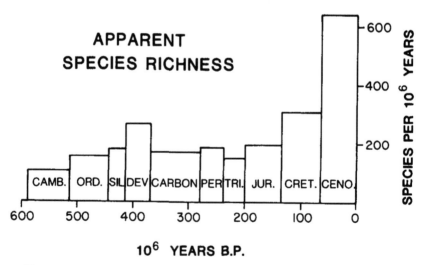

Figure 1. Numbers of described species of marine invertebrates through time. Numbers of species are normalized to period length. Data from Raup (1976a).

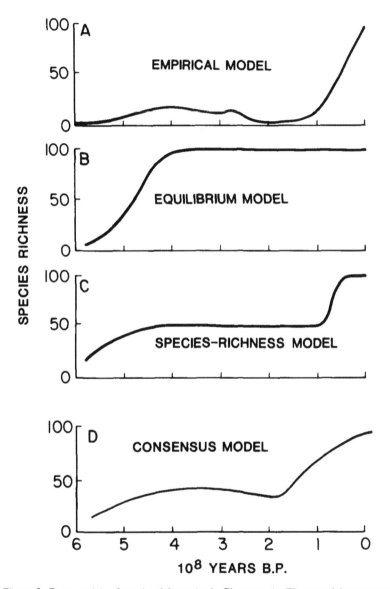

Figure 2. Four models of species richness in the Phanerozoic. These models represent changes in standing species richness relative to the Recent. The models are from the following sources: A, Valentine, 1970; B, Gould et al. 1977; C, Bambach, 1977; D, modified from Sepkoski et al., 1981.

had occurred, there must have been a tremendous increase in actual species richness at the end of the Mesozoic.

A similar pattern was obtained through an alternative analysis by Valentine et al. (1978). In this second study, Valentine et al. estimated how many species would be contained within high and low diversity provinces. By calculating how the numbers of provinces varied through time, and factoring in variations in within-habitat species richness (from Bambach, 1977), they were able to infer past trends in global species richness. Their results closely matched the Empirical model proposed earlier by Valentine.

Gould et al. (1977) proposed a second model, the Equilibrium model (Figure 2b), for the history of species richness. Gould et al. found that the center of gravity of artificial clades was always near the clade's midpoint when their simulated universe was at equilibrium. However, when new lineages were being added to the universe the clade's center of gravity tended to fall before the clade's midpoint. Applying this statistic to the distribution of actual supraspecific taxa within higher taxa, they found that the center of gravity of real clades was close to the clade's midpoint for post-Ordovician clades. Earlier clades tended to have their center of gravity prior to their midpoint. On the basis of this analysis they concluded that global species richness had been approximately constant since the Ordovician.

The Equilibrium model was supported by Sepkoski's (1978) analysis of variation in the numbers of marine metazoan orders through time. Sepkoski showed that the number of orders of marine Metazoa has remained approximately constant since the mid-Ordovician. Using simulations and qualitative arguments he explicitly contended that species richness would have paralleled the numbers of orders through time. If these arguments are valid, actual species richness must have been approximately constant through time.

The third model of Phanerozoic species richness, the Species Richness model (Figure 2c), was proposed by Bambach (1977). Bambach compiled lists of species found at different times in three generalized marine benthic habitats: nearshore stressed, variable nearshore and open marine environments. He found that species richness in nearshore stressed environments has remained approximately constant through the Phanerozoic but approximately doubled in nearshore and open marine environments around the end of the Mesozoic. Other factors being equal, global species richness should also have doubled near the end of the Mesozoic. Of course, other factors (e.g., provinciality) were not equal but, with the exception of Valentine et al. (1978), no attempt has been made to integrate Bambach's species richness data with other factors that might impact global species richness. It appears that Bambach's data contradict the Equilibrium model, but may be compatible with the Empirical model.

In a recent paper, four of the debate's protagonists (Sepkoski et al. 1981) joined to propose a new model of species richness, here termed the Consensus model (Figure 2d). Sepkoski et al. pointed out that five independent or partially independent diversity data sets (trace fossil diversity, numbers of described species, within-habitat species richness, numbers of genera and numbers of families) all show an identical pattern. Even when the autocorrelation with time contained in each of the data sets was removed the patterns in each of the five sets were very similar. Sepkoski et al. interpreted this pattern to reflect a primary signal of global species richness. Whether the pattern documented in the Consensus model reflects a primary signal or a pervasive filter, such as sampling bias, remains to be determined. Regardless, the Consensus model represents withdrawal of support for the Equilibrium model by two of the latter model's ardent supporters.

As noted above, the foregoing models are not based directly upon data on global species richness, and their accuracy depends upon the degree that species-level phenomena are reflected by the various methods devised to infer global species richness. The fact that such disparate results are obtained with alternative methods of inferring species richness seems to speak to this problem. The remaining unsolved problem is to deduce actual species richness from the data on apparent species richness without recourse to additional assumptions of uncertain validity (e.g., numbers of orders reflect trends in apparent species richness). We need to develop a method of removing sampling bias from trends in apparent species richness through time.

COMPENSATING FOR SAMPLING BIAS

The standard technique used to compensate for sampling bias in ecological and paleoecological data is rarefaction (for a review see Tipper, 1979). Unfortunately, application of rarefaction would require data on the numbers of individuals of every species that have been collected. This information is obviously not available and, consequently, rarefaction cannot be applied.

An alternative solution is possible. If we can infer the frequency distribution of species in a given geological interval, and if measures of sampling intensity are available, then it is possible to estimate how many additional or fewer species might be recovered by an increase or decrease in sampling intensity. This concept is illustrated in Figure 3. Sampling effort and the shape of the sampling curve (which is determined by the numbers of species and their relative abundance) combine to determine how many species will be recovered. When the species-abundance distributions are similar, the height of the sampling curve above the abscissa is controlled by the number of species in the distribution. For example, there are five sampling curves illustrated in Figure 3; each curve represents a lognormal distribution but curves a through e diverge because the total numbers of species present in each distribution vary. This raises an important point. If sampling intensity and the numbers of described species are known, and if the species abundance distribution of species in each geological period are lognormal, then it should be possible to determine which curve is being sampled and, indirectly, the total number of species that existed in each geological period.

The crucial aspect in the foregoing method is the nature of the species-abundance distribution. Numbers of described species are already available (Raup, 1976a), as are three estimates of sampling intensity: sediment volume, sediment area, and Paleontologist Interest Units (P.I.U.) (Raup, 1976b). The problem is tractable if use of the lognormal (or another) distribution can be justified.

The most general and commonly occurring species-abundance distribution in nature is the lognormal distribution (McNaughton and Wolf, 1970; May, 1975; Whittaker, 1975; Pielou, 1975). The lognormal distribution is not derived from specific biological properties or characteristics of communities,

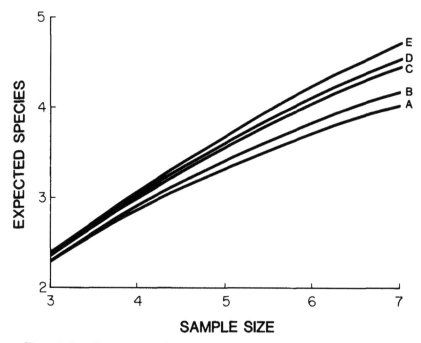

Figure 3. Sampling curves for five lognormally distributed populations. Both sample size and numbers of species have been converted to logarithms. The population sizes for the five curves are: A, 5×10^4 species; B, 10^5 species; C, 5×10^5 species; D, 10^6 species; and +, 5×10^6 species. The expected number of species for each sample and population size was calculated using the sampling model developed by Signor (1978).

but is a statistical model based on the assumption that the relative abundances of different species in a community are determined by a large number of independent factors (May, 1975; Pielou, 1975). The species-abundance distributions of organisms in diverse communities or where more than one community is sampled together is usually lognormal (May, 1975; Pielou, 1975; Whittaker, 1975; see Koch, 1978 for an example in fossil communities). The following analysis is based upon the assumption that the lognormal distribution adequately characterizes the global diversity of marine invertebrates.

The lognormal distribution is a continuous curve but it is easier to work with a discrete version, in which the species are divided up into frequency classes (or 'octaves') where a species in an octave is on average twice as abundant as those species in the next lower octave (see Signor, 1978). The distribution is defined by the equation:

$$Y_R = Y_o e^{-(aR)^2} \qquad (1)$$

where Y_o is the number of species in the modal (central) octave, Y_R is the number of species in the Rth octave and a is a constant for each distribution. "a" can be estimated by the following relationship (when Y_T, the total number of species is very large: May, 1975):

$$a \sim \frac{\ln Z}{2 \cdot \sqrt{\ln Y_T}} \qquad (2)$$

Finally, Y_o can be calculated using the equation:

$$Y_o = \frac{a Y_T}{\sqrt{\pi}} \qquad (3)$$

Since the relative abundances of species in each octave are known it is possible to estimate the number of individuals that would be drawn from an octave in a given sample. The number of species n represented among the individuals drawn from each octave can be calculated from the Poisson relationship:

$$n = Y_R - Y_R e^{-s/Y_R} \qquad (4)$$

The number of species that would be recovered in a sample of a given size can be calculated using equations 1 − 4, assuming the sampling intensity and the total number of species present in the distribution are known.

Raup (1976b) proposed that sediment volume (data from Gregor, 1970) and sediment area (data from Blatt and Jones, 1975) are two valid estimates of sampling intensity (Figure 4). Sheehan (1977) suggested a third alternative, Paleontologist Interest Units, which are estimates of the distribution of

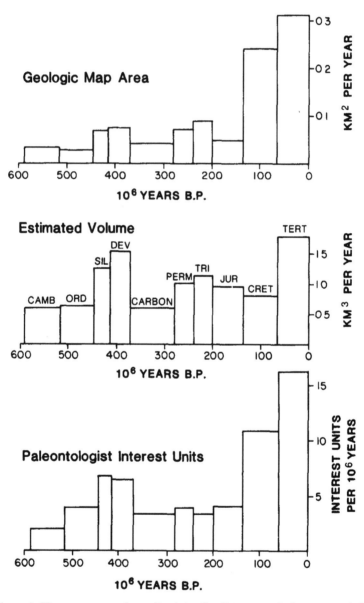

Figure 4. Three measures of sampling intensity. Data on geologic map area is from Blatt and Jones (1975). Data on estimated sediment volume are from Gregor (1970), and data on Paleontologist Interest Units are from Sheehan (1977).

taxonomists' efforts among the different geologic periods. All three are relative measures of sampling intensity and must be converted to absolute, quantitative estimates of sampling intensity. In an earlier paper (Signor, 1978), I demonstrated this can be done by calibrating the Cenozoic sample to "discover" approximately 43,000 species, the number of species estimated by Raup (1976a) to have been described from the Cenozoic. Sample sizes for the remaining periods are then set proportionately to sediment volume, area, or P.I.U.s.

Unfortunately, we do not know how many species existed during the course of the Cenozoic and this figure is necessary if sampling intensity is to be calibrated to this interval. Total Cenozoic species richness can be estimated if three variables are known: the number of Recent fossilizable marine species, the pattern of diversity change through the Cenozoic and mean species duration. Following my earlier treatments of this problem (Signor, 1978, 1982), I used two estimates of standing species richness and three estimates of mean species duration, combined into six data sets (Table 1). Each set of calculations will be done under three additional assumptions: Cenozoic standing diversity was constant, diversity increased by a factor of two and diversity increased by a factor of five. If results obtained with the sampling model are robust then variations in the number of species existing in the Cenozoic will not greatly affect the outcome. But if the results are not robust then the sampling model must be rejected as an inadequate tool for inferring past patterns of global species richness.

Once the sample size for each geologic period is known it is possible to calculate how many species were present in the interval of time, or in the pool that was sampled (Signor, 1982). This is done by taking advantage of the fact that a lognormal distribution for a given number of species defines a single sampling curve. Since the numbers of described species (from Raup 1976a)

Table 1. Data used to calculate total Cenozoic species richness. Estimates of mean durations in millions of years are from the following sources: (2.75) Simpson (1952); (6.5) Stanley (1975); (12) Teichert (1957). Data on Recent standing diversity of readily preservable marine invertebrates are from Teichert (1957) (169,700) and Valentine (1970) (100,775).

Data Set	Mean Species Duration	Standing Diversity
1	2.75	100,775
2	6.5	100,775
3	12	100,775
4	2.75	169,700
5	6.5	169,700
6	12	169,700

and sample size are known it is possible to estimate (using an iterative solution) which sampling curve the data define, which in turn yields the number of species present in the period. The computer program written to produce the solution calculates how many species would be discovered given some number of species were available for sampling. If the number of species discovered is higher or lower than the number actually described from that period, the program adds species or subtracts species from the total number present and then recalculates how many species would be discovered by sampling the new total. This process continues until the number of species "discovered" matches the number actually described from the various periods (± 100). These calculations provide an independent estimate of species richness through the Phanerozoic.

ESTIMATING APPARENT DIVERSITY UNDER DIFFERENT PATTERNS OF SAMPLING

A second, closely related question is how patterns of apparent species richness would change if sampling was equal in each of the geologic periods. This is analogous to the approach utilized in rarefaction analysis, where several samples are adjusted to a common size. This pattern can be estimated by calculating how many species were present in each period and then applying constant sampling to each. Here, I have set sampling intensity for each period equal to the Cenozoic sampling intensity. This procedure should produce results comparable to those which might be obtained through rarefaction (although with rarefaction it is always necessary to rarify to the smallest sample size because rarefaction curves cannot be extended to larger samples).

A final question that may be examined with the sampling model is how increased sampling will, in the future, modify trends in apparent species richness. By doubling or tripling the estimated sampling intensity we can project future patterns of apparent species richness, assuming that patterns of sampling remain constant. This assumption seems reasonable, as there is no reason to expect large shifts in the distribution of paleontological interest. Therefore, we can estimate what the fossil record may look like to succeeding generations of paleontologists.

RESULTS

Figures 5, 6 and 7 present the estimated trends in numbers of fossilizable marine invertebrates through time, calculated assuming Cenozoic diversity was either constant or increasing. In each case the sampling model predicts that diversity was low during the Paleozoic and Mesozoic and then increased dramatically to modern levels. The results are consistent regardless of variations in estimated Cenozoic species richness. These results clearly support the Empirical model of species richness (Figure 2a), and simultaneously contradict the Equilibrium model. While differing in detail from the Consensus model, the results obtained here are quite similar in several respects to the data presented by Sepkoski et al. (1981). Both models predict low Cambrian diversity followed by increasing diversity to a Cenozoic high. The major difference between the Consensus model and the results presented here is that the Consensus model predicts much higher overall levels of diversity in the Paleozoic and Mesozoic.

Figures 5, 6 and 7 do not necessarily represent changes in standing species richness. The figures represent species per million years, normalized to the number present in the Cenozoic. Only if mean species durations are constant in time will this pattern approximate standing species richness. This assumption may be valid for the late Paleozoic, Mesozoic and Cenozoic but is probably not true for the Cambrian, which is dominated by trilobites (Raup, 1976a). Trilobites have relatively short species durations (Stanley, 1979). Thus, Cambrian standing richness was probably lower than suggested by figures presented here.

When estimated sampling intensity is modified to simulate equal sampling of each geologic period a different pattern emerges (Figure 8). This pattern resembles Raup's (1976a) estimate of the numbers of species described from each geologic period, but suggests more species would be recovered from the Paleozoic and Mesozoic. This result is not surprising as sampling is being increased in every period except the Cenozoic. The differences between Figure 8 and Figures 5, 6 and 7 do not reflect inconsistencies in the sampling model but, rather, the different kind of question posed by the two approaches.

When sampling intensity is doubled to simulate future trends in numbers of described species, yet a third pattern is found (Figure 9). The Cenozoic peak is much higher relative to the nine Paleozoic and Mesozoic periods. This change probably reflects the fact that the return on sampling declines monotonically with increasing sampling, and the point of diminishing returns is reached more quickly in sampling smaller numbers of species (see Figure 3). Regardless, the important point is that additional sampling of the fossil record will produce no appreciable changes in patterns of apparent species richness.

DISCUSSION

Which of the three measures of sampling is the most accurate? I would agree with Sheehan (1977) that paleontologist interest units are probably the more

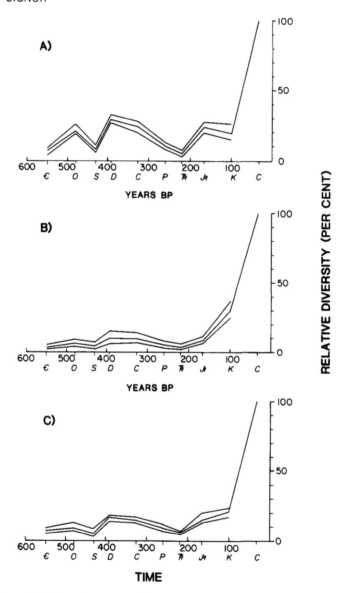

Figure 5. Estimated patterns of species richness through time, assuming Cenozoic species richness was constant and equal to Recent standing species richness. A is calculated assuming that sampling is proportional to sediment area, B assumes sampling is proportional to sediment volume, and C assumes sampling is proportional to Paleontologist Interest Units. The three lines represent the high, low and average predictions obtained using the six data sets in Table 1. No range is given for the Cenozoic because all estimates are based, in part, on estimated Cenozoic species richness.

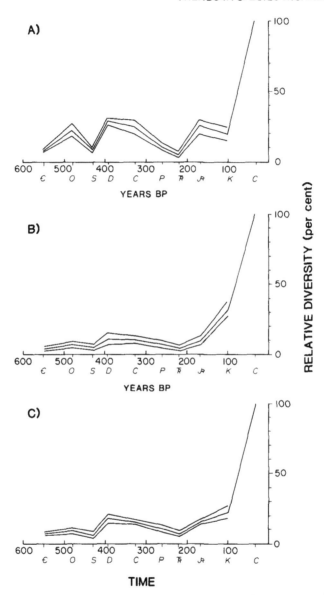

Figure 6. Estimated patterns of species richness through time, assuming Cenozoic species richness increased by a factor of two. See Figure 5 for further explanation.

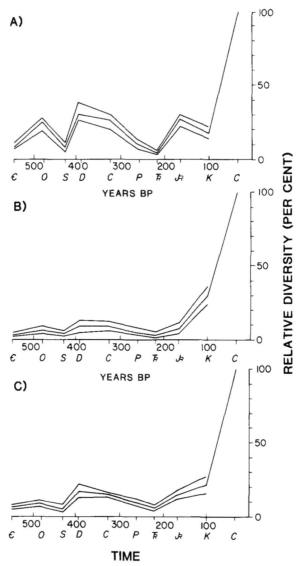

Figure 7. Estimated patterns of species richness through time, assuming Cenozoic species richness increased by a factor of five. See Figure 5 for further explanation.

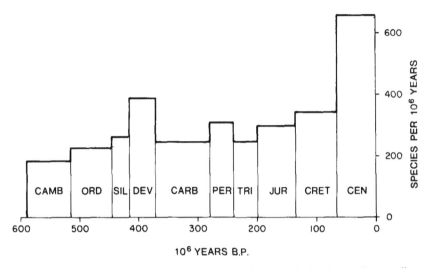

Figure 8. Estimated pattern of apparent species richness calculated assuming sampling is constant in all periods and equal to Cenozoic levels of sampling. Apparent species richness is normalized to period length.

valid measure of sampling intensity. Hence, the results presented in Figures 5c, 6c and 7c are probably most realistic. Since all three postulate a tremendous Cenozoic increase in diversity, the results calculated assuming a five-fold increase in Cenozoic diversity (Figure 7c) seem the best model for species richness in the Phanerozoic.

The diversity values calculated under the assumption that sampling is proportional to sediment volume are virtually identical to those calculated from P.I.U.s, but those calculated from sediment area suggest relatively higher diversity in the Paleozoic. The extremely low Cretaceous diversities, predicted by assuming sampling is proportional to sediment area, probably reflect a bias related to the large areas of Cretaceous sediments located in the continental interiors. Although more species have been described from the Cretaceous than any other period except the Tertiary (Raup, 1976a), the number of species is still small in comparison to the relative areal extent of Cretaceous sediments. Similarly, the small areal extent of most Paleozoic systems, in comparison to the number of species known from each period, would make the estimated total diversity appear high. Also, the sediment area data includes areas of non-marine sediment, which may be a significant bias (Flessa and Sepkoski, 1978). Cretaceous sediments, in particular, contain large quantities of non-fossiliferous intramontain deposits which probably bias the data set. Both effects can be interpreted as an aberration reflecting the inadequacy of sediment area as a measure of sampling intensity.

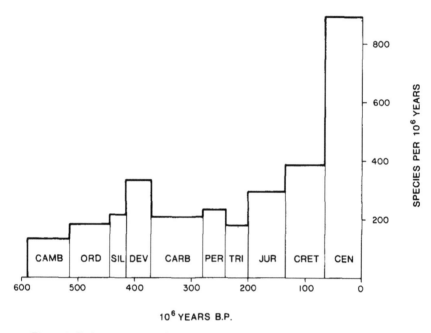

Figure 9. Estimated pattern of apparent species richness calculated assuming a doubling of sampling intensity, with no change in the distribution of sampling effort. Apparent species richness is normalized to period length.

One factor that has not been considered in the foregoing analysis is how possible temporal variations in diagenesis may impact the results. It is possible that diagenesis could be significantly more severe in older sediments which would, in effect, reduce the return on sampling effort. If this is the case, species richness in older periods could be greater than predicted here.

WHY DID DIVERSITY INCREASE?

Several authors have addressed the question of what processes control global species richness. Valentine (1970), Valentine and Moores (1972), Valentine et al. (1978), and Schopf (1979) have argued that the changing configurations of continents upon the earth, driven by plate tectonics, has produced varying levels of provinciality through the Phanerozoic. When provinciality is high, global diversity will also be high without any changes in species packing. While we need better data, provinciality appears to be higher now than at any other time in the earth's history (Valentine and Moores, 1972; Valentine et al., 1978; Schopf, 1979).

Other factors may also have been important. Bambach (1977) argued that within-habitat species richness increased dramatically in nearshore and open marine environments at the end of the Mesozoic. As Valentine et al. (1978) pointed out, this increase in species packing combined with variations in provinciality may be responsible for the Permo-Triassic mass extinctions. Conceivably, more subtle variations in shelf area over a longer term could also affect standing species richness (see Wise and Schopf, 1981). Finally, Ausich and Bottjer (1982) have suggested that variations in tiering, or vertical partitioning of space by suspension feeders, has played a role in the history of species richness. Although their estimates of variations in tiering do not resemble the results presented here, it is possible that tiering may have played some part in altering within-habitat species richness. The nature of this factor and the pattern of its effect remains to be established. While I agree with Schopf (1979) and Wise and Schopf (1981) that provinciality is the most significant factor controlling species richness in the Phanerozoic, it is clear that several factors are important and that further work will be required to sort out the relative impact of the various processes controlling global species richness.

HOW GOOD IS THE FOSSIL RECORD?

One measure of the adequacy of the fossil record is the percent of species that existed in the past and have been recovered from the fossil record and described. This can be estimated by comparing the patterns of species richness predicted here to the numbers of described species published by Raup (1976a). A reasonable estimate of past species richness (calculated assuming sampling was proportional to P.I.U., using data set 2 [Table 1] and assuming a five-fold Cenozoic diversity increase) is about ten times larger than the number of described species calculated by Raup. This suggests that the fossil record may actually be slightly better than many paleontologists have suspected (e.g., Durham, 1967), but is within the range suggested by Valentine (1970).

Comparing different periods reveals an interesting pattern. We have apparently discovered a higher percentage of Cambrian species than Cenozoic species (Table 2). In fact, we have discovered a percentage of Cenozoic species smaller than the percent recovered from any other period. This result is contrary to intuition, because the Cenozoic has been sampled more intensively than any other interval. Nevertheless, there is a reasonable explanation for the pattern, closely related to the apparent cause of the post-Mesozoic diversity increase. In the Paleozoic, during a time of relatively low provinciality, species would have tended to have wider ranges than later, during times of high provinciality. Therefore, paleontologists only need sample a small geographic range to pick up most common species from times of low provinciality. In contrast, when provinciality is high paleontologists must sample many

Table 2. Percent of fossilizable marine invertebrates recovered from the fossil record. Numbers of described species are from Raup (1976a). Estimated richness was calculated using data set 2 (Table 1), assuming sampling was proportional to P.I.U.s and Cenozoic diversity increased by a factor of five. Accuracy implied by the number of significant figures in left two columns is not justified and are given only because they represent actual program input and output.

	Species Described	Estimated Richness	Percent Described
CAMBRIAN	8,260	44,000	19
ORDOVICIAN	10,916	59,500	18
SILURIAN	5,519	18,500	30
DEVONIAN	12,381	77,500	16
CARBONIFEROUS	15,259	117,500	13
PERMIAN	7,592	39,000	19
TRIASSIC	6,006	27,000	22
JURASSIC	13,039	86,500	15
CRETACEOUS	22,166	133,500	17
CENOZOIC	43,056	605,000	7
TOTALS:	144,194	1,208,000	12

geographic areas to discover even the relatively common species. This latter case would seem to correspond to the situation in the Cenozoic, and would explain the low percentage of species known from that era.

Identical results would be obtained if the numbers of species described from each geologic period were compared to the Empirical model of species richness. The estimated high recovery rate of Paleozoic species is not dependent upon the sampling model developed here, but is a direct result of low diversity in the Paleozoic and Mesozoic.

DO HIGHER TAXA REFLECT SPECIES LEVEL EVENTS?

Traditionally, paleontologists have attempted to circumvent the problem of sampling bias at the species level by using higher taxa to infer species level evolutionary events. Sepkoski (1978) explicitly defended this technique; other studies have incorporated this approach as an implicit assumption (e.g., Gould et al., 1977). The results presented in this paper are the first estimate of global species richness calculated independently of the number of higher taxa, and afford the opportunity to test the idea that counts of higher taxa reflect species richness.

The numbers of orders, families, genera and species through time are illustrated in Figure 10. It is fairly obvious from even a cursory inspection that the higher the taxonomic level the more disparate the patterns of taxonomic richness. Even the lowest supraspecific level, the genus, does not accurately reflect trends in species richness. These dissimilarities suggest that higher taxa should not be trusted to reveal trends in species richness.

This conclusion is supported by Raup's (1979) analysis of the Permo-Triassic mass extinction. Raup estimated that a 17 percent decrease in the number of orders would translate into a 96 percent decrease in the number of species. Similarly, a 52 percent decline in families would also translate into a 96 percent decrease in species richness. Raup's results show that the relationship between numbers of species is not straightforward; the richness of higher taxa is well damped against variations in species richness.

This conclusion does not imply that variations in numbers of higher taxa are uninteresting or not useful. Higher taxa may still reflect other factors, such as the presence of important evolutionary innovations within a clade which permit new modes of life ("adaptive zones" *sensu* Simpson, 1953). Analysis of taxic richness and taxonomic frequency rates will undoubtedly continue to play a role in future paleontological studies.

CONCLUSIONS

Global levels of fossilizable marine invertebrate species richness were much lower in the Paleozoic, perhaps as low as one tenth Cenozoic levels. In the Paleozoic, diversity was highest in the Devonian and Carboniferous. Mesozoic diversity levels were highest in the Jurassic and Cretaceous. This pattern of diversity closely resembles the Empirical model of species richness proposed by Valentine (1970) and Valentine et al. (1978) and, to a lesser extent, the Consensus model of Sepkoski et al. (1981). While it remains uncertain which processes are responsible for this pattern, it is clear that variations in levels of provinciality played a major role.

Approximately ten percent of the fossilizable marine invertebrates that have existed in the past have been discovered and described. Further sampling of the fossil record will result in recovery of more species but will not significantly modify present patterns of variation in the number of described species through time if patterns of sampling do not change. The estimated trends in actual species richness through time do not match trends in the number of higher taxa. This poor match suggests that higher taxa are insensitive indicators of species level trends in time.

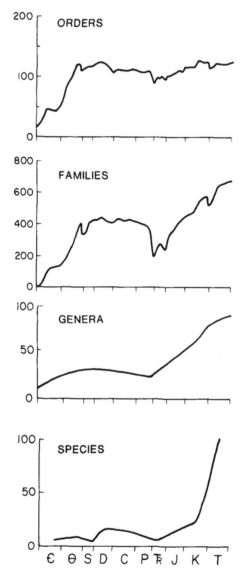

Figure 10. Trends in taxonomic richness through time. Note that the species richness curve is not congruent with the order, family and generic curves. Data on orders are from Sepkoski (1978), family data are from Sepkoski (1979), generic data are from Sepkoski et al. (1981), and species data were calculated using data set 2 (Table 1) and assuming sampling was proportional to P.I.U.s and that Cenozoic species richness increased by a factor of five. Scales for genera and species are percent of Recent richness.

REFERENCES

Ausich, W. I., and Bottjer, D. J., 1982. Tiering in suspension-feeding communities on soft substrata throughout the Phanerozoic. Science 216:173-174.

Bambach, R. K., 1977. Species richness in marine benthic habitats through the Phanerozic. Paleobiology 3:152-167.

Blatt, H., and Jones, R. L., 1975. Proportions of exposed igneous, metamorphic, and sedimentary rocks. Geol. Soc. Am. Bull. 86:1085-1088.

Cooper, G. A., and Williams, A., 1952. Significance of the stratigraphic distribution of brachiopods. J. Paleontol. 26:326-337.

Durham, J. W., 1967. The incompleteness of our knowledge of the fossil record. J. Paleontol. 41:559-565.

Flessa, K. W., and Sepkoski, J. J., Jr., 1978. On the relationship between Phanerozoic diversity and changes in habitable area. Paleobiology 4:359-366.

Gould, S. J., Raup, D. M., Sepkoski, J. J., Jr., Schopf, T. J. M., and Simberloff, D. S., 1977. The shape of evolution: A comparison of real and random clades. Paleobiology 3:23-40.

Gregor, B., 1970. Denudation of the continents. Nature 228:273-275.

Gregory, J. T., 1955. Vertebrates in the geologic time scale. G.S.A. Spec. Pap. 62:593-608.

Koch, C. F., 1978. Bias in the published fossil record. Paleobiology 4:367-372.

May, R. M., 1975. Patterns of species abundance and diversity. In Cody, M. L., and Diamond, J. M. (eds.), Ecology and evolution of communities, Cambridge: Belknap Press, 81-120.

McNaughton, S. J., and Wolf, L. L., 1970. Dominance and the niche in ecological systems. Science 167:131-139.

Pielou, E. C., 1975. Ecological diversity. New York: John Wiley and Sons, 165.

Raup, D. M., 1972. Taxonomic diversity during the Phanerozoic. Science 177:1065-1071.

Raup, D. M., 1976a. Species diversity in the Phanerozoic: A tabulation. Paleobiology 2:279-288.

Raup, D. M., 1976b. Species diversity in the Phanerozoic: An interpretation. Paleobiology 2:289-297.

Raup, D. M., 1977. Species diversity in the Phanerozoic: Systematists follow the fossils. Paleobiology 3:328-329.

Raup, D. M., 1979. Size of the Permo-Triassic bottleneck and its evolutionary implications. Science 206:217-218.

Schopf, T. J. M., 1974. Permo-Triassic extinctions: Relation to sea-floor spreading. J. Geol. 82:129-139.

Schopf, T. J. M., 1979. The role of biogeographic provinces in regulating marine faunal diversity through geologic time. In Gray, J., and Boucot, A. J.

(eds.), Historical biogeography, plate tectonics, and the changing environment, Corvallis: Oregon State Univ. Press, 449-457.

Sepkoski, J. J., Jr., 1978. A kinetic model of Phanerozoic taxonomic diversity I. Analysis of marine orders. Paleobiology 4:223-251.

Sepkoski, J. J., Jr., 1979. A kinetic model of Phanerozoic taxonomic diversity II. Early Phanerozoic families and multiple equilibria. Paleobiology 5:222-251.

Sepkoski, J. J., Jr., Bambach, R. K., Raup, D. M., and Valentine, J. W., 1981. Phanerozoic marine diversity and the fossil record. Nature 293:435-537.

Sheehan, P. M., 1977. Species diversity in the Phanerozoic: A reflection of labor by systematists? Paleobiology 3:325-328.

Signor, P. W., 1978. Species richness in the Phanerozoic: An investigation of sampling effects. Paleobiology 4:394-406.

Signor, P. W., 1982. Species richness in the Phanerozoic: Compensating for sampling bias. Geology 10:625-628.

Simpson, G. G., 1952. How many species? Evolution 6:342.

Simpson, G. G., 1953. The major features of evolution. New York: Columbia Univ. Press, 434.

Simpson, G. G., 1960. The history of life. In Tax, S. (ed.), Evolution after Darwin, Chicago: Univ. of Chicago Press, 117-180.

Stanley, S. M., 1975. A theory of evolution above the species level. Nat. Acad. Sci. Proc. 72:646-650.

Stanley, S. M., 1979. Macroevolution. San Francisco: Freeman, 332.

Stanton, R. J., and Evans, I., 1972. Community structure and sampling requirements in paleoecology. J. Paleontol. 46:845-858.

Teichert, C., 1957. How many fossil species? J. Paleontol. 31:967-969.

Tipper, J. C., 1979. Rarefaction and rarefiction—the use and abuse of a method in paleoecology. Paleobiology 5:423-434.

Valentine, J. W., 1970. How many marine invertebrate fossil species? A new approximation. J. Paleontol. 44:410-415.

Valentine, J. W., Foin, T. C., and Peart, D., 1978. A provincial model of Phanerozoic marine diversity. Paleobiology 4:55-66.

Valentine, J. W., and Moores, E. M., 1972. Plate-tectonic regulation of faunal diversity and sea level: A model. Nature 228:657-659.

Whittaker, R. H., 1975. Communities and ecosystems. Second ed. New York: MacMillan, 385.

Williams, A., 1957. Evolutionary rates of brachiopods. Geol. Mag. 94:201-211.

Wise, K. P., and Schopf, T. J. M., 1981. Was marine faunal diversity in the Pleistocene affected by changes in sea level? Paleobiology 7:394-399.

Part II

PATTERNS OF FAUNAL EXPANSION, PARTITIONING AND REPLACEMENT

Waxing and waning clades must eventually be studied in ecological contexts if we are ever to understand the processes and events which have created the clade profiles. Three different approaches to this task are presented in this section. Sepkoski and Miller examine the patterns of major marine faunas and show that during the Paleozoic, each successive fauna first appears onshore of the others and then extends its dominance into offshore environments over millions of years. The cause of this unexpected but large-scale pattern, not yet understood, is clearly a process of major significance in the history of marine life. Bambach compares the ecological roles represented during the successive dominance of the major Phanerozoic marine faunas and shows that many adaptive zones simply lie vacant for hundreds of millions of years. Commonly the eventual filling of vacant zones is associated with the origin of middle- to upper-level taxa. Ausich and Bottjer examine the occupation pattern along a single environmental dimension—tiering—in some detail. They find that times of expansion and partitioning of tiers correspond well with times of general diversity increase; tiering is obviously a significant mode of accommodation of species diversity within marine ecosystems.

Chapter 5

EVOLUTIONARY FAUNAS AND
THE DISTRIBUTION OF PALEOZOIC
MARINE COMMUNITIES IN SPACE AND TIME

J. JOHN SEPKOSKI, Jr. and ARNOLD I. MILLER

Department of the Geophysical Sciences, University of Chicago

INTRODUCTION

The concept of evolutionary faunas and floras provides a useful vehicle for describing and analyzing major changes in the composition of the Earth's biotas through time (Sepkoski, 1981a; Niklas et al., this volume). Much of the variation in the global marine biota through the whole of the Phanerozoic can be summarized in terms of just three major evolutionary faunas: a Cambrian Fauna, consisting largely of trilobites, inarticulate brachiopods, monoplacopherans, hyolithids, and eocrinoids; a later Paleozoic Fauna, composed primarily of articulate brachiopods, crinoids, corals, and stenolaemate bryozoans; and a Mesozoic-Cenozoic, or "Modern," Fauna, consisting mostly of gastropods, bivalves, bony fishes, malacostracan crustaceans, and echinoids (see also Sepkoski, 1981b; Sepkoski and Sheehan, 1983; Ausich and Bottjer, this volume; Bambach, this volume). As illustrated in Figure 1, all of these faunas originated early in the Phanerozoic but then diversified at different rates, with each fauna attaining a successively higher maximum diversity and appearing to displace the fauna before it (Sepkoski, 1979; Kitchell and Carr, this volume). The Cambrian Fauna, which had the lowest diversity, expanded rapidly during the Cambrian Period but then slowly declined following the

Phanerozoic Diversity Patterns·
Profiles in Macroevolution

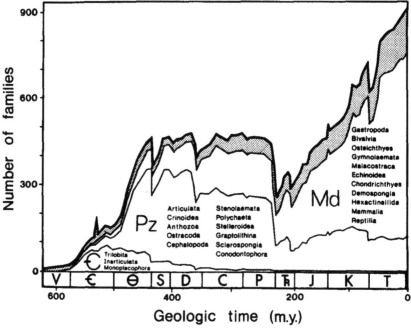

Geologic time (m.y.)

Figure 1. The diversity of animal families in the oceans through the Phanerozoic, with the three global evolutionary faunas delineated. The uppermost curve shows the total number of animal families that have been described from each stage of the Phanerozoic marine fossil record. The stippled area represents the contribution to total diversity made by families of rarely fossilized soft-bodied and lightly sclerotized animals. The fields below the stippled area show the diversities of the three evolutionary faunas: the Cambrian Fauna ("ε"), Paleozoic Fauna ("Pz"), and Modern Fauna ("Md"). The principal classes in each fauna are listed within the fields in the order of their importance (i.e., maximum diversity). This figure is similar to Figure 6 in Sepkoski (1979), but the evolutionary faunas are delineated by their actual diversities rather than by projections of factor loadings. Data are from Sepkoski (1982).

onset of the Ordovician. The Paleozoic Fauna, with a familial diversity three times greater than that of the Cambrian, diversified during the Ordovician Period and dominated the marine biota until its massive decline during the Late Permian mass extinctions. Finally, the Modern Fauna, with a maximum familial diversity nearly twice that of the Paleozoic Fauna, increased in diversity slowly from the Cambrian to the Permian Periods and then rapidly rose to dominance after the Permian extinctions; it has remained the dominant marine fauna to the present.

These broad patterns are derived from data on marine diversity summed over the entire world and therefore do not immediately relate to changes occurring in any particular ecological community or local environmental setting. However, there have been several suggestions that a parallel relationship did in fact exist between turnover in global evolutionary faunas and distributional changes in local benthic communities. Berry (1972, 1974), for example, found that trilobite-rich communities, characteristic of the Cambrian Fauna, became restricted to offshore environments with expansion of the typically Paleozoic brachiopod-rich communities during the Ordovician. Bretsky (1968, 1969) observed that mollusc-rich assemblages, broadly suggestive of Mesozoic-Cenozoic communities, were concentrated in nearshore environments through the Paleozoic Era, while brachiopod-rich communities were concentrated primarily offshore. Recently, Sepkoski and Sheehan (1983) analyzed more than 100 Cambrian and Ordovician level-bottom communities and concluded that representatives of the three evolutionary faunas were indeed segregated into different portions of the shelf environment during the early part of the Paleozoic Era.

In this chapter, we present a further analysis of the composition and distribution of marine communities[1] during the Paleozoic Era. We wish to document that:

1. there is a definite correspondence between the development of global evolutionary faunas and the evolution of marine benthic communities throughout the Paleozoic Era;
2. the development of evolutionary faunas was characterized by strong spatio-temporal patterns, with faunal transitions taking place diachronously across shelf environments;
3. the originations and expansions of evolutionary faunas probably reflected evolutionary processes occurring continuously in local environments and were not necessarily direct by-products of large-scale environmental perturbations such as those presumably associated with mass extinctions.

In the first part of this chapter we discuss the nature of and problems associated with data on Paleozoic benthic communities. We then present results of a Q-mode factor analysis of these data and consider how these relate to global evolutionary faunas. Finally, we conclude with some brief considerations of problems in community evolution and paleoenvironmental reconstruction.

---

[1] The term "community" is used throughout this chapter to mean the assemblage of organisms that lived together in a particular place or were fossilized together in a particular facies. No organizational properties are necessarily implied. Mostly, we use "communities" as operational units of sampling of the biota in a particular environment. Similarly, we use the term "community evolution" to mean simply evolutionary change in the composition of the biota in a particular environment and do not imply any special evolutionary processes unless explicitly stated otherwise.

PALEOZOIC COMMUNITY DATA

The analytic design for this study of Paleozoic communities was simple in conception but became complicated in execution. Basically, we endeavored to collect information on a large number of Paleozoic level-bottom communities, then search for statistical patterns of faunal variation among these communities, and finally investigate how these patterns relate to the distribution of the communities in space and time. Many of the details of the analysis follow the procedures used by Sepkoski and Sheehan (1983).

Our initial data base consisted of information on the age, sedimentary environment, and faunal composition of more than 300 communities, gleaned largely from the published literature (see Appendix). Actually, these "communities" represented a diverse array of paleoecologic communities and assemblages as well as biostratigraphic faunules and biofacies, all of which shared the quality of being samples of the total fossil content of some restricted stratigraphic and environmental interval. The assortment of faunal units was collected in order to obtain as broad an environmental and temporal sampling as possible. This sampling was limited to North America (exclusive of Acadia), however, so that large-scale provincial overprints on local ecological variation would be minimized. In assembling the data base, we began with the set of Cambro-Ordovician communities compiled by Sepkoski and Sheehan (1983) and then added approximately 200 more faunal units of Silurian through Permian age.

A simplified representation of the distribution of the communities in space and time is given by the "time-environment diagram" in Figure 2. Each numbered box in this diagram shows the age (vertical dimension) and approximate environment (horizontal dimension) of a single analyzed community. The environmental framework utilized in this figure is a very simple shelf-slope gradient, caricatured at the top of Figure 2. This framework is, of course, a great oversimplification of the actual environmental situations of the communities, reducing the complex nature of marine benthic environments to the simple dimensions of depth and distance from shore. It ignores differences between

---

Figure 2. A "time-environment diagram" showing the temporal and environmental distribution of communities analyzed in this study. Each stippled box represents one community. Its vertical placement in the diagram indicates the community's age, and its horizontal position indicates the community's approximate environmental range. (The environmental framework is illustrated by the simple shelf-slope model at the top of the diagram.) The timescale is shown in the column to the right of the diagram, with systems and series of the Paleozoic Era delineated. (Note that this column is not scaled to absolute time.) The literature sources of the information on the communities are listed by community number in the appendix to this paper.

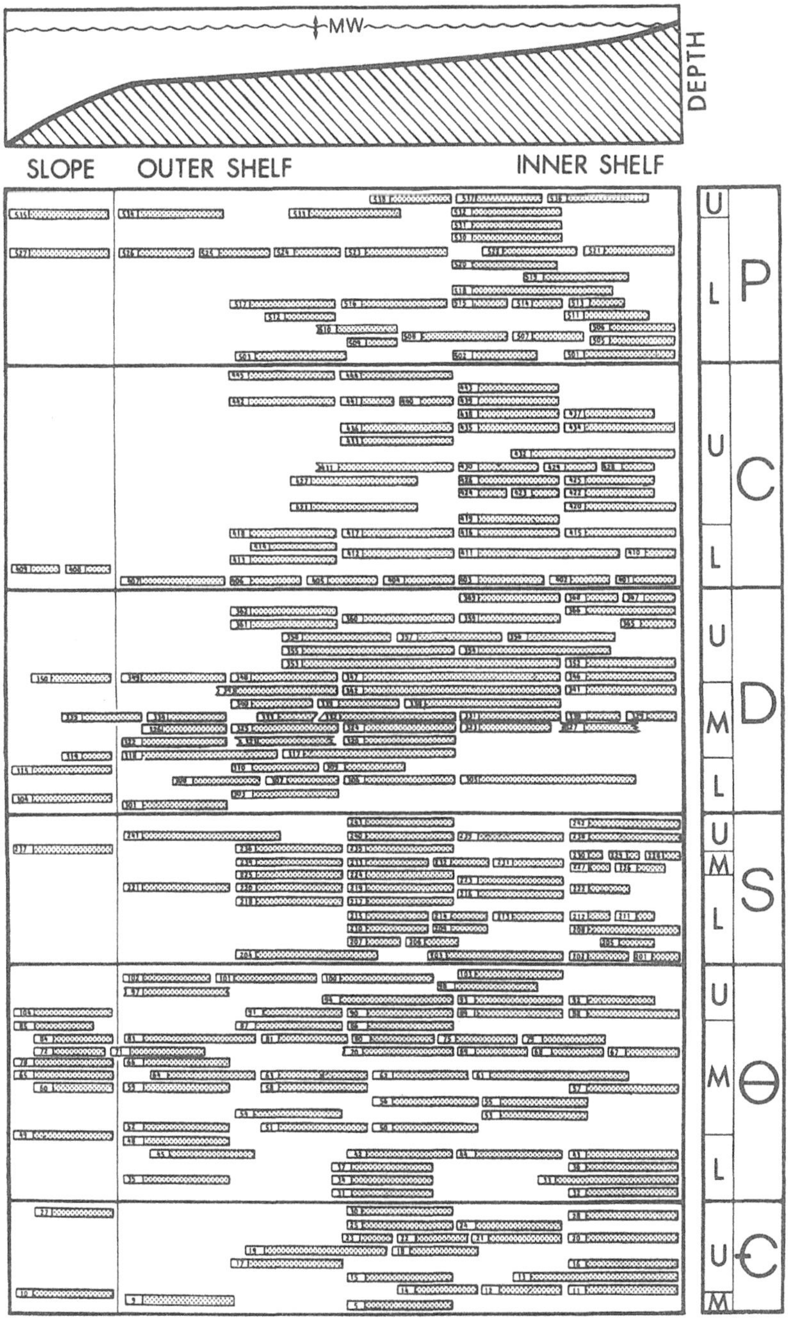
MW
DEPTH
SLOPE
OUTER SHELF
INNER SHELF
P
C
D
S
Θ
Є

carbonate and terrigenous environments, soft and firm substrates, hyper- and hyposaline conditions, epeiric and ocean-facing seas, and tropical and temperate climatic settings. Thus, the diagram provides only a first approximation of the environmental differences among the communities and represents no more than a starting point for interpretation of faunal variation among them.

The positioning of communities in the time-environment diagram involved a combination of rigorous guidelines and educated guesses. In general, we attempted to use lithologic, and not paleontologic, indicators in making environmental decisions, and we relied heavily (although critically) upon the interpretations of the primary authors of the community data. Communities from very shallow-water facies with features such as desiccation cracks, flaser bedding, and stromatolites were placed along the righthand margin of the time-environment diagram. Communities from deepwater black shales and/or turbidite sequences were placed along the lefthand margin. (This positioning involved some uncertainty, since it is not always clear whether a black shale reflects deposition in a deepwater basin or a shallow-water lagoon; see Heckel, 1977.) Communities from environments close to wavebase were placed near the center of the shelf gradient. (Thus, this gradient is not precisely linear with respect to depth or distance from shore.) Environments in the central position include offshore shoals and delta-front sands and are represented by the equivalents of Ziegler et al.'s (1968) *Pentamerus* Community and Boucot's (1975) Benthic Assemblage 3. To the right of the center we placed communities from delta platforms, nearshore lagoons, etc., and to the left, communities from deeper water, open-shelf facies. Obviously, there is considerable uncertainty in our placement of many communities in Figure 2, especially since lithologic and other paleoenvironmental information is often scant or ambiguous in the paleontologic literature.

We were not able to obtain a completely even coverage of environments throughout the Paleozoic Era. As shown in Figure 2, outer shelf and slope environments are poorly represented, especially in the Carboniferous and Permian, whereas inner to mid shelf environments are often densely represented. This problem reduces the resolution of the data set and may impede recognition of regular patterns of faunal variation; however, there is no reason to believe that this difficulty should lead to the recognition of non-existent trends. Similarly, we do not believe that there are systematic biases in the proportions of carbonate vs. terrigenous, open-ocean vs. cratonic, or tropical vs. temperate facies through the data set. Thus, trends in faunal composition among the sampled communities should reflect evolutionary patterns and not long-term environmental changes.

The faunal compositions of the communities were compared to one another by examining the importances (that is, internal diversities) of the orders within them. Orders were chosen as an appropriate taxonomic level for analysis because they are neither too restricted with respect to time nor too widely distributed with respect to environment. Most lower taxa have geologic durations that are considerably shorter than the length of the Paleozoic Era so that analyses based on them tend to cluster communities by age rather than environment. On the other hand, taxa above the ordinal level often have environmental ranges that encompass the entire shelf so that analyses based on them tend to miss significant changes in the compositions of certain kinds of communities over the course of time.

Ordinal importances were measured by counting the number of genera within each order in each community. This metric permits us to see where each order was attaining its maximum, or minimum, diversity through the Paleozoic. It also reflects to some extent the ecological dominance of the orders within the communities since the richness of subtaxa (species, genera, etc.) within higher taxa correlate with the relative abundances of those subtaxa in many communities. (That is, communities with similar relative abundances of species also tend to have similar diversities of species within higher taxa.) An example of this correlation is illustrated by the scattergram in Figure 3. This figure shows the relationship between statistical similarities computed for ordinal importances and for species abundances in a data set on molluscan species sampled at 37 stations in shallow water off St. Croix (Miller, 1981; Miller and Bambach, 1981). The similarities were computed by first constructing two data arrays, one of the number of species within each order at each sampling station and the other of the number of individuals within each species at the same stations. Between-station similarity matrices were calculated for each array using the Cosine θ Coefficient (or "Coefficient of Proportional Similarity") of Imbrie and Purdy (1962), and the two similarity matrices were plotted and correlated, using the "matrix correlation" technique of Sneath and Sokal (1973:280 ff). The correlation coefficient for the two similarity matrices is 0.592 which, although not high, is sufficient for multivariate analyses to provide comparable ordinations of stations for both the species abundances and the ordinal importances.

In the analysis of Paleozoic communities, genera rather than species had to be counted because of inconsistencies in the delineation of fossil species among the communities and because the genus is commonly used as the operational taxonomic unit in paleoecologic studies. Even with this concession, several exceptions to the use of genera and orders had to be made:

1. No microfossils were counted. Only a few researchers routinely include both micro- and macrofossils in their faunal lists, and thus inclusion of scattered microfossils in our data set would have induced artificial sampling variation. For operational purposes, we defined a microfossil as any

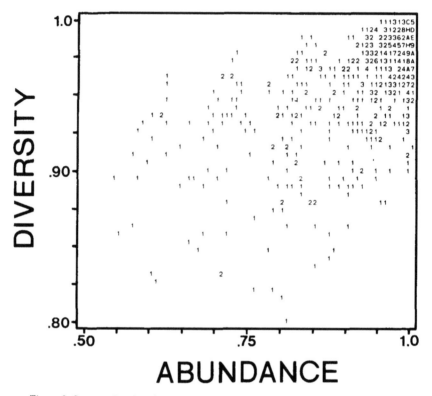

Figure 3. Scatter plot showing the relationship between similarity coefficients computed for the abundance of individuals within species (x-axis) and for the diversity of speccies within orders (y-axis). The data are from a census of dead molluscan shells at 37 stations along a shallow-water depth gradient off the coast of St. Croix, U.S. Virgin Islands (Miller, 1981). Cosine θ similarity coefficients (Imbrie and Purdy, 1962) were computed for all 666 pairs of stations based first on the abundance of individuals within species and then on the diversity (richness) of species within orders. The scatter plot shows these two sets of coefficients plotted against each other, with multiple overlapping points denoted by numbers and letters (A = 10 points, B = 11, etc.). The product-moment correlation for the two sets is 0.592, indicating the two types of data reflect basically similar patterns in the environmental distributions of the molluscs.

taxon with adult individuals smaller than 5 mm in maximum dimension. This criterion excluded all conodonts and radiolarians and most ostracodes (except leperditicopids) and foraminifers (except fusulinids and a few other subgroups).

2. Some classes and subclasses were used instead of orders as variables. Some kinds of fossils are so seldomly identified to the ordinal level that any attempt to tabulate their diversity within orders was impractical. For such

groups, we counted the number of identified taxa within the class or sub-class. Affected taxa include the Hexactinellida, Demospongia, Scapho-poda, Nautiloidea, Ammonoidea, Hyolitha, Crinoidea, Eocrinoidea, and Echinoidea (see Table 1).

3. Crinoids were treated as a special variable. Most published faunal lists mere-ly note the presence of "crinoid debris," or the like, without any identifi-cation of the groups present. Some faunal lists do include large numbers of crinoid taxa, however, as a result of unique preservation and/or special effort by the workers. In such cases, we tabulated the number of identified *orders* within the Crinoidea (as opposed to the mere presence of Crinoidea) in an attempt to prevent the affected communities from behaving as statistical outliers. This arbitrary compromise undoubtedly underestimates the importance of crinoids in Paleozoic communities, but it does seem to minimize nonbiological variation resulting from preservation or sampling.

A total of 80 orders and substituted higher taxa were present within our ini-tial data set. However, because of the limitations of the computer programs available to us, we had to reduce this number to 50 (Table 1). We therefore eliminated those orders that occurred in fewer than 10 communities. This manipulation resulted in some communities (particularly those in the Lower and Middle Cambrian) losing a considerable portion of their generic diversity; we therefore chose to remove all communities that lost more than a third of their genera. This resulted in a final data set that consisted of counts of gen-era within 50 orders distributed among 280 communities.

Patterns of faunal variation among the communities were investigated using Q-mode factor analysis with Varimax rotation. This multivariate statistical technique has three main uses, as outlined by Jöreskog et al. (1976:88):

(1) To find the minimum number. . .of "end-member" assemblages of which the observed objects [communities, in this case] may be con-sidered combinations.

(2) To specify the compositions of the end-members in relation to the. . . constituents [that is, orders] .

(3) To describe each object in terms of the end-members; that is, to "un-mix" the objects into their end-member components.

In terms of the present study, the analytic properties of Q-mode factor analy-sis permitted us first to test whether three end-members, or "community types," are adequate for describing the basic composition of most Paleozoic marine communities; then to assess how the taxonomic compositions of these end-members are related to the three global evolutionary faunas; and finally to investigate how the end-members are distributed in space and time. All computations in this analysis were performed with the CABFAC computer program of Klovan and Imbrie (1971).

Table 1. Taxa used in the factor analysis of Paleozoic marine communities. The 50 orders and higher taxa actually used as variables are listed in italics.

PROTOZOA
  Cl. Rhizopodea
    Foraminferida

PORIFERA
  Cl. *Hexactinellida*
  Cl. *Demospongia*
  Cl. Sclerospongia
    Stromatoporoidea

COELENTERATA
  Cl. Scyphozoa
    Conulariida
  Cl. Anthozoa
    Tabulata
    Rugosa

MOLLUSCA
  Cl. Monoplacophora
    Cyrtonellida
  Cl. Gastropoda
    Bellerophontida
    Archaeogastropoda
    Mesogastropoda
  Cl. *Scaphopoda*
  Cl. Bivalvia
    Nuculoida
    Mytiloida
    Arcoida
    Pterioida
    Modiomorphoida
    Trigonioida
    Veneroida
    Pholadomyoida
  Cl. Cephalopoda
    Nautiloidea
    Ammonoidea
  Cl. Cricoconarida
    Tentaculitida
    Dacryoeonarida
  Cl. *Hyolitha*

ANNELIDA
  Cl. Polychaeta
    Serpulimorpha

ARTHROPODA
  Cl. Trilobita
    Agnostida
    Ptychopariida
    Proetida
    Phacopida
  Cl. Ostracoda
    Leperditicopida

BRYOZOA
  Cl. Stenolaemata
    Cyclostomata
    Cystoporata
    Trepostomata
    Cryptostomata

BRACHIOPODA
  Cl. Inarticulata
    Lingulida
    Acrotretida
  Cl. Articulata
    Orthida
    Strophomenida
    Pentamerida
    Rhynchonellida
    Spiriferida
    Atrypida
    Terebratulida

ECHINODERMATA
  Cl. *Crinoidea*
  Cl. *Cystoidea*
  Cl. *Eocrinoidea*
  Cl. *Echinoidea*

HEMICHORDATA
  Cl. Graptolithina
    Dendroidea
    Graptoloidea

FACTOR ANALYSIS OF PALEOZOIC COMMUNITY DATA

EIGENVALUES

The data set on Paleozoic communities proved to be very heterogeneous, as might be expected from the diverse array of faunal units and sampling techniques represented in the data, the arbitrary decisions made when scoring the variables, and the inherent variability of faunal distributions in the complex array of sampled environments. The data appear far from random, however, as seen by the character of the eigenvalues, scores, and loadings resulting from the factor analysis. The magnitudes of the eigenvalues from the first 10 unrotated factors (or "principal vectors") are illustrated in the "scree graph" in Figure 4. This graph was constructed by taking the relative eigenvalue (that is,

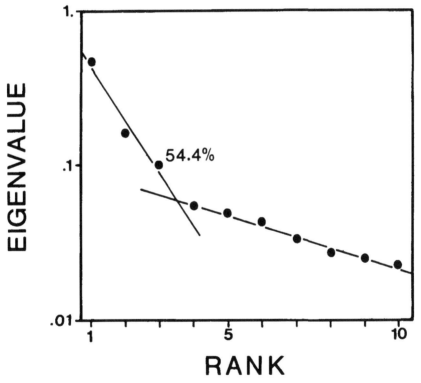

Figure 4. A scree graph showing the relationship between eigenvalues and factor ranks in the Q-mode factor analysis of Paleozoic community data. The relative eigenvalues for the first ten unrotated factors are plotted on a logarithmic axis against the ranks of their respective factors. The lines fitted to the points show a change in rate of decay of the eigenvalues between the first three factors and the following seven. The first three factors together account for 54.4% of the community data.

proportion of encompassed data) associated with each factor and plotting it logarithmically against the factor's rank. Such plots permit rapid visual assessment of how the amount of data encompassed by each factor declines as more factors are extracted in the analysis; this, in turn, permits easy identification of changes in the rate of decline which may signal changes in the kind of information encompassed by the factors (see Sepkoski, 1981a).

In Figure 4 the rate of decline seems to change markedly after three factors, as indicated by interpretive lines fitted to the eigenvalues. The first three factors, encompassing 54.4% of the data, exhibit a rapid decline in eigenvalues as opposed to the more gradual and continual decline of eigenvalues associated with factors 4 through 10. A very similar pattern emerged when the entire data set was logarithmically transformed; the first three factors, encompassing 53.5% of the data, displayed a steep decline in eigenvalues whereas the higher-numbered factors had a much more gradual decline (with a small break between factors 7 and 8). This structure suggests that the three largest factors in the data set are incorporating a different and perhaps more general kind of information than are the minor, higher-numbered factors.

TAXONOMIC COMPOSITION OF FACTORS

The ordinal-level compositions of the first three factors, following varimax rotation and rearrangement into stratigraphic sequence, are shown by the factor scores in Table 2. As evident, the compositions of the factors for the Paleozoic community data closely parallel the class-level compositions of the evolutionary faunas (Figure 1). Factor 1, which accounts for 13.4% of the data, is dominated by members of the Cambrian Fauna, including trilobites (Ptychopariida, Phacopida, Agnostida, and Proetida), inarticulate brachiopods (Lingulida and Acrotretida), monoplacophorans (Cyrtonellida), eocrinoids, and hyolithids. The only order that is "misclassified" on this factor is the Graptoloidea; its moderate score on Factor 1 results from the preservation of graptolites with trilobites and inarticulate brachiopods in deepwater environments during the Ordovician (see below). The moderate score of the Phacopida on Factor 2, on the other hand, reflects the common occurrence of these trilobites with members of the Paleozoic evolutionary fauna (especially orthid brachiopods) in shelf environments of the Ordovician through Devonian.

Factor 2 is the largest of the three rotated factors, accounting for 26.4% of the data. It receives high scores only from members of the Paleozoic evolutionary fauna. The five highest scores all come from orders of articulate brachiopods: the Strophomenida, Spiriferida, Orthida, Atrypida, and Rhynchonellida; pentamerids and terebratulids (not listed in Table 2) also have their highest scores on this factor. The substantial scores of orthids and pentamerids on Factor 1 reflect the high diversities of these brachiopods during the

Table 2.   Sorted scores for the first three rotated factors in the Q-mode
analysis of Paleozoic community data. Only scores with absolute
values greater than 0.04 are listed. (Lower scores are of negligible
importance in the factor structure.) Relative eigenvalues for the
factors are given at the bottom of the table. (Note that the
absolute order of the factors has been altered to parallel their
temporal sequence.)

TAXA	FACTORS		
	1	2	3
Ptychopariida	.91	-.07	–
Lingulida	.20	–	.04
Phacopida	.12	.10	–
Acrotretida	.12	–	.04
Agnostida	.10	–	–
Graptoloidea	.09	–	–
Proetida	.06	–	.05
Cyrtonellida	.06	–	–
Eocrinoidea	.04	–	–
Hyolitha	.04	–	–
Strophomenida	–	.70	.06
Spiriferida	-.08	.51	–
Orthida	.24	.27	-.08
Atrypida	–	.17	-.04
Rhynchonellida	–	.17	.06
Rugosa	–	.16	–
Crinoidea	–	.14	–
Cryptostomata	–	.12	.10
Tabulata	–	.11	–
Pentamerida	.06	.07	-.04
Archaeogastropoda	.04	-.08	.74
Pterioida	-.05	.07	.37
Nuculoida	–	–	.32
Bellerophontida	–	-.04	.24
Mesogastropoda	–	–	.18
Nautiloidea	.06	–	.17
Veneroida	–	–	.11
Scaphopoda	–	–	.10
Pholadomyoida	–	–	.09
Modiomorphoida	–	–	.08
Trepostomata	–	.07	.08
Arcoida	–	–	.05
Serpulimorpha	–	–	.05
Trigonioida	–	–	.05
EIGENVALUES	13.4%	26.4%	14.6%

Ordovician Period when the Cambrian evolutionary fauna was still moderately important. Other typically Paleozoic taxa with substantial scores on Factor 2 include rugose and tabulate anthozoans, cryptostome bryozoans (also trepostome bryozoans, listed under Factor 3), and crinoids.

The third factor, which encompasses 14.6% of the data, receives high scores mostly from members of the Modern Fauna. These include orders of gastropods (Archaeogastropoda, Bellerophontida, and Mesogastropoda), bivalves (Pterioida, Nuculoida, Veneroida, Pholadomyoida, Modiomorphoida, Arcoida, and Trigonioida), and scaphopods. Nautiloids also have their highest score on Factor 3, even though as cephalopods they are best treated as members of the Paleozoic Fauna (Figure 1). This "misclassification" primarily reflects the occurrence of diverse nautiloids in shallow-water, gastropod-rich communities in the Lower Ordovician. Similarly, the moderate scores of the benthic Trepostomata (stenolaemate bryozoans) and Serpulimorpha (polychaetes) on Factor 3 indicate further associations of some elements of the Paleozoic and Modern evolutionary faunas in shallow-water communities of the later Paleozoic Era (see below).

DISTRIBUTION OF FACTORS IN SPACE AND TIME

The statistical sorting of the Cambrian, Paleozoic, and Modern evolutionary faunas onto separate factors would not be surprising if the analyzed communities covered the whole of the Phanerozoic; the sorting would then merely reflect the sequential dominance of the faunas through time. But the data analyzed here pre-date the rise to dominance of the Modern Fauna (Figure 1). Therefore the segregation of this fauna onto a distinct factor must indicate that it was separated from the Paleozoic Fauna in space as well as time.

This conclusion can be demonstrated quite clearly by contouring the factor loadings on the time-environment diagram, as shown in Figure 5. Each panel in this figure represents a simplified version of the time-environment diagram in Figure 2, with the box for each community having been replaced by a dot located at the box's center. The contours have magnitudes of 0.33 and 0.67, with stippling covering the area within the higher contour. Prior to contouring, the communality of each community on the three-factor solution was restored to unity and the absolute values of its loadings were transformed to sum to 1.0. This manipulation in effect eliminates the variability not encompassed by the first three factors and thus clarifies which of the general factors are most important in each community.

The contoured loadings show that Factor 1 is the most temporally restricted of the three factors. It receives uniformly high loadings from all Cambrian points in our data set, reflecting the dominance of trilobites and inarticulate brachiopods in nearly all Cambrian communities regardless of environment.

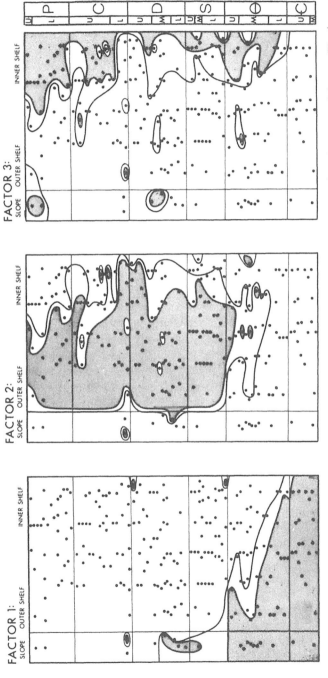

Figure 5. Contoured time-environment diagrams showing the loadings of the communities on the first three rotated factors. The dots, which represent the communities, have been placed at the centers of the boxes illustrated in Figure 2. The contours have values of 0.33 and 0.67, with the stippling covering areas within the 0.67 contour. The contours show that the communities with high loadings on the factors are segregated into well-defined fields in each time-environment diagram, indicating non-random distributions of the taxonomic orders in space and time. See text for further explanation.

This dominance was relinquished during the Ordovician with the rise of the Paleozoic and Modern Faunas. The distribution of loadings on Factor 1 shows that this turnover had a regular spatial pattern. In the Lower Ordovician, communities from most shelf environments in the data set still load highly on Factor 1; the several very nearshore communities, however, display only moderate to low loadings, reflecting the partial changeover to mollusc-dominated communities in these environments (represented in part by Boucot's [1983] Ecologic-Evolutionary Unit III). Through the Middle Ordovician, communities with high loadings on Factor 1 become progressively restricted to deeper, more offshore environments, so that by the top of the Ordovician only communities from outermost shelf, slope, and basinal environments continue to load heavily. This pattern reflects the diachronous onshore-offshore replacement of the Cambrian evolutionary fauna during the Ordovician radiations, discussed by Berry (1972, 1974) and Sepkoski and Sheehan (1983) (see also Lockley, 1983); it is this environmental constriction which may be largely responsible for the slow decline in diversity of that Fauna after the close of the Cambrian Period (see Figure 1).

The apparent persistence of high loadings on Factor 1 in "slope" environments beyond the Ordovician may be more an artifact of the methodology than a reality of the data. All post-Ordovician deepwater ("slope") communities have low communalities (i.e. are poorly fit) in the three-factor solution. The one Carboniferous point that loads highly on Factor 1 is a low-diversity community with several genera of inarticulate brachiopods, whereas the Silurian and Devonian points are mostly graptolite assemblages with scattered sponges, cricoconarids, proetids, and articulate and inarticulate brachiopods. These latter communities load highly on Factor 1 only because graptolites co-occur with trilobites and inarticulate brachiopods in the better-fitted deepwater communities of the Ordovician. Still, the representation in Figure 5 may not be entirely wrong. There is some evidence, especially in European sections, that trilobites, inarticulate brachiopods, and even hyolithids maintained their highest diversities or greatest abundances in outer shelf, slope, and basinal facies throughout the Silurian and Devonian Periods (Marr and Nicholson, 1888; Elles, 1939; McKerrow, 1978; Goldring and Langenstrassen, 1979; Mikulic and Watkins, 1981). This suggests that some remnants of the Cambrian Fauna did indeed persist as a unit in deepwater environments beyond the Ordovician.

Factor 2, which encompasses many elements of the Paleozoic evolutionary fauna, receives high loadings from mid to outer shelf communities of the Silurian through Permian (Figure 5). This factor dominates more communities in the data set than do either of the other two. However, the dominance of this factor is not evenly distributed in time. In addition to the obvious dearth of high loadings in the Cambrian and Ordovician, there appears to be a trend toward fewer high loadings in the Carboniferous and Permian; 53% of sampled Carboniferous shelf communities and only 33% of Permian shelf communities have high loadings on Factor 2 in contrast to the 70% of Siluro-Devonian shelf communities with their highest loadings on this factor. Although this statistical pattern may partially reflect the uneven sampling of shelf environments in the upper Paleozoic, it seems mostly to correspond to the slow decline of brachiopod-bryozoan-pelmatozoan communities and expansion of bivalve-gastropod assemblages in the Carboniferous and Permian, noted by Boucot (1983) in his Ecologic-Evolutionary Unit VIII. Note, however, that Figure 5 indicates that this change did not involve a gradual, even transformation of all shelf communities; rather, the late Paleozoic change in community compositions appears to have involved an onshore-offshore contraction (albeit irregular) of brachiopod-rich communities, just as the previous Ordovician turnover involved an onshore-offshore contraction of trilobite-rich communities.

The relatively small number of high loadings that Factor 2 receives from shelf communities in the Ordovician is somewhat anomalous. In their analysis of Cambro-Ordovician communities, Sepkoski and Sheehan (1983) obtained a factor similar in composition to Factor 2 here which received high loadings from Middle and Upper Ordovician communities in mid to outer shelf environments; they interpreted this factor as encompassing the brachiopod-rich communities of the Paleozoic evolutionary fauna which expanded diachronously across the shelf at the expense of the Cambrian-type trilobite-rich communities (clearly seen in Factor 1 here). The small number of Ordovician communities that load highly on Factor 2 in the present study seems to reflect the somewhat transitional nature of Middle and Upper Ordovician brachiopod-rich communities. Factor 2 is dominated by strophomenid and spiriferid brachiopods (Table 2), reflecting the great diversity of these two groups in mid to upper Paleozoic communities. Ordovician brachiopod-rich communities, however, are dominated by orthids and contain virtually no spiriferids (excluding atrypids from this group). This difference, along with the moderate score that orthids contribute to Factor 1, causes Factor 2 to receive relatively low loadings from the Ordovician brachiopod-rich communities in the three-factor solution.

The small number of deepwater, "slope" communities with high loadings on Factor 2 also seems somewhat anomalous, particularly in comparison with Factor 1. The several post-Ordovician deepwater points in our data set constitute a heterogeneous set of low-diversity assemblages with various inarticulate brachiopods, articulate brachiopods (especially rhynchonellids and strophomenids), bivalves (especially praecardioids and pterioids), cephalopods, hexactinellid sponges, cricoconarids, and graptolites, as discussed above. The heterogeneity of these assemblages probably reflects in large part a variable admixing of benthic and pelagic organisms (including organisms attached to floating algae; see Watkins and Berry, 1977); variable removal of calcareous fossils by post-mortem dissolution may also be important. All of the post-Ordovician deepwater communities have low communalities in the three-factor solution and, because of their generally low diversities (which causes the presence or absence of a single species to affect a community's overall composition greatly), they have maximum loadings scattered over all three factors. These problems are evident in the factor analysis only in Silurian and younger deepwater communities because both deepwater benthic and pelagic organisms (other than graptolites) were largely derived from the same Cambrian evolutionary fauna during the Cambrian and Ordovician Periods (see for example Robison, 1972; Bergstrom, 1973; Taylor, 1977).

The third factor, which is dominated by molluscan orders, receives high loadings from communities in nearshore environments throughout the post-Cambrian portion of the Paleozoic Era (Figure 5). As suggested above, the domain of this factor is completely contemporaneous with that of Factor 2, indicating a significant environmental segregation of early members of the Modern evolutionary fauna from most elements of the Paleozoic Fauna. The oldest communities that have moderate loadings on Factor 3 are assemblages with gastropods, nautiloids, orthids, and trilobites from very shallow-water facies in the lower Ordovician; these communities also have moderate loadings on Factor 1, reflecting a gradation with the underlying Cambrian Fauna. Above the Lower Ordovician, Factor 3 dominates inner shelf communities, reflecting the disappearance of most trilobites and orthids and the diversification of bivalves in inner shelf environments. (The small number of Devonian communities with high loadings on Factor 3 and the excursion of Factor 2 into inner shelf communities in the Upper Devonian result from a general paucity of archaeogastropods described from these communities in North America and may represent an artifact of sampling.) In the upper Paleozoic there appears to be a significant, if somewhat irregular, expansion of Factor 3 into mid shelf environments, corresponding to the slight offshore contraction of brachiopod-rich communities as well as to the further diversification of gastropods and bivalves and increasing abundance of scaphopods and echinoids in shallow-water environments.

The spatial and temporal domain of Factor 3 is much the same as Bretsky's (1968, 1969) "Linguloid-molluscan association" which he recognized in nearshore facies throughout much of the Paleozoic Era (see also Anderson, 1971; Boucot, 1975; McKerrow, 1978; Steele-Petrovic, 1979). Several refinements can be made to Bretsky's scheme on the basis of the factor analysis:

1. The very shallow-water assemblages dominated by *Lingula* with occasional other non-molluscan taxa (e.g. leperditicopid ostracodes and rhynchonellid

brachiopods) are compositionally distinct from adjacent molluscan communities. (The low-diversity lingulid assemblages tend to have their highest loadings on either Factor 1 or Factor 2, depending on the relative diversity of articulate brachiopods.)

2. The nearshore molluscan communities are not static through the Paleozoic Era but rather become increasingly widespread in shallow-water environments and expand somewhat into offshore environments in the Carboniferous and Permian, paralleling the late Paleozoic diversification of families seen in the global Modern Fauna (Figure 1).

The offshore expansion of molluscan communities might best be considered a statistical trend rather than an orderly pattern, as indicated by the irregularities in the contours for Factor 3. These irregularities probably reflect both inaccuracies in the positioning of communities in the time-environment diagram and real complexities in the distribution of Paleozoic molluscs (especially as related to their affinities for hyper- and hyposaline environments), as well as intergradations between nearshore mollusc-rich assemblages and more offshore brachiopod-rich assemblages (reflected in the comparatively wide areas between the 0.33 and 0.67 contours for Factor 3 in some systems). The irregularity in the distribution of Factor 3 is made all the more evident by the apparent slowness of the offshore expansion of the Paleozoic molluscs. This contrasts with the comparatively rapid offshore restriction of the Cambrian Fauna (Factor 1) during the Ordovician; this onshore-offshore change took only several tens of million years to complete and is therefore reflected in only a few communities in our sample, greatly limiting the potential for irregularity in the contours. The molluscan communities took hundreds of million years to expand and evidently expanded at slightly different rates in different nearshore environments.

"MINOR" FACTORS

The Modern evolutionary fauna, as represented by Factor 3, appears to be relatively homogeneous in terms of taxonomic composition at the ordinal level; most of the change within it between the Ordovician and Permian involves addition and diversification of higher taxa. This is not true of the more offshore Paleozoic evolutionary fauna, which underwent considerably more turnover at low taxonomic levels during the same time interval (cf. Bretsky and Lorenz, 1970, 1971). Some of this turnover is reflected in factor analyses based on more than three principal vectors. For example, a seven-factor solution (warranted by the small break in eigenvalues between the seventh and eighth principal vectors) produces four "minor" factors, each of which accounts for approximately 5% of the data following rotation. (The total proportion of data encompassed by the seven-factor solution is approximately 70%.) Three of the minor factors encompass stratigraphic variation within the Paleozoic Fauna. These are:

1. a factor receiving high scores from orthid and atrypid brachiopods and phacopid trilobites and differentiating Middle and Upper Ordovician shelf assemblages from the preceding Cambrian-type shelf communities of the Lower Ordovician (cf. Boucot's [1983] Ecologic-Evolutionary Unit IV);
2. a factor contrasting the diversity of rugosans, tabulates, spiriferids, and rhynchonellids with that of strophomenids and pterioids and distinguishing mid to outer shelf communities of the Silurian through Middle Devonian from those preceding (cf. Boucot's Ecologic-Evolutionary Unit VI);
3. a factor contrasting the diversity of strophomenids, crinoids, and cryptostome and trepostome bryozoans with that of spiriferid brachiopods and largely distinguishing Upper Carboniferous and Permian offshore shelf communities from those of the Devonian (cf. Boucot's Ecologic-Evolutionary Unit VIII).

The fourth minor factor contrasts the diversity of pterioid and modiomorphoid bivalves and rhynchonellid brachiopods with that of archaeogastropods and serves largely to differentiate inner shelf communities of the Middle and Upper Devonian from the more offshore brachiopod-rich communities of the Paleozoic Fauna; this final factor may be reflecting more the vagaries of sampling (as previously noted) than the realities of faunal turnover on the Paleozoic shelf.

All of these minor factors, however, seem to be merely "variations upon themes" within the data set. Even after rotation, the seven-factor solution retains three major factors that are almost identical to those in the three-factor solution. Thus, the three major community types seem to embody the major themes in faunal distribution on the Paleozoic shelf, just as the three evolutionary faunas embody the principal themes in faunal diversity in the Phanerozoic oceans.

DISCUSSION

The principal conclusions that can be drawn from the factor analysis of Paleozoic level-bottom communities are summarized in the three points below:

1. The three global evolutionary faunas of the Phanerozoic oceans are clearly manifested in local community compositions. The first three factors in the analysis separately encompass trilobite-rich, brachiopod-rich, and mollusc-rich communities.
2. These three community types, and hence the evolutionary faunas they reflect, occupied different environments during the post-Cambrian portion

of the Paleozoic Era, with mollusc-rich communities concentrated in near-shore environments, brachiopod-rich communities in more offshore shelf environments, and trilobite-rich communities in deepwater environments (particularly during the Ordovician).

3. Changes in dominance of the global evolutionary faunas involved onshore-offshore expansion of new community types, as seen during the Ordovician Period when surviving elements of the Cambrian Fauna became restricted to deepwater environments by expansion of the Paleozoic and Modern Faunas, and as seen again in the later Paleozoic Era when the Paleozoic Fauna slowly became restricted to middle and outer shelf environments by onshore expansion of the Modern Fauna.

These three points are embodied in the summary diagram in Figure 6. This final time-environment diagram shows the relationships among the three major community types in space and time as inferred from the contoured factor loadings in Figure 5. Again, each community type reflects a different evolutionary fauna: trilobite-rich communities constitute the Cambrian Fauna, (articulate) brachiopod-rich communities constitute the Paleozoic Fauna, and mollusc-rich communities constitute the Modern Fauna. The stippling of the boundaries between the fields emphasizes the intergradation of the various community types as reflected in the factor analysis. Finally, the inclusion of ranges of several dominant taxa in very shallow and deep environments signifies that not all communities are easily or accurately represented by this simple three-fauna scheme.

Figure 6 and its underlying analysis provide simply a geometric representation of the distribution of evolutionary faunas in space and time; they do not indicate necessary causes for this distribution. However, the existence of an environmental geometry does limit the possible spectrum of causes. The fact that contemporaneous evolutionary faunas inhabited different marine environments suggests that environmental factors formed the template for assembly and development of the faunas (see also Sepkoski and Sheehan, 1983). On the other hand, the long timescales involved in the onshore-offshore changes in the faunas, taking tens to hundreds of million years, imply that it was not progressive change in the physical environment that governed the success of the faunas; we are aware, for example, of no aspect of the physical environment that changed monotonically from the Ordovician to the Permian and could reasonably have caused the offshore expansion of the mollusc-rich communities. In the same vein, the gradual, if somewhat irregular, nature of the onshore-offshore changes suggests that major episodic events, such as mass extinctions, were not governing influences, at least not over the course of the entire Paleozoic Era (see also Kitchell and Carr, this volume). Large-scale episodic changes should have produced abrupt, stepwise expansions or

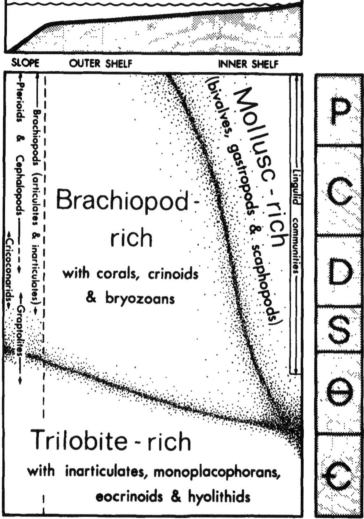

Figure 6. Summary of the results of the Q-mode factor analysis of Paleozoic level-bottom communities. The fields labelled trilobite-rich, brachiopod-rich, and mollusc-rich in this time-environment diagram delineate the distributions of the three major community types indicated by the factor analysis. The boundaries between these fields are stippled in order to emphasize the intergradation of the community types. The taxa listed at the margins of the diagram are the principal members of communities in very shallow and deepwater environments that do not fit easily into any of the major community types; the arrows show the approximate geologic ranges of these taxa within these environments in the analyzed data set. The vertical dimension in this figure is scaled to absolute time.

contractions in the environmental boundaries between the evolutionary faunas and more dramatic turnovers within the faunas. Therefore we conclude that persistent or mildly fluctuating features of the physical environment, and biological responses to these features, were responsible for the assembly and expansion of the evolutionary faunas.

The appearance of both the Paleozoic and the Modern evolutionary fauna in the nearshore zone suggests that there is something unique about shallow-water environments that promotes the origin of evolutionary novelties or the assembly of novel community types. The most distinctive ecological features of shallow-water, nearshore environments are the frequent disturbances and the stressful or rigorous ambient conditions (Brongersma-Sanders, 1957; Sanders, 1968; Valentine, 1973). Several authors have argued recently that biological responses to persistent stress and disturbance may be conducive, in the long run, to the evolution of novel taxa and communities. Their arguments are summarized very briefly below:

1. Valentine and Jablonski (1983) and Jablonski and Bottjer (1983) have hypothesized that nearshore species, which are frequently widely dispersed but essentially panmictic as a result of their planktotrophic larval stage, may occasionally produce small peripheral isolates that undergo genetic revolutions, or transiliences (Templeton, 1980), which lead to rapid shifts in morphology or physiology. These processes may promote the appearance of evolutionary novelties in nearshore regions, whereas vicariance events, clinal speciation, etc., which affect offshore species that reproduce by direct development, will tend to produce ecological vicars rather than unique or novel adaptive types.

2. Sepkoski and Sheehan (1983) have speculated that frequent defaunation and recolonization in disturbed, nearshore areas might aid the establishment of new species which normally would not be able to invade stable communities and thus would be liable to rapid extinction as a consequence of small population sizes. Frequent recolonization might also induce constant recombination of established species, eventually leading to the chance assembly of new community types composed of species that are well adapted to fluctuating resources and are maximally co-adapted (or "mutually accommodated") to each other. These qualities might permit such novel, nearshore communities to persist somewhat longer than offshore communities and to expand slowly through a combination of ecological efficiency, extinction resistance, and competitive superiority of many of their constituent species.

3. Steele-Petrovic (1979) has argued that stress and disturbance in nearshore environments require unique adaptations for efficient utilization of fluctuating resources. Consequently, such adaptations might make nearshore

taxa (especially filter-feeding bivalves) physiologically more efficient than, and therefore competitively superior to, offshore taxa (especially articulate brachiopods); offshore replacement may be slow, however, because of the difficulty that competitors, even superior competitors, have in invading stable, diverse communities (MacArthur, 1972). (Steele-Petrovic''s argument is supported by the observation that the initial habitat of the nearshore molluscan fauna—somewhat hypersaline, stromatolitic carbonate lithotopes of the earliest Ordovician or even latest Cambrian—seem not to have been densely inhabited during the Cambrian [Garrett, 1970; Kepper, 1974; Mazzullo and Friedman, 1977]; thus, molluscan communities became established in an essentially vacant environment which required a peculiar suite of adaptations that members of the Cambrian Fauna never achieved.)

We are not in a position at present to differentiate among these three hypotheses for the nearshore origin of evolutionary faunas. However, in view of the complexity of ecological interactions in nearshore communities and the long timescales involved in faunal turnover, we suspect all may have an element of truth.

In addition to providing some insight into substantive issues concerning the evolution of global faunas, the factor analysis of Paleozoic communities also suggests some methodological problems involved in the paleoecologic reconstruction of ancient environments. Sedimentary environments and their contained animal communities are frequently interpreted in part on the basis of "taxonomic uniformitarianism." As Dodd and Stanton (1981:17) note, this approach involves "strict substantive application of the principle of uniformitarianism, that is, the ecology of present organisms is the key to past organisms." The analysis presented here, however, supports the oft-voiced concern that mean environmental preferences of the members of various higher taxa can shift in time. Distributions of trilobites or inarticulate brachiopods in the Upper Ordovician may not reflect the same environmental constraints as in the Cambrian, and abundances of molluscs in Permian facies may mean something quite different from abundances in Ordovician facies. In view of the progressive changes in the environmental distributions of major community types through the Paleozoic Era, we urge care in comparing temporally distant sedimentary facies on the basis of their contained faunas.

SUMMARY AND CONCLUSIONS

1. There is a strong correspondence between global evolutionary faunas and local marine communities, with trilobite-rich, brachiopod-rich, and

mollusc-rich communities, corresponding to the Cambrian, Paleozoic, and Modern Faunas, occupying different portions of the shelf environment through the Paleozoic Era. Changes in dominance among the evolutionary faunas appears to have corresponded to onshore-offshore expansions of the respective community types.

2. Temporal changes in the environmental distributions of the community types occurred independently of either progressive change or large-scale episodic fluctuations in the global physical environment through the Paleozoic Era. Instead, the faunal changes appear to have resulted from assembly of novel communities in nearshore areas and, perhaps, from subsequent competitive replacement of offshore taxa; however, precise mechanisms for these changes remain to be clarified.

3. Uncritical use of taxonomic uniformitarianism in the reconstruction of paleoenvironments may lead to inaccurate results unless habitat shifts among members of higher taxa are accommodated.

4. Considerably more work is required in compiling community data, refining present environmental interpretations, and analyzing patterns and rates of change in faunal distributions before we will have a precise description of habitat shifts through time or a reasonable understanding of the processes governing the appearance and expansion of global evolutionary faunas.

ACKNOWLEDGMENTS

We thank R. Ludvigsen, G. R. McGhee, Jr., P. M. Sheehan, and A. M. Ziegler for their invaluable help and advice in compiling data. We also thank P. W. Bretsky, K. W. Flessa, and J. W. Valentine for critical comments on early manuscript drafts. Research for this paper, which represents a progress report on an ongoing study of the nature and rate of faunal turnover in fossil communities, received partial support from NSF Grant DEB 81-08890 to JJS.

REFERENCES

Anderson, E. J., 1971. Environmental models for Paleozoic communities. Lethaia 4:287-302.

Ausich, W. I., and Bottjer, D. J., 1985. Phanerozoic tiering in suspension-feeding communities on soft substrata: Implications for diversity. In Valentine, J. W. (ed.), Phanerozoic diversity patterns: Profiles in macroevolution, Princeton, N.J.: Princeton Univ. Press and Amer. Assoc. Adv. Sci. (this volume).

Bambach, R. K., 1985. Classes and adaptive variety: The ecology of diversification in marine faunas through the Phanerozoic. In Valentine, J.W. (ed.), Phanerozoic diversity patterns: Profiles in macroevolution, Princeton, N.J.: Princeton Univ. Press and Amer. Assoc. Adv. Sci. (this volume).

Bergstrom, J., 1973. Organization, life, and systematics of trilobites. Fossils and Strata 2:1-69.

Berry, W. B. N., 1972. Early Ordovician bathyurid province lithofacies, biofacies, and correlations—their relationship to a proto-Atlantic Ocean. Lethaia 5:69-84.

Berry, W. B. N., 1974. Types of early Paleozoic faunal replacements in North America: Their relationship to environmental change. J. Geol. 82:371-382.

Boucot, A. J., 1975. Evolution and extinction rate controls. Amsterdam: Elsevier, 427.

Boucot, A. J., 1983. Does evolution take place in an ecological vacuum? II. J. Paleontol. 57:1-30.

Bretsky, P. W., 1968. Evolution of Paleozoic marine invertebrate communities. Science 159:1231-1233.

Bretsky, P. W., 1969. Evolution of Paleozoic benthic marine invertebrate communities. Palaeogeogr., Palaeoclimat., Palaeoecol. 6:45-59.

Bretsky, P. W., and Lorenz, D. M., 1970. An essay on genetic-adaptive strategies and mass extinctions. Geol. Soc. Am. Bull. 81:2449-2456.

Bretsky, P. W., and Lorenz, D. M., 1971. Adaptive response to environmental stability: A unifying concept in paleoecology. N. Am. Paleontol. Conv., 1969, Proc., Pt. E., 522-550.

Brongersma-Sanders, M. 1957. Mass mortality in the sea. Geol. Soc. Am. Mem. 67(1):941-1010.

Dodd, J. R., and Stanton, R. J., 1981. Paleoecology, concepts and applications. Somerset, N.J.: Wiley, 544.

Elles, G. L., 1939. Factors controlling graptolite succession and assemblages. Geol. Mag. 76:181-187.

Garrett, P. 1970. Phanerozoic stromatolites: Noncompetitive ecologic restrictions by grazing and burrowing animals. Science 169:171-173.

Goldring, R., and Langenstrassen, F., 1979. Open shelf and near-shore clastic facies in the Devonian. In House, M. R., Scrutton, C. T., and Bassett, M. G. (eds.), The Devonian system, London: Palaeontol. Assoc., Spec. Pap. Palaeontol. (23):81-98.

Heckel, P. H., 1977. Origin of phosphatic black shale facies in Pennsylvanian cyclothems of mid-continent North America. Am. Assoc. Petrol. Geol. Bull. 61:1045-1068.

Imbrie, J. and Purdy, E. G., 1962. Classification of modern Bahamian carbonate sediments. In Ham, W. E. (ed.), Classification of carbonate rocks, Tulsa, Okla.: Am. Assoc. Petrol. Geol. Mem. 1:253-272.

Jablonski, D. and Bottjer, D. J., 1983. Soft-bottom epifaunal suspension-feeding assemblages in the Late Cretaceous: Implications for the evolution of benthic paleocommunities. In Tevesz, M. J. S., and McCall, P. L. (eds.),

Biotic interactions in recent and fossil benthic communities, New York: Plenum Publ. Co., 747-812.

Jöreskog, K. G., Klovan, J. E., and Reyment, R. A., 1976. Geological factor analysis. Amsterdam: Elsevier, 178.

Kepper, J. C., 1974. Antipathetic relation between Cambrian trilobites and stromatolites. Am. Assoc. Petrol. Geol. Bull. 58:141-143.

Kitchell, J. A., and Carr, T. R., 1985. Nonequilibrium model of diversification: Faunal turnover dynamics. In Valentine, J. W. (ed.) Phanerozoic diversity patterns: Profiles in macroevolution, Princeton, N.J.: Princeton Univ. Press and Amer. Assoc. Adv. Sci. (this volume).

Klovan, J. E., and Imbrie, J., 1971. An algorithm and FORTRAN-IV program for large-scale Q-mode factor analysis and calculation of factor scores. Math. Geol. 3:61-77.

Lockley, M. G., 1983. A review of brachiopod dominated palaeocommunities from the type Ordovician. Palaeontology 26:111-145.

MacArthur, R. H., 1972. Geographical ecology. New York: Harper & Row, Publishers, 269.

Marr, J. E., and Nicholson, H. A., 1888. The Stockdale shales. Quart. J. Geol. Soc. London 44:654-734.

Mazzullo, S. J., and Friedman, G. M., 1977. Competitive algal colonization of peritidal flats in a schizohaline environment: The Lower Ordovician of New York. J. Sed. Pet. 47:398-410.

McKerrow, W. S., ed., 1978. The ecology of fossils. Cambridge, Mass.: M.I.T. Press, 384.

Mikulic, D. G., and Watkins, R., 1981. Trilobite ecology in the Ludlow Series of the Welsh Borderland. In Gray, J., Boucot, A. J., and Berry, W. B. N. (eds.), Communities of the past, Stroudsburg, Pa.: Hutchinson Ross Publ. Co., 101-117.

Miller, A. I., 1981. Gradients in nearshore marine molluscan assemblages: Smuggler's Cove, U.S. Virgin Islands. Unpubl. M.S. thesis, Virginia Polytechnic Institute and State University, Blacksburg, Va., 155.

Miller, A. I., and Bambach, R. K., 1981. Gradients in nearshore molluscan marine assemblages and their implications for paleoenvironmental analysis. Geol. Soc. Am. Abst. Prog. 13:30.

Niklas, K. J., Tiffney, B. H., and Knoll, A. H., 1985. Patterns in vascular land plant diversification: An analysis at the species level. In Valentine, J. W. (ed.), Phanerozoic diversity patterns: Profiles in macroevolution, Princeton, N.J.: Princeton Univ. Press and Amer. Assoc. Adv. Sci. (this volume).

Robison, R. A., 1972. Mode of life of agnostid trilobites. Int. Geol. Congr., 24th, Montreal, Sect. 7:33-40.

Sanders, H. L., 1968. Marine benthic diversity: A comparative study. Am. Nat. 102:243-282.

Sepkoski, J. J., Jr., 1979. A kinetic model of Phanerozoic taxonomic diversity: II. Early Phanerozoic families and multiple equilibria. Paleobiology 5:222-251.

Sepkoski, J. J., Jr., 1981a. A factor analytic description of the Phanerozoic marine fossil record. Paleobiology 7:36-53.

Sepkoski, J. J., Jr., 1981b. The uniqueness of the Cambrian fauna. In Taylor, M. E. (ed.), Short papers for the second international symposium on the Cambrian System, U.S. Geol. Surv. Open-File Rpt. 81-743:203-207.

Sepkoski, J. J., Jr., 1982. A compendium of fossil marine families. Milwaukee Public Museum Contr. Biol. Geol. (51), 125.

Sepkoski, J. J., Jr., and Sheehan, P. M., 1983. Diversification, faunal change, and community replacement during the Ordovician radiations. In Tevesz, M. J. S., and McCall, P. L. (eds.), Biotic interactions in recent and fossil benthic communities, New York: Plenum Publ. Co., 673-717.

Sneath, P. H. A., and Sokal, R. R., 1973. Numerical taxonomy. San Francisco: W. H. Freeman and Co., 573.

Steele-Petrović, M., 1979. The physiological differences between articulate brachiopods and filter-feeding bivalves as a factor in the evolution of marine level-bottom communities. Palaeontol. 22:101-134.

Taylor, M. E., 1977. Late Cambrian of western North America: Trilobite biofacies, environmental significance, and biostratigraphic implications. In Kauffman, E. G., and Hazel, J. E. (eds.), Concepts and methods of biostratigraphy, Stroudsburg, Pa.: Dowden, Hutchinson & Ross, Inc., 397-426.

Templeton, A. R., 1980. Modes of speciation and inferences based on genetic distances. Evolution 34:719-729.

Valentine, J. W., 1973. Evolutionary paleoecology of the marine biosphere. Englewood Cliffs, N.J.: Prentice-Hall, Inc., 511.

Valentine, J. W., and Jablonski, D., 1983. Speciation in the shallow sea: General patterns and biogeographic controls. In Sims, R. W., Price, J. H., and Whalley, P. E. S. (eds.), Evolution, time and space: The emergence of the biosphere, London: Academic Press, 203-228.

Watkins, R., and Berry, W. B. N., 1977. Ecology of a Late Silurian fauna of graptolites and associated organisms. Lethaia 10:267-286.

Ziegler, A. M., Cocks, L. R. M., and Bambach, R. K., 1968. The composition and structure of Lower Silurian marine communities. Lethaia 1:1-27.

APPENDIX

SOURCES OF INFORMATION ON NORTH AMERICAN
PALEOZOIC COMMUNITIES ANALYZED IN THIS STUDY

The numbers correspond to the numbered communities illustrated in Figure 2. (Italicized numbers indicate the several communities left off the figure because of space limitations.) The literature references listed for each set of communities are the sources of the faunal lists that were analyzed. Note that the numbering and information sources for communities 5 to 102 are identical to those in Sepkoski and Sheehan (1983).

Communities	References
5	Willoughby R., 1976. Lower and Middle Cambrian fossils from the Shady Formation, Austinville, Virginia. Geol. Soc. Amer. Abstr. Prog. 8:301-302.
9	Bright, R. C., 1956. A paleoecologic and biometric study of the Middle Cambrian trilobite *Elrathia kingii* (Meek). J. Paleontol. 33:83-98.
	Robison, R. A., 1971. Additional Middle Cambrian trilobites from the Wheeler Shale of Utah. J. Paleontol. 45: 796-804.
10	Conway Morris, S., 1979. The Burgess Shale (Middle Cambrian) fauna. Ann. Rev. Ecol. Syst. 10:327-349.
	Fritz, W. H., 1969. Geological setting of the Burgess Shale. N. Am. Paleont. Conv. 1969, Proc. B:1155-1170.
11,22	Kulik, J. W., 1965. Stratigraphy of the Deadwood Formation, Black Hills, South Dakota and Wyoming. Unpubl. M.S. thesis, South Dakota School of Mines and Technology, Rapid City, S.D.
12,13,14,15	Lochman, C., and Duncan, D., 1944. Early Upper Cambrian faunas of central Montana. Geol. Soc. Amer. Spec. Pap. 54, 181.
	Lochman, C., and Hu, C.-H., 1961. Upper Cambrian faunas from the northwest Wind River Mountains, Wyoming, Pt. II. J. Paleontol. 35:125-246.
	Sepkoski, J. J., Jr., 1977. Dresbachian (Upper Cambrian) stratigraphy in Montana, Wyoming, and South Dakota. Unpubl. Ph.D. dissert., Harvard University, Cambridge, Mass.
16	Stinchcomb, B. L., 1975. Paleoecology of two new species of Late Cambrian *Hypseloconus* (Monoplacophora) from Missouri. J. Paleontol. 49:416-421
17	Palmer, A. R., 1954. The faunas of the Riley Formation in central Texas. J. Paleontol. 28:709-786.

18,19	McBride, D. J., 1976. Outer shelf communities and trophic groups in the Upper Cambrian of the Great Basin. In Robison, R. A., and Rowell, A. J. (eds.), Paleontology and depositional environments: Cambrian of western North America, Brigham Young Univ. Geol. Studies 23(2):139-152.
20,21,23,24	Grant, R. E., 1965. Faunas and stratigraphy of the Snowy Range Formation (Upper Cambrian) in southwestern Montana and northwestern Wyoming. Geol. Soc. Amer. Mem. 96, 171.
25,26,30,31, 34,37,42,44	Lochman-Balk, C., and Wilson, J. L., 1967. Stratigraphy of Upper Cambrian-Lower Ordovician subsurface sequence in Williston Basin. Amer. Assoc. Petrol. Geol. Bull. 51:883-917.
27	Cook, H. E., and Taylor, M. E., 1975. Early Paleozoic continental margin sedimentation, trilobite facies and the thermocline western United States. Geology 3:559-562.
28	Fisher, D. W., 1956. The Cambrian System of New York State. In Rogers, J. (ed.), El Sistema Cámbrico, XX Int. Geol. Congr., Mexico, Proc. 321-351.
29	Taylor, M. E., and Halley, R. B., 1974. Systematics, environment, and biogeography of some Late Cambrian and Early Ordovician trilobites from eastern New York State. U.S. Geol. Surv. Prof. Pap. 834:1-38.
32,36,43	Mazzullo, S. J., and Friedman, G. M., 1977. Competitive algal colonization of peritidal flats in a schizohaline environment: The Lower Ordovician of New York. J. Sed. Pet. 47:398-410.
33	Sando, W. J., 1957. Beekmantown Group (Lower Ordovician) of Maryland. Geol. Soc. Amer. Mem. 68, 159.
35,45	Ross, R. J., Jr., 1970. Ordovician brachiopods, trilobites, and stratigraphy in eastern and central Nevada. U.S. Geol. Surv. Prof. Pap. 639:1-103.
48	Ross, R. J., Jr., 1951. Stratigraphy of the Garden City Formation in northeastern Utah and its trilobite faunas. Peabody Mus. Nat. Hist. Bull. 6, 161.
	Ross, R. J., Jr., 1968. Brachiopods from the upper part of the Garden City Formation (Ordovician) north-central Utah. U.S. Geol. Surv. Prof. Pap. 593H:H1-H13.
49,78,243, 303,433	Kay, M. and Crawford, J. P., 1964. Paleozoic facies from the miogeosynclinal to the eugeosynclinal belt in thrust slices, central Nevada. Geol. Soc. Amer. Bull. 75:425-454.
50,51,53,55, 56,57,58	Berry, W. B. N., Lawson, D. A., and Yancey, E. S., 1979. Species-diversity patterns in some Middle Ordovician communities from California-Nevada. Palaeogeogr., Palaeoclimat., Palaeoecol. 26:99-116.

Ross, R. J., Jr., 1967. Some Middle Ordovician brachiopods and trilobites from the Basin Ranges, western United States. U.S. Geol. Surv. Prof. Pap. 523D:D1-D43.

52 Kay, M., 1962. Classification of Ordovician shelly and graptolite sequences from central Nevada. Geol. Soc. Amer. Bull. 73:1421-1430.

McKee, E. H., Norford, B. S., and Ross, R. J., Jr., 1972. Correlation of the Ordovician shelly facies *Orthidiella* Zone with zones of the graptolitic facies, Toquima Range, Nevada, and North White River Region, British Columbia. U.S. Geol. Surv. Prof. Pap. 800C:C145-C156.

54 Ross, R. J., Jr. 1949. Stratigraphy and trilobite faunal zones of the Garden City Formation, northeastern Utah. Am. J. Sci. 247:472-491.

59,60 Cooper, G. A., 1956. Chazyan and related brachiopods. Smithson. Misc. Coll. (127), 1245.

Walker, K. R., Broadhead, T. W., and Keller, F. B., 1980. Middle Ordovician carbonate shelf to deep water basin deposition in the southern Appalachians. Univ. Tenn., Dept. Geol. Sci., Studies in Geol. (4), 120.

61,62,63,64, 65 Hayes, B. J. R., 1980. A cluster analysis interpretation of Middle Ordovician biofacies, southern MacKenzie Mountains. Can. J. Earth Sci. 17:1377-1388.

Ludvigsen, R., 1979. A trilobite zonation of Middle Ordovician rocks, southwestern district of Mackenzie. Geol. Surv. Can. Bull. 312:1-98.

66 Ross, R., J., Jr., and Shaw, F. C., 1972. Distribution of the Middle Ordovician Copenhagen Formation and its trilobites in Nevada. U.S. Geol. Surv. Prof. Pap. 749.

67,68,69,70 Walker, K. R., 1972. Community ecology of the Middle Ordovician Black River Group of New York State. Geol. Soc. Amer. Bull. 83:2499-2524.

71,72 Cooper, B. N., and Cooper, G. A., 1946. Lower Middle Ordovician stratigraphy of the Shenandoah Valley, Virginia. Geol. Soc. Amer. Bull. 57:35-113.

75,79,80,81, 83,84 Titus, R., and Cameron, B., 1976. Fossil communities of the lower Trenton Group (Middle Ordovician) of central and northwestern New York State. J. Paleontol. 50:1209-1225.

85 Cisne, J. L., 1973. Beecher's trilobite bed revisited: Ecology of an Ordovician deep water fauna. Peabody Mus. Nat. Hist., Yale Univ. Postilla (160), 25.

86,87,88,89 90,91 Bretsky, P. W., 1969. Central Appalachian Late Ordovician communities. Geol. Soc. Amer. Bull. 80:193-212.

Bretsky, P. W., 1970. Upper Ordovician ecology of the

central Appalachians. Peabody Mus. Nat. Hist. Bull. 34: 1-150.

92,93,94   Bretsky, P. W., 1970. Late Ordovician benthic marine communities in north-central New York. N.Y. State Mus. and Sci. Surv. Bull. 414:1-34.

97   Bayer, T. N., 1965. The Maquoketa Formation in Minnesota and an analysis of its benthonic communities. Unpubl. Ph.D. dissert., Univ. of Minnesota, Minneapolis, Minn.

Bayer, T. N., 1967. Repetitive benthonic community in the Maquoketa Formation (Ordovician) of Minnesota. J. Paleontol. 41:417-422.

98,100,101, 103   Fox, W. T., 1968. Quantitative paleoecologic analysis of fossil communities in the Richmond Group. J. Geol. 76: 613-640.

MacDaniel, R. P., 1976. Upper Ordovician sedimentary and benthic community patterns of the Cincinnati Arch area. Unpubl. Ph.D. dissert., Univ. of Chicago, Chicago.

Richards, R. P., 1972. Autecology of Richmondian brachiopods (Late Ordovician) of Indiana and Ohio. J. Paleontol. 46:386-405.

Walker, K. R., 1972. Trophic analysis: A method for studying the function of ancient communities. J. Paleontol. 46: 82-93.

102   Ross, R. J., Jr., Nolan, T. B., and Harris, A. G., 1979. The Upper Ordovician and Silurian Hanson Creek Formation of central Nevada. U.S. Geol. Surv. Prof. Pap. 1126-C.

104   Ludvigsen, R., 1981. Biostratigraphic significance of Middle Ordovician trilobites from the Road River Formation, northern Cordillera. Geol. Assoc. Can. Abstr.G:A-36.

201,202,203, 204,205,211, 212,213,215, 217,220,221, 222,223,226, 227,228,229, 230,231,232, 233,234   Bolton, T. E., 1957. Silurian stratigraphy and paleontology of the Niagaran Escarpment in Ontario. Geol. Surv. Can. Mem. 289, 145.

Fisher, D. W., 1953. Additions to the stratigraphy and paleontology of the lower Clinton of western New York. Buffalo Soc. Nat. Sci. Bull. 21(2):26-36.

Fisher, D. W., 1954. Stratigraphy of Medina Group, New York and Ontario. Amer. Assoc. Petrol. Geol. Bull. 38: 1979-1996.

Gillette, T., 1947. The Clinton of western and central New York. N.Y. State Mus. and Sci. Serv. Bull. 341, 191.

Schuchert, C., 1914. Notes on Arctic Paleozoic fossils. Am. J. Sci. 37-83:467-477.

Zenger, D. H., 1971. Uppermost Clinton of western and central New York. N.Y. State Mus. and Sci. Serv. Bull. 417, 58.

Ziegler, A. M., Newall, G., Halleck, M. S., and Bambach, R. K., 1971. Repeated community-sediment patterns in the Silurian of the northern Appalachian Basin. Geol. Soc. Amer. Abstr. Prog. 3:760-761.

206,207,214, 224   Copper, P., 1978. Paleoenvironments and paleocommunities in the Ordovician-Silurian sequence of Manitoulin Island, Michigan Basin. Geol. Soc. Amer. Spec. Pap. 3: 47-61.

208,209,210   Johnson, M. E., and Campbell, G. T., 1980. Recurrent carbonate environments in the Lower Silurian of northern Michigan and their inter-regional correlation. J. Paleontol. 54:1041-1057.

216   Miller, R. L., and Brosgé, W. P., 1954. Geology and oil resources of the Jonesville District, Lee County, Virginia. U.S. Geol. Surv. Bull. 990, 240.

218,219   Johnson, M. E., 1980. Paleoecological structure in Early Silurian platform seas of the North American midcontinent. Palaeogeogr., Palaeoclimat., Palaeoecol. 30:191-216.

225   Boucot, A. J., and Thompson, J. B., Jr., 1963. Metamorphosed Silurian brachiopods from New Hampshire. Geol. Soc. Amer. Bull. 74:1313-1334.

235,236   Amsden, T. W., 1975. Hunton Group (Late Ordovician, Silurian, and Early Devonian) in the Anadarko Basin of Oklahoma. Okla. Geol. Surv. Bull. 121, 214.

Amsden, T. W., 1981. Biostratigraphic and paleoenvironmental relations: A Late Silurian example. In Dutro, J.T., and Boardman, R. S. (eds.), Lophophorates: Notes for a shortcourse, Univ. Tenn. Stud. Geol. 5:154-169.

Campbell, K. S. W., 1967. Trilobites of the Henryhouse Formation (Silurian) in Oklahoma. Okla. Geol. Surv. Bull. 115, 68.

Decker, C. E., 1935. Graptolites from the Silurian of Oklahoma. J. Paleontol. 9:434-446.

Lundin, R. F., 1965. Ostracodes of the Henryhouse Formation (Silurian) in Oklahoma. Okla. Geol. Surv. Bull. 108, 104.

Strimple, H. L., 1963. Crinoids of the Hunton Group (Devonian-Silurian) of Oklahoma. Okla. Geol. Surv. Bull. 100, 169.

Sutherland, P. K., 1965. Rugose corals of the Henryhouse Formation (Silurian) in Oklahoma. Okla. Geol. Surv. Bull. 109, 92.

237   Churkin, M., and Brabb, E., 1965. Ordovician, Silurian, and Devonian biostratigraphy of east-central Alaska. Amer. Assoc. Petrol. Geol. 49:172-185.

238,239,240   Warshauer, S. M., and Smosna, R., 1977. Paleoecologic controls of the ostracode communities in the Tonoloway Limestone (Silurian; Pridoli) of the Central Appalachians. In Loffler, H., and Danielopolis, D. (eds.), Aspects of ecology and zoogeography of Recent and fossil Ostracoda, The Hague: Junk, 475-485.

241   Johnson, J., G., Boucot, A. J., and Murphy, M.A., 1967. Lower Devonian faunal succession in central Nevada. In Oswald, D. H. (ed.), International symposium on the Devonian System, Calgary, Alberta, 1967, Alberta Soc. Petrol. Geol. 679-691.

242,*244*,327   Buehler, E. J., and Tesmer, I. H., 1963. Geology of Erie County, New York. Buffalo Soc. Nat. Sci. Bull. 21(3), 118.

301,*302*,309 310   Johnson, J. G., 1965. Lower Devonian stratigraphy and correlation, northern Simpson Park Range, Nevada. Can. Petrol. Geol. Bull. 13:365-381.

Johnson, J. G., 1974. Early Devonian brachiopod biofacies of western and Arctic North America. J. Paleontol. 48: 809-819.

Johnson, J. G., Boucot, A. J., and Murphy, M. A., 1967. Lower Devonian faunal system in central Nevada. In Oswald, D. H. (ed.), International symposium on the Devonian System, Calgary, Canada, 1967, Alberta Soc. Petrol. Geol. 679-691.

304,315,*316*, 334,335   Lenz, A. C., 1976. Lower Devonian brachiopod communities of the northern Canadian Cordillera. Lethaia 9:19-27.

Lenz, A. C., and Pedder, A. E. H., 1972. Lower and middle Paleozoic sediments and paleontology of Royal Creek and Peel River, Yukon, and Powell Creek, Northwest Territories. XXIV Int. Geol. Cong. Field Excursion Guidebook A14.

305,306,307, 308   Laporte, L. F., 1969. Recognition of a transgressive carbonate sea within an epeiric sea: Helderberg Group (Lower Devonian) of New York State. In Friedman, G. M. (ed.), Depositional environments in carbonate rocks, a symposium, Soc. Econ. Paleontol. Mineral, Spec. Publ. (14):98-118.

Rickard, L. V., 1962. Late Cayugan (Upper Silurian) and Helderbergian (Lower Devonian) stratigraphy in New York. N.Y. State Mus. and Sci. Serv. Bull. 386, 157.

Rickard, L. V., and Zenger, D. H., 1964. Stratigraphy and paleontology of the Richfield Springs and Cooperstown Quadrangles, New York. N.Y. State Mus. and Sci. Serv. Bull. 396, 101.

311,312,313, 314   Lespérance, P. J., and Sheehan, P. M., 1975. Middle Gaspé limestone communities on the Forillon Peninsula, Quebec,

	Canada (Siegenian, Lower Devonian). Palaeogeogr., Palaeoclimat., Palaeoecol. 17:309-326.
317,318,319, 320	Koch, W. F., II, 1981. Brachiopod community paleoecology, paleobiogeography, and depositional topography of the Devonian Onondaga Limestone and correlative strata in eastern North America. Lethaia 14:83-104.
321	Johnson, J. G., and Flory, R. A., 1972. A Rasenriff fauna from the Middle Devonian of Nevada. J. Paleontol. 46: 892-899.
322	Chatterton, B. D. E., and Perry, D. G., 1978. An early Eifelian invertebrate faunule, Whittaker Anticline, northwestern Canada. J. Paleontol. 52:28-39.
323,324,325, 326,*328*	Feldman, H. R., 1980. Level-bottom brachiopod communities in the Middle Devonian of New York. Lethaia 13:27-46.
329,330,331, 332,*336,337,* *370*	McCollum, L. B., 1980. Distribution of marine faunal assemblages in a Middle Devonian stratified basin, lower Ludlowville Formation, New York. Unpubl. Ph.D. dissert., State Univ. N.Y., Binghamton.
333	Beerbower, J. R., Bray, R. E., Buehler, E. J., and Jordan, F. W., 1969. "Paleomicroecologic" study of Mid-Devonian marine assemblages (Hamilton Group) in western New York. In Devonian marine assemblages of New York, North American Paleontological Convention, 1969, Field Trip Guide (5):48-77.
338,339,340	Mazzullo, S. J., 1973. Deltaic depositional environments in the Hamilton Group (Middle Devonian), southeastern New York State. J. Sed. Pet. 43:1061-1071.
341,342,343	Ellison, R. L., 1965. Stratigraphy and paleontology of the Mohantango Formation in south-central Pennsylvania. Penn. Geol. Surv. Bull. G48, 298.
344,345	Klovan, J. E., 1964. Facies analysis of the Redwater Reef Complex, Alberta, Canada. Can. Petrol. Geol. Bull. 12: 1-100.
346,347,348, 349,350,*351*	Thayer, C. W., 1974. Marine paleoecology in the Upper Devonian of New York. Lethaia 7:121-156.
352,353,354, 355	Bowen, Z. P., Rhoads, D. C., and McAlester, A. L., 1974. Marine benthic communities in the Upper Devonian of New York. Lethaia 7:93-120.
356,357,358, *372,373*	McGhee, G. R., and Sutton, R. G., 1981. Late Devonian marine ecology and zoogeography of the central Appalachians and New York. Lethaia 14:27-43.
359,360,361, 362	McGhee, G. R., 1976. Late Devonian benthic marine communities of the central Appalachian Allegheny Front. Lethaia 9:111-136.
365	Sanford, B. V., and Norris, A. M., 1975. Devonian

stratigraphy of the Hudson Platform—Pt. 1: Stratigraphy and economic geology. Can. Geol. Surv. Mem. 379, 124.

366   Feldman, R. M., and McKenzie, S., 1981. *Echinocaris multispinosis*, a new echinocarid (Phyllocarida) from the Chagrin Formation (Late Devonian) of Ohio. J. Paleontol. 55:383-388.

367,368,369   Gutschick, R. C., McLane, M., and Rodriguez, J., 1976. Summary of Late Devonian-Early Mississippian biostratigraphic framework in western Montana. Mont. Bur. Mines Geol. Spec. Pub. 73:91-124.

Gutschick, R. C., Suttner, L. J., and Switek, M. J., 1962. Biostratigraphy of the transitional Devonian-Mississippian Sappington Formation of southwest Montana. Billings Geol. Soc. 13th. Field Conf. Guidebook, 79-89.

Rodriguez, J., and Gutschick, R. C., 1967. Brachiopods from the Sappington Formation (Devonian-Mississippian) of western Montana. J. Paleontol. 41:364-384.

401,402,403,   Ausich, W. I., Kammer, T. W., and Lane, N. G., 1979. Fossil
404,405,406,   communities of the Borden (Mississippian) Delta in Indi-
407   ana and northern Kentucky. Paleontol. J. 53:1182-1197.

Lane, N. G., 1973. Paleontology and paleoecology of the Crawfordsville site (upper Osagian), Indiana. Calif. Univ. Publ. Geol. Sci. 99, 141.

408,409   Sandburg, C. A., and Gutschick, R. C., 1979. Guide to conodont biostratigraphy of Upper Devonian and Mississippian rocks along the Wasatch Front and Cordilleran Hingeline, Utah. In Sandburg, C. A., and Clark, D. L. (eds.), Conodont biostratigraphy of the Great Basin and Rocky Mountains, Brigham Young Univ. Geol. Stud. 26(3):107-134.

Sandburg, C. A., and Gutschick, R. C., 1980. Sedimentation and biostratigraphy of Osagean and Meramecian starved basin and foreslope, western United States. In Fouch, T. D., and Magathan, E. R. (eds.), Paleozoic paleogeography of the west-central United States, Denver Colo., Soc. Econ. Paleontol. Mineral. Rocky Mtn. Sect. 129-147.

410,411,412   Broadhead, T. W., 1976. Depositional systems and marine
413.414   benthic communities in the Floyd Shale, Upper Mississippian, northwest Georgia. In Scott, R. W., and West, R. R. (eds.), Structure and classification of paleocommunities, Stroudsburg, Pa.: Dowden, Hutchinson, and Ross, Inc., 263-278.

415,416,417,   Watkins, R., 1973. Carboniferous faunal associations and
418,419,444,   stratigraphy, Shasta County, northern California. Amer.
445   Assoc. Petrol. Geol. Bull. 57:1743-1764.

420,422,423, 424   West, R. R., 1972. Relationship between community analysis and depositional environments: An example from the North American Carboniferous. XXIV Int. Geol. Cong., Proc., Sec. 7:130-146.

421   Brew, D. C., and Beus, S. S., 1976. A Middle Pennsylvanian fauna from the Naco Formation near the Kohl Ranch, central Arizona. J. Paleontol. 50:888-906.

425,426,427   Johnson, R. G., 1962. Interspecific associations in Pennsylvanian fossil assemblages. J. Geol. 70:32-55.

428,429,430, 431   Melton, R. A., 1972. Paleoecology and paleoenvironment of the upper Honaker Trail Fm. near Moab, Utah. Brigham Young Univ. Geol. Stud. 19:45-88.

432   Richardson, E., 1978. Mazon Creek faunal list. Unpubl. ms.

434   Morris, R. W., Rollins, H. B., and Shaak, G. D., 1973. A new ophiuroid from the Brush Creek Shale (Conemaugh Group, Pennsylvanian) of western Pennsylvania. J. Paleontol. 47:473-478.

435,436   Shaak, G. D., 1975. Diversity and community structure of the Brush Creek marine interval (Conemaugh Group, Upper Pennsylvanian), in the Appalachian Basin of western Pennsylvania. Florida State Mus. Bio. Sci. Bull. 19:69-133.

437,438   Hickey, D. R., and Younker, J. L., 1981. Structure and composition of a Pennsylvanian benthic community. J. Paleontol. 55:1-12.

439,440,441, 442   MacLeod, N., 1980. The paleoecology of the Wolf Mountain Shale: Community structure and trophic analysis. Unpubl. M.S. thesis, Southern Methodist Univ., Dallas, Tx.

  MacLeod, N., 1982. Upper Pennsylvanian peri-tidal benthic marine communities from the Wolf Mountain Formation (Canyon Group), north-central Texas. Soc. Econ. Paleontol. Mineral., Permian Basin Sec., Publ. 82-21:167-178.

443   Fagerstrom, J. A., 1964. Fossil communities in paleoecology: Their recognition and significance. Geol. Soc. Amer. Bull. 75:1197-1216.

501   West, R. R., 1976. Comparison of seven Lingulid communities. In Scott, R. W., and West, R. R. (eds.), Structure and classification of paleocommunities, Stroudsburg, Pa.: Dowden, Hutchinson, and Ross, Inc. 171-192.

502   Parker, W. C., 1983. Fossil ecological succession in Paleozoic level bottom brachiopod-bryozoan communities. Unpubl. Ph.D. dissert., Univ. of Chicago, Chicago.

503   Laporte, L. F., 1962. Paleoecology of the Cottonwood Limestone (Permian), northern mid-continent. Geol. Soc. Amer. Bull. 73:521-544.

505,506,507, 508,509,510   Terrell, F. M., 1972. Lateral facies and paleoecology of the Permian Elephant Canyon Formation, Grand County,

	Utah. Brigham Young Univ. Geol. Stud. 19:3-44.
511,512,513, 514,515,516, 517	Stevens, C. H., 1965. Pre-Kaibab Permian stratigraphy and history of Butte Basin, Nevada and Utah. Amer. Assoc. Petrol. Geol. Bull. 49:139-156.
	Stevens, C. H., 1966. Paleoecologic implications of Early Permian fossil communities in eastern Nevada and western Utah. Geol. Soc. Amer. Bull. 77:1121-1130.
518	Plas, L. P., Jr., 1972. Upper Wolfcampian (?) Mollusca from the Arrow Canyon Range, Clark County, Nevada. J. Paleontol. 46:249-260.
519,520,522, 523,524,525, 526,527	Yancey, T. E., and Stevens, C. H., 1981. Early Permian fossil communities in northeast Nevada and northwest Utah. In Gray, J., Boucot, A. J., and Berry, W. B. N. (eds.), Communities of the past, Stroudsburg, Pa.: Dowden, Hutchinson, and Ross, Inc.,243-270.
521	Winters, S. S., 1963. Supai Formation (Permian) of eastern Arizona. Geol. Soc. Amer. Mem. 89, 99.
529	Mudge, M. R., and Yochelson, E. L., 1962. Stratigraphy and paleontology of the uppermost Pennsylvanian and lowermost Permian rocks in Kansas. U.S. Geol. Surv. Prof. Pap. 323.
530	Nicol, D., 1944. Paleoecology of three faunules in the Permian Kaibab Formation at Flagstaff, Arizona. J. Paleontol. 18:553-557.
531	Clifton, R. L., 1942. Invertebrate faunas from the Blaine and the Dog Creek Formations of the Permian Leonard Series. J. Paleontol. 16:685-699.
532	Newell, N. D., 1940. Invertebrate fauna of the Late Permian Whitehorse Sandstone. Geol. Soc. Amer. Bull. 51:261-336.
533,534,535	Newell, N. D., Rigby, J. K., Fischer, A. G., Whitman, A. J., Hickox, J. E., and Bradley, J. S., 1953. The Permian Reef Complex of the Guadalupe Mountains Region, Texas and New Mexico. San Francisco: W. H. Freeman and Co., 236.
536,537,538	Walter, J. C., 1953. Paleontology of the Rustler Formation, Culberson County, Texas. J. Paleontol. 27:679-702.
539,540	Mayon, T. V., 1967. Paleontology of the Permian Loray Formation in White Pine County, Nevada. Brigham Young Univ. Geol. Stud. 14:101-122.

Chapter 6

CLASSES AND ADAPTIVE VARIETY: THE ECOLOGY OF DIVERSIFICATION IN MARINE FAUNAS THROUGH THE PHANEROZOIC

RICHARD K. BAMBACH

Department of Geological Sciences, Virginia Polytechnic Institute and State University, Blacksburg

PATTERNS OF DIVERSITY CHANGE

Step-like, rather than gradual or continuous, increase in diversity character-izes the history of the marine biosphere (Sepkoski et al., 1981). The pattern is one of low initial diversity in the Cambrian Period, higher but not persis-tently increasing diversity through the rest of the Paleozoic Era, a drop in diversity at the Paleozoic-Mesozoic boundary, increasing diversity through the Mesozoic Era and into the Cenozoic Era and a higher level of diversity during the Late Cenozoic than at any previous time. A variety of comprehen-sive studies on diversity, each using a different data source, reflect this pat-tern (Figure 1).

Four aspects of this pattern demonstrate that a relationship exists between diversity change in the marine biosphere and change in the ecologic structure of the fauna:

1. The relationship between the turnover of diversity dominance by various marine classes and the steps in total diversity increase noted by Sepkoski (1981) has major ecologic ramifications (see also Sepkoski and Sheehan, 1983; Bambach, 1983; Sepkoski and Miller, this volume). The Cambrian

fauna (Figure 2—first five classes at the top left, dominated by Trilobita) was comprised primarily of sedentary or creeping epifaunal surface deposit feeders, grazers or suspension feeders. The replacing fauna that dominated the rest of the Paleozoic included organisms with a broad range of solitary and colonial epifaunal habits (Anthozoa, Bryozoa, Brachiopoda, and Crinoidea and the other classes on the top row of Figure 2). Tiering above the substratum was added to the structure of the marine ecosystem by these epifaunal groups (Ausich and Bottjer, this volume) just as the Cephalopoda, Polychaeta, Bivalvia and Gastropoda added tropic variety and the increased utilization of shallow infaunal ecospace. The more diverse fauna of the Mesozoic and Cenozoic (the second row of Figure 2) includes classes with highly varied life styles ranging from deep infaunal to active nektonic habits. These classes also display a full spectrum of trophic activities including a dramatic elaboration of predatory specializations. The general increase in adaptive variety with each change of faunal dominance suggests that ecosystem organization has changed as diversity has increased.

2. As discussed by Valentine (1969, 1973, 1980), the number of higher level taxa (phyla and classes) has not increased despite turnover in faunal dominance and increase in diversity during the Phanerozoic (Figure 3). The total number of classes remained at about 50 throughout the Phanerozoic. The number of classes with very high diversity (over 30 families) also stayed nearly constant. During the same span of time the number of small classes (fewer than 5 families) steadily declined, as did the number of new classes originating in each time interval. The latter decrease was particularly great in the Mesozoic and Cenozoic although diversity reached all time high levels. Increases in diversity, especially in the Mesozoic and Cenozoic, were accommodated primarily within existing class level taxa.

3. The increase in species richness within habitats noted by Bambach (1977) and implicit in the trace fossil data of Seilacher (1974) and Crimes (1974) shows that diversity increase through the Phanerozoic includes packing more species into communities. Increase in diversity is not achieved simply by forming more numerous ecologic or biogeographic units, although that also may occur.

4. Adaptive variety within communities has increased with the increase in species richness of communities (Bambach, 1983). Increased species packing within communities has been accompanied by a broader utilization of ecospace, not by packing more species into the same realized ecospace. The ecologic structure of marine communities has been altered as diversity has increased.

These four phenomena indicate that the ecologic structure of the marine biosphere has changed as diversity has changed. This has influenced the biosphere at all levels from the interaction of species within communities to the overall utilization of ecospace by the entire marine fauna. These changes in ecologic properties occurred without increasing the number of diverse class-level taxa. The ecologic change in the marine fauna must be related to turnover in dominant taxa and change within those taxa.

This paper examines the changes in adaptive variety of the dominant class-level taxa through the Phanerozoic and develops the argument that there is a relationship between change in adaptive variety and change in diversity. The subject of the study is the pattern of increasing, constant and waning diversity observed within class and ordinal level clades. These patterns are compared with the adaptive variety represented by the members of each clade. Distinctive structural features that each taxon utilizes in achieving its range of life habits are noted, as is the role these features play in expanding the adaptive range of those taxa that increase in diversity.

The following empirical study of the style of diversification within subordinal to class level taxa reveals that a link exists at one level or another between the increase of diversity and the evolution of new structural attributes that permit greater adaptive variety. The evolutionary fate of low level taxa (species, genera and families) may best be described in stochastic terms because there are so many variables that can influence their success through time, most of which can not be specified because of the veil of time and diagenesis. Diversity change in the total biosphere is best described with generalized mathematical models because it is the summation of so many separate clade histories (Raup et al., 1973; Gould et al., 1977; Raup, 1977; Sepkoski, 1978, 1979; Carr and Kitchell, 1980; Kitchell and Carr, this volume; Walker, this volume). However, deterministic explanation, relating structural properties of organisms to their adaptive variety and ecologic success, can be applied to the study of ordinal and class level diversity history. Stanley et al. (1981) and Thomas and Foin (1982) have also raised questions concerning purely stochastic methods of the study of diversity. As pointed out by Schopf (1979), stochastic and deterministic viewpoints each have appropriate applications, depending on the level of analysis under study.

ANALYTIC SCHEME

If there is no strong relationship between diversification and ecologic properties of organisms we should find that some classes that contain taxa that develop new modes of life and invade new ecospace may still decline in diversity whereas other classes that contain taxa that remain structurally unchanged and do not invade new habitats will increase in diversity nonetheless. If, on

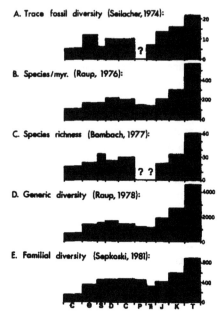

Figure 1. Five compilations of marine diversity through the Phanerozoic.

A. Variety of trace fossils in flysch deposits (tabulated by Seilacher, 1974).

B. Estimate of relative species variety on a worldwide basis (Raup, 1976).

C. Median species richness within communities (Bambach, 1977).

D. Total generic diversity reported in the *Treatise on Invertebrate Paleontology* (Raup, 1978).

E. Family diversity from various sources (Sepkoski, 1981).

Figure after Sepkoski et al. (1981).

Figure 2. (Facing page). Diversity of families within classes of marine animals. Classes arranged in order of earliest time of maximum diversity to illustrate the turnover in diversity dominance through Phanerozoic time. *pC* refers to Vendian time, P to the Paleozoic Era; M the Mesozoic Era and *C* the Cenozoic Era. The third row contains those class level taxa with little total familial diversity or a poor fossil record. Figure modified from Sepkoski (1981). Compilation includes only families with a fossil record and omits recent families with no fossil record. This figure serves as source for Figures 4, 6, 8, 9, 13, 18, 23 and 27.

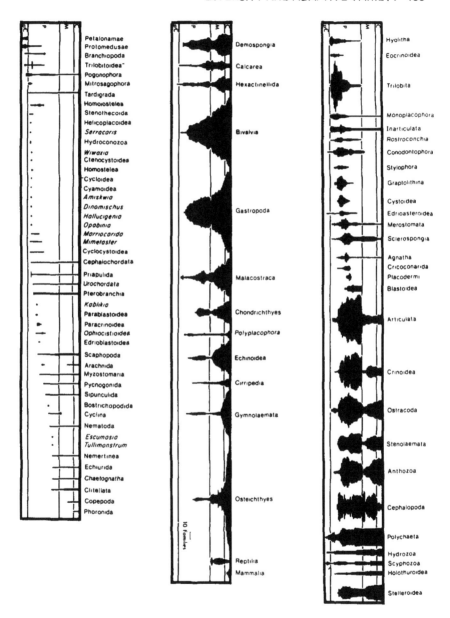

CLASS LEVEL TAXA

	NUMBER OF FAMILIES IN CLASS				TOTAL DIVERSE CLASSES	SMALL CLASSES 1-4 fam.	TOTAL CLASSES	NEW CLASSES
	5-9	10-19	20-29	30+				
CENOZOIC	5	7	5	7	24	24	48	2
MESOZOIC	2	2	8	7	19	29	48	2
L. PALEOZOIC	6	3	4	6	19	33	52	6
M. PALEOZOIC	8	10	4	7	29	37	66	28
E. PALEOZOIC	4	2		1	7	44	51	41
PRE-CAMBRIAN						11	11	11

Figure 3. The number of class level taxa for different segments of the Phanerozoic tabulated by number of families within classes. Right hand column (New Classes) gives the number of classes originating during each time segment. E. Paleozoic includes the Cambrian and Early Ordovician, M. Paleozoic includes Mid Ordovician through Devonian and L. Paleozoic includes the Carboniferous and Permian. Figure from Bambach, 1983.

the other hand, diversification is related to ecologic factors, then only those classes that include taxa that acquire new structures that permit them to invade additional adaptive zones should increase in diversity through time. This paper examines these possibilities.

The primary subject of this paper is the diversity history of classes. This avoids the problem of taxonomic decisions interfering with the perception of the relationship between diversity within a group and its ecologic variety. The diversity histories of classes are phylogenetically independent of each other. For example, the decline of the Trilobita was not because organisms arose from trilobites that were removed from the class Trilobita and called new classes. Trilobites really did decline and Malacostraca really have diversified since the Cambrian. The Malacostraca are not the modified descendents of trilobites representing continued high diversity within that lineage group. Therefore, it is reasonable to compare classes and their ecologic variety and it is possible to relate their diversity histories to the range of ecospace they exploit without fearing that we have lost members from some classes and erected others just because of the methodology of taxonomy.

At lower taxonomic levels the development of features that permit the exploitation of new ecospace usually does lead to the designation of new taxa for those organisms with the new features. For example, the flattening of the test and shift in position of the periproct in the irregular echinoids permits them to live infaunally whereas the regular echinoids are epifaunal. These modifications are among those that have justified erecting two super-orders for the irregular echinoids, separating them taxonomically from the other echinoids (which are also assigned to several other orders and superorders). These groupings within classes make it possible to identify the particular lower taxa that contribute to the diversity changes of a class and document the relationship those taxa have to the spectrum of ecospace utilization by the whole class.

In this study the term ecospace is used for the theoretical ecologic hypervolume occupied by a group of organisms. Dimensions of this theoretical hypervolume are general features such as mode of life and feeding type. Organisms with a wide variety of modes of life and feeding types exploit more ecospace than those with more limited habits. For example, bivalve mollusks, which include epifaunal and infaunal, mobile and sedentary, and deposit feeding as well as suspension feeding forms, exploit more ecospace than articulate brachiopods, which are all epifaunal, sedentary, suspension feeders.

This study examines patterns of total diversity for various taxa rather than changes in diversity within geographic areas or habitats. One of the major conclusions that can be drawn from the results reported in this paper is that diversity change at intermediate taxonomic levels is related to change in utilization of ecospace rather than just to change in geographic area inhabited by a group. For example, bivalve mollusks have increased in diversity because of the development of deep infaunal modes of life, not because they have expanded their habitat range from nearshore shelf environments until they have even invaded the deep sea. The bivalve families present in the deep sea, however, are all represented in shelf environments and many were present in shelf settings in the Paleozoic.

Examination of Figure 2 reveals that there are several consistent forms to family diversity in classes through the Phanerozoic. Those groups in the bottom row of Figure 2 are classes with either poor representation in the fossil record or only low diversity or short duration. They are represented by "dots" or "lines" only. The record is obscure as to their complete diversity history. The classes in the top two rows of Figure 2 have four basic forms: (1) bottom-heavy, either with or without a later "tail," (2) irregular or fluctuating, ranging from irregular to "dumbbell" shaped, (3) relatively smooth and even column, and (4) top-heavy, often with a rather long narrow early shaft. Few class level clades have a diamond or symmetrical form of expansion and contraction. The four common forms for these clades represent: (1) for bottom heavy clades an early, rapid diversification followed by unrecovered loss in diversity even though a few forms may hang on for a long time creating the "tail," (2) for irregular shaped clades a rapid diversification followed by diversity contractions and expansions that recover some or all of the earlier diversity, (3) for even columns an early diversification followed by maintenance of diversity without much fluctuation, and (4) for top-heavy clades a large expansion in relatively recent geologic time, often after a lengthy interval of relatively low diversity for the group. The 16 clades at the left of the top row of Figure 2 are "bottom-heavy," the next six after the Blastoidea are irregular, the Blastoidea and the five on the right hand side of the top row are even columns and all clades in the second row, except the Reptilia, are "top-heavy."

The following sections examine these various clades, grouped into phylum groupings and also subdivided into subclasses, orders or suborders. The goal is to establish whether changes in diversity of class level clades are related to particular included taxa and, if so, to see if there are ecologic reasons for those taxa to diversify. Therefore, the ecologic properties of the constituent taxa will also be noted.

The data used in the following sections were drawn from a variety of sources. The purpose of the study is to identify pattern rather than to determine the exact number of currently known taxa. Because all of the sources are comprehensive reviews the data are adequate even though some are out of date for current detailed taxonomy. For example, the original volume of the *Treatise on Invertebrate Paleontology* on the Coelenterata from 1956 was used for data on all corals (Figure 7) rather than the new revision (1981) because the original work includes the Scleractinia which have not been revised in the new work. The interest here is in comparing the time of occurrence and proportional diversity of all three groups. In 1956 the *Treatise* reported 333 genera of Rugosa and 108 genera of Tabulata. It lists 770 genera of Rugosa and 334 genera of Tabulata in 1981. Despite much valuable new information accumulated in the past quarter century, there has been no drastic change in proportional diversity among the major taxa.

Diversity within taxa below the class level is given using various taxonomic levels. Genera are used in nine examples, families in six, and superfamilies in one case. Although this breaks from consistent use of family diversity it does not obscure the relative fluctuations in diversity within groups. In fact, one benefit of using different levels of taxonomic resolution is the demonstration of the similarity of diversity change at all scales of taxonomic resolution. The proportional variation between generic and familial diversity fluctuation is relatively small in most cases.

1. PORIFERA (Figures 4 and 5)

All classes of the Porifera originated in the Vendian or Cambrian (Figure 4). The Sclerospongia reached their peak of family diversity during the Middle Paleozoic (Silurian and Devonian), whereas the Demospongia, Calcarea and Hexactinellida did not achieve their maximum diversity levels until the Mesozoic. The Demospongia and Hexactinellida have maintained their high diversity levels through the Cenozoic.

The Sclerospongia secrete a massive calcareous skeleton with the living

PORIFERA

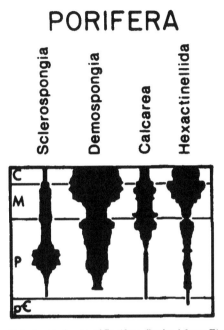

Figure 4. Family diversity in classes of Porifera. Derived from Figure 2.

tissue confined to a thin external veneer (Wendt in Hartment et al., 1980). They were the capping organisms in many middle Paleozoic reef structures, overgrowing the tabulate and rugose corals as the reef built up into agitated near-surface waters. Their diversity decline parallels the collapse of the mid-Paleozoic reef ecosystem (Heckel, 1974), but they have been able to persist in a variety of settings, although not as community dominants. Today they are members of cryptic and deeper water communities in modern scleractinian-dominated reefs. The sponge classes that have diversity increase later than the Sclerospongia are all characterized by a three-dimensional elaboration of living tissue with the skeleton serving as a supporting mesh-like framework rather than only as a foundation underneath the living sponge. The proportion of living material to inorganic skeleton is considerably greater, too.

In each class of the sponges other than the Sclerospongia there is an order or group of orders that originate in the Vendian or Cambrian and persist through the Paleozoic or through the entire Phanerozoic without much apparent change in diversity (the Heteractinellida in the class Calcarea, Demospongia with simple megascleres lacking tetraxons and regional differentiation of placement of megasclere types (the orders Haplosclerida, Hadromerida and Halichondria) and Hexactinellida without fused spicules (the orders

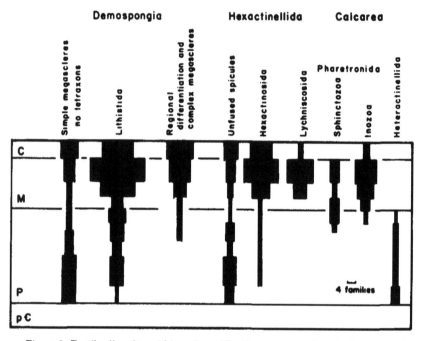

Figure 5. Family diversity within orders of Porifera as reported in the *Treatise on Invertebrate Paleontology* (Part E, 1955).

Lyssacinosida, Reticulosida and Amphidiscosida (Figure 5). This apparent lack of diversity change may be at least partly an artifact of the fossil record because the skeletons of the sponge groups in question easily disarticulate after death unless the organism is rapidly buried; they are not easily preserved. However, the groups of Demospongia that have spicule types segregated into particular regions of the sponge and those with complex spicule forms such as tetraxons do show diversification through the Mesozoic and Cenozoic. This grouping also includes the largest and most structurally diverse order of living Demosponges, the Poecilosclerida (Bergquist, 1978). Because the sponges in this group are also capable of disaggregation after death and yet show a markedly different time of origin and a different diversity pattern (matched by the recent fauna) than the other groups of sponges with unfused spicules, it is likely that the clade shapes for these groups do reflect their general pattern of diversity history even though the patterns may be somewhat degraded by preservational biases. The group of Demospongia with simple megascleres, the Hexactinellida with unfused spicules and the Heteractinellida all seemed to have had an early Paleozoic origin and then simply maintained their diversity afterwards at approximately that early Paleozoic level. These groups are responsible for the long interval of low diversity in the Hexactinellida, Calcarea and, in part, the Demospongia (Figure 4) but do not contribute to the later expansion of diversity in those classes.

The sponges that have the most influence on diversity increase are the Lithistida, Hexactinosida, Lynchniscosida, and Pharetronida (Figure 5). In all cases they have skeletons of interlocking or fused materials and therefore they have the most complete fossil record among the sponges. The Lithistida are a polyphyletic group (Finks, 1967; Hartman in Hartman et al., 1980) but are kept as an order because they all bear multi-branched siliceous spicules called desmas. Several lineages of lithistids arose in the Cambro-Ordovician and maintained their initial diversity into the late Paleozoic at which time they became extinct but were replaced by other lineages, most of which have more complex (tetraxon) spicules and complex spicular relationships including firm linkage of desmas. It is these more recently evolved groups plus several groups that do not originate until the Mesozoic that undergo diversity expansion in the Mesozoic. In the Hexactinellida it is the two orders with fused spicules (Hexactinosida and Lychniscosida) that create the Mesozoic diversity increase in that class (and the Lychniscosida originate only in the Mesozoic). Among the Calcarea it is the development of the Sphinctozoa in the Carboniferous and the radiation of the Inozoa during the Mesozoic that create the diversity expansion in that class. These are the rigid Pharetronida. Those Demospongia with a more regular degree of spicular organization and more complex spicular structure also contribute to the diversity expansion of that class. They did not arise until the late Paleozoic and have increased in diversity during the Mesozoic and Cenozoic. They include the orders Poecilosclerida, Astrophorida, Spirophorida and Homosclerophorida.

In each class of the sponges some groups were established in the early Paleozoic. They maintained their diversity through the rest of the Paleozoic but it is only newly developed groups with fused, rigid skeletons or more complex spicules or spicular organization that contribute to the Late Paleozoic and Mesozoic diversity expansions in each class of the Porifera. The Sclerospongia, the class that had its diversity peak the earliest, has remained a part of the marine fauna but never expanded in diversity after the sponges with three-dimensional development of living tissue and higher biomass to skeletal proportions began to diversify. Because of their strength and erect growth forms, sponges with interlocking or fused skeletons or forms with complex spicular organization or structure have been able to invade a variety of habitats for which the other sponges were not suited. Those other sponges have been able to continue life in the settings that they initially inhabited and they still maintain their former diversity.

2. COELENTERATA (Figures 6, 7, and 8)

Among the Coelenterata, the Hydrozoa, Scyphozoa and several extinct groups originated in the Late Precambrian. Anthozoa do not appear until the early Paleozoic (Figure 6). Both the Hydrozoa and Scyphozoa maintain a rather constant family diversity throughout the Phanerozoic. The modest increase in each class is probably related to the somewhat better record of rarely preserved organisms in younger rocks. Although poor preservation of these generally non-skeletal organisms may have obscured their complete diversity pattern (as can happen for any such organisms; see also discussion of sponges above and worm-like phyla below), it is interesting that major bulges in the diversity curves do not appear at the times of fortunate preservation of soft-bodied organisms (lagerstätten) such as the Middle Cambrian (Burgess Shale), Late Carboniferous (Francis Creek Shale) or Late Jurassic (Solenhofen Limestone). Such deposits can reveal groups for which the general fossil record may not reflect actual diversity (see Trilobitoidea and Priapulida in the Middle Cambrian on bottom column of Figure 2). The only one of these deposits with a visible effect for the Coelenterata is the small bulge for the Scyphozoa in the middle of the Mesozoic caused by the Solenhofen.

The Anthozoa have a different diversity pattern. Their fossil record is rich because the corals have readily preserved skeletons. The Anthozoa lost diversity at the close of the Paleozoic and diversified again in the Mesozoic,

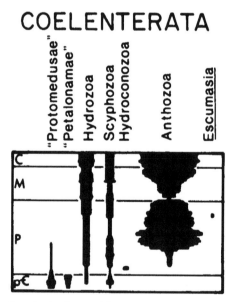

Figure 6. Family diversity in classes of Coelenterata. Derived from Figure 2.

reaching a family diversity in the Cenozoic equal to but not exceeding their Paleozoic maximum (Figure 6). The Ordovician diversity expansion of the Anthozoa is the result of the origin and diversification of both the Tabulata and Rugosa and the diversity collapse reflects the extinction of these two sub-classes of corals (Figure 7). The recovery of Paleozoic levels of diversity resulted from the diversification of the modern subclass of corals, the Scleractinia. Figure 7 illustrates that the diversity of genera of Scleractinia in the Cretaceous and again in the Cenozoic is about that of the Paleozoic maximum of the Tabulata and Rugosa combined, just as their familial diversities are similar (Figure 6).

The Scleractinia have replaced the Paleozoic corals and they equal but do not exceed them in diversity. The Scleractinia differ from Paleozoic corals in several ways but they have not developed any structural attributes that permit them to live in any environments that Paleozoic corals did not also inhabit. The fauna that interacts with the Scleractinia is markedly different from the Paleozoic fauna, however, and the new features of the Scleractinia such as the ability to firmly cement or attach the entire base of large colonies (because of the coenosarc and its skeletal secretions) and the apparent more rapid growth rates (ascribed to both their less massive, more porous and open skeletal structure and the physiological properties of hermatypic corals) may have been necessary to permit the second full diversification of the group in the Mesozoic.

Figure 8 illustrates the generic diversity of the orders of the Tabulata and of suborders of the Rugosa grouped by time of origin (pre- and post-Middle Silurian) and growth habit (predominantly solitary versus compound or mixed compound and solitary). All tabulates were compound (colonial). The diversity of the Tabulata declined severely in the Late Devonian at which time the Sarcinulida and Heliolitida became extinct and the Favositida dropped to

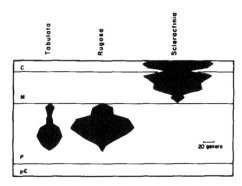

Figure 7. Generic diversity within the orders of corals as reported in the *Treatise on Invertebrate Paleontology* (Part F, 1956).

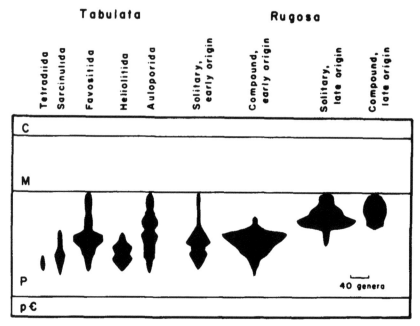

Figure 8. Generic diversity within orders of the Tabulata and early and late originating groups of orders of the Rugosa as reported in the *Treatise on Invertebrate Paleontology* (Part F, Supplement 2, 1981). Solitary form groups of the Rugosa originating prior to the Late Silurian are the Streptelasmatina (Middle Ordovician–Middle Devonian), Metriophyllina (Middle Ordovician–Late Permian), and Lycophyllina (Early Silurian–Late Devonian). Compound or mixed solitary and compound groups of the Rugosa originating prior to the Late Silurian are the Cystiphllida (Middle Ordovician–Late Devonian), Calostylina (Middle Devonian), Arachnophyllina (Early Silurian–Middle Devonian), Ketophylina (Late Ordovician–Late Late Carboniferous), Ptenophyllina (Early Silurian–Late Devonian), Columnarina (Middle Silurian–Late Devonian), and Cyathophyllina (Early Silurian–Late Early Carboniferous). The solitary form groups of the Rugosa originating after the Middle Silurian are the Stereolasmatina (Early Devonian–Late Permian), Plerophyllina (Late Silurian–Late Permian), Caniniina (Early Early Carboniferous–Late Permian), Aulophyllina (Late Devonian–Early Permian) and Heterocorallia (Middle Devonian–Early Late Carboniferous). The compound or mixed compound and solitary form groups of the Rugosa originating after the Middle Silurian are the Lithostrotionina (Early Early Carboniferous–Late Permian) and Lonsdaleiina (Early Early Carboniferous–Late Permian).

about 25% of their former diversity. Only the Auloporida, in which the coral-lites are at least proximally recumbent and often cemented to the substratum and in which the corallum is composed of separate or anastomosing but part-ly separated corallites, maintain their earlier Devonian diversity into the Late Paleozoic. Among the Rugosa all the suborders that originated in or prior to the Middle Silurian also suffered a collapse in diversity in the Late Devonian. The Rugosa as a whole (Figure 7) do not show this sharp change because two suborders that originated in the Late Silurian and Early Devonian plus one suborder starting in the Late Devonian and three new suborders first appear-ing in the Early Early Carboniferous all diversify in the Carboniferous. The whole clade declines in diversity only when several of these late-originating and late-diversifying suborders decrease in diversity in the Late Carbonifer-ous.

Prior to the Late Devonian-Early Carboniferous replacement of diverse suborders in the Rugosa the suborders with predominantly solitary coralla had been less than half as diverse as those suborders comprised of colonial or mixed colonial and solitary forms. After the Late Devonian-Early Carbon-iferous turnover the newly diversified suborders of predominantly solitary form had three to four times greater generic diversity than the suborders containing compound and mixed compound and solitary forms. The Late Carboniferous decrease in diversity in the Rugosa occurred in predominantly solitary-form suborders but they remained almost as diverse as the compound-form suborders as late as the Late Permian just prior to the extinction of all Rugosa.

The major diversity change in the Paleozoic corals occurred in the Late De-vonian. All the types of massive compound corals, both tabulate and rugose, declined severely at that time, as did the older suborders of solitary rugosans. The Tabulata never recovered although the group with the firmest attachment potential and the least massive colony form (the Auloporida) maintained their former diversity. The Rugosa recovered their former diversity as newly developed taxa diversified in the Carboniferous but the compound taxa never reached half the diversity of pre-Late Devonian compound Rugosa. These fea-tures all relate to the disappearance of coral-dominated reef systems in the Late Devonian (Heckel, 1974). Coral diversity only returned to Middle Paleo-zoic levels with the development of Scleractinian reef systems in the Meso-zoic.

3. LOPHOPHORATE PHYLA (Figure 9)

There are three lophopore-bearing phyla, the solitary Brachiopoda (classes In-articulata and Articulata), the colonial Bryozoa (classes Stenolaemata, Gymnolaemata and Phylactolaemata, the latter with no fossil record) and the

LOPHOPHORATE PHYLA

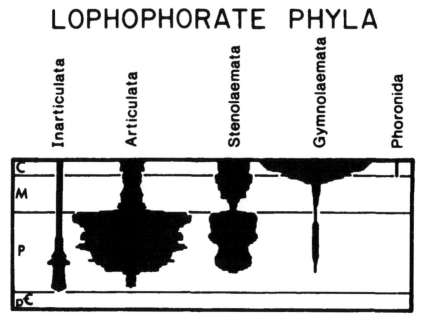

Figure 9. Family diversity in classes of the lophophorate phyla Brachiopoda (Inarticulata and Articulata), Bryozoa (Stenolaemata and Gymnolaemata) and Phoronida. Derived from Figure 2.

Phoronida (Figure 9). The two classes of Brachiopoda diversified most extensively in the Paleozoic with the Inarticulata reaching their diversity maximum in the Cambrian and the Articulata maintaining high diversity from the Ordovician to Permian. Both groups persist into the Recent but at reduced diversity. The two classes of Bryozoa with an extensive fossil record originate in the Ordovician. The Stenolaemata maintain a high diversity through the Paleozoic, collapse at the end of the Permian and rediversify to near Paleozoic levels in the Mesozoic and Cenozoic. The Gymnolaemata increase enormously in diversity in the Cretaceous and Cenozoic. The Phoronida are soft bodied and only have a very poor Cenozoic record. The diversity histories of each of the skeletal classes of lophophorates illustrate features that reflect changing diversity within various orders and suborders rather than coherent diversity change across whole classes.

A. Brachiopoda (Figures 10 and 11). The Inarticulata diversified into five orders (Lingulida, Acrotretida, Obolellida, Paternida and Kutorginida) in the Early Cambrian. Three of these orders became extinct by the Middle

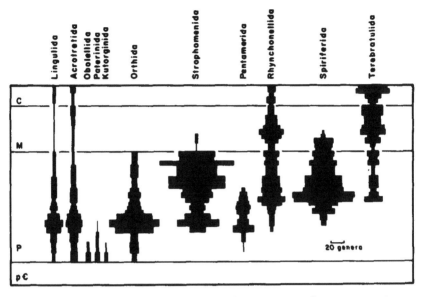

Figure 10. Generic diversity within orders of Brachiopoda. After *Treatise on Inverte-brate Paleontology* (Part H, 1965).

Ordovician whereas the Lingulida (including infaunal forms) and the Acrotretida (attached forms), after a modest Ordovician increase in generic diversity, declined slightly and then remained at nearly constant low levels of diversity thereafter. They have neither expanded nor declined much for 400 million years (if they had declined very much they would have become extinct). The Articulata are represented in the Cambrian by two orders (Orthida and Pentamerida). These diversify in the Ordovician, as do the newly originated Strophomenida, to establish the general high level of articulate brachiopod diversity for the rest of the Paleozoic. These three orders are not responsible for maintaining that diversity, however. The Orthida contract in diversity after the Ordovician and the Pentamerida become extinct by the end of the Devonian. Of these three dominant Early Paleozoic orders, only the Strophomenida increase in diversity in the later Paleozoic. The maintenance of high articulate brachiopod diversity through the Middle and Late Paleozoic is a result of the diversification of three new orders, the Rhynchonellida, Spiriferida, and Terebratulida.

Discussions by Williams and Rowell (1965) and Williams and Hurst (1977) point out morphological and developmental relationships that can be used to justify grouping the Orthida with the Pentamerida and grouping the Rhynchonellida, Spiriferida and Terebratulida together for purposes of examining overall diversity changes in the Articulata. These features include, besides different times of origin, the general lack of restriction of the pedicle opening, general lack of calcified support for the lophophore (except for brachiophores) and lack of evidence for mantle reversal during development in the Orthida and Pentamerida whereas the Rhynchonellida, Spiriferida and Terebratulida all have deltidial plates to restrict or modify the pedicle opening, calcified lophophore support structures (crura, spiralia or loops), and pedicle base insertion and musculature of a highly developed style that would only develop with mantle reversal in the developing larvae (Williams and Rowell, 1965:H194).

The Strophomenida, on the other hand, are a diverse group in which the suborders have distinctive attributes and different diversity histories. The Strophomenidina are predominantly concavo-convex to resupinate, often free-lying unattached forms with long hinge lines. The Chonetidina also are long hinged concavo-convex shells but they have a spine row along the margin of the pedicle valve interarea and other distinctive features. The Productidina are generally deep-bodied plano-convex shells without pedicle but with spines on the pedicle valve and often with other, sometimes bizarre, features such as

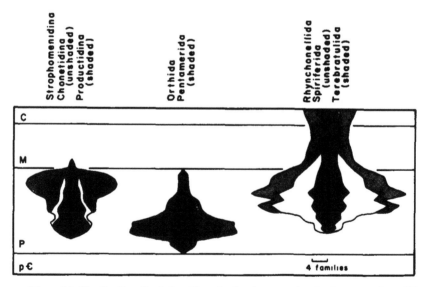

Figure 11. Family diversity in brachiopod suborders or orders in three groupings: (1) stromphomenide suborders, (2) the two orders of primarily inpunctate shells lacking deltidial plates (Orthida and Penamerida) and (3) the three orders with true deltidial plates, probable mantle reversal and calcified lophephere supports (Rhynchonellida, Spiriferida and Terebratulida). Compiled from the *Treatise on Invertebrate Paleontology* (Part H, 1965).

geniculated valves with trails. The Oldhaminidina are a small group of Late Paleozoic brachiopods with weird irregular shells.

Figure 11 illustrates the diversity histories of the three groupings of articulate brachiopods discussed above. The Orthida and Pentamerida have parallel histories. Both reach high diversity in the Ordovician and both decline in diversity in the Middle Paleozoic. Within the strophomenids, the Strophomenidina diversify in the Ordovician and contract in diversity in the Middle Paleozoic. The Chonetidina simply persist at moderate diversity from the Silurian through the Carboniferous without much diversity change. The Productidina, however, increase the total diversity of the Strophomenida considerably Their bizarre forms and varied structures permitted them to occupy many life habits that were not open to the more conservative and restricted range of morphologies present in the Strophomenidina and Chonetidina. The Productidina not only recovered the diversity lost as the Strophomenidina dwindled but diversified more than earlier, less varied strophomenids. The Rhynchonellida, Spiriferida and Terebratulida maintained and even increased the diversity of pedunculate brachiopods established by the Orthida and Pentamerida. It is interesting that these three orders have features that are not possessed by the earlier diversified pedunculate orders. The Spiriferida diversified the most and also had the widest range of shell morphologies and lophophore supports. Indeed, it is the Spiriferida with their broad range of morphologies adapted to various substrata (Copper, 1966, 1967) that add diversity to the Middle Paleozoic articulates as a whole. The Rhynchonellida and Terebratulida only supply enough new diversity to regain that lost by the decline of the Orthida and Pentamerida.

The earliest forms to develop in each group in Figure 11 (the Orthida, the Strophomenidina and the Rhynchonellida) also persist the longest. In the case of the Rhynchonellida, Spiriferida and Terebratulida it is also of interest that two of these orders survive to the Recent whereas all other orders of the Articulata are extinct (the difficult group Thecideidina is included in the Terebratulida in this paper). It is in this group of orders that the support for the feeding organ, the lophophore, is most developed and it is in these orders that the pedicle insertion and attachment is elaborated. This style of pedicle is strong (Thayer, 1975), capable of varied growth and function (Richardson, 1981), including active positioning of the animal in currents in some forms (LaBarbera, 1977). A last point is that the proportional changes in diversity within groups are similar at the generic (Figure 10) and familial (Figure 11) levels.

B. Bryozoa (Figure 12). The diversity history of the Stenolaemata is one of diversification by several groups during the Paleozoic, extinction of most of them at the end of the Paleozoic, and recovery of diversity during the

Mesozoic by diversification in a new group. The added diversity for the Gymnolaemata in the Mesozoic and Cenozoic is related to the diversification of a new group with markedly different features (the Cheilostomata) whereas the Paleozoic gymnolaemate group simply persists with little diversity change.

In Figure 12 the Paleozoic diversification of the Stenolaemata is recorded by the expansion of the Trepostomata, Cryptostomata and Cyclostomata, all in the Ordovician. The Trepostomata contract in diversity after the Ordovician but persist until the end of the Paleozoic. The Cryptostomata increase some in diversity in the later Paleozoic but this simply recovers the total stenolaemate diversity attained in the Ordovician with no added expansion. The Cyclostomata have a relatively modest but fairly stable diversity from the Ordovician to Permian, a contraction in diversity at the end of the Paleozoic (when the other two stenolaemate orders become extinct) and a Mesozoic increase in diversity. The Cyclostomata as reported in the *Treatise on Invertebrate Paleontology* (Bassler, 1953) have been subject to much later debate. Astrova (1966) proposed that the Ceramoporoidea and Fistuliporoidea and related forms be grouped in a new order, the Cystoporata. Ryland (1970) reviewed the evidence and accepted that scheme. If this is done, then 49 of the 62 genera of Cyclostomata of Paleozoic age listed in the *Treatise* would be placed in the extinct order Cystoporata, leaving only 13 genera of Paleozoic

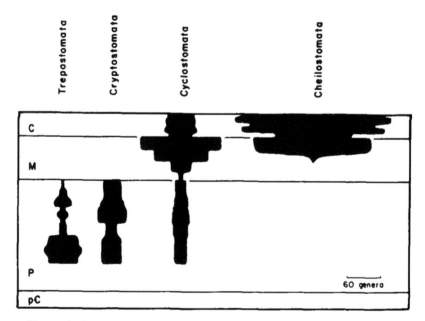

Figure 12. Generic diversity within orders of Bryozoa well represented in the fossil record. After Müller (1958) in Schopf (1977).

age assigned to the Cyclostomata. Schopf (1977) argues that most bryozoan genera cited as having long geologic histories actually do not persist as claimed. He cites studies by Brood in 1972 and 1975 (not seen by me) that seriously question whether several of the other genera of Paleozoic cyclostomes have the characteristics universally found in Mesozoic and Cenozoic Cyclostomata. It appears that the Cyclostomata of the Mesozoic and Cenozoic may be a newly developed Mesozoic group and not one with a long Paleozoic history. If there are any true Paleozoic Cyclostomata they are restricted to only a few genera and they do not seem to possess the porous walls characteristic of the diverse post-Paleozoic families of the order. Therefore, it seems as if there is a nearly complete turnover in the Stenolaemata at the end of the Paleozoic. The variety of orders present in the Paleozoic is replaced by the single order Cyclostomata in the Mesozoic and Cenozoic. That order diversifies in the Mesozoic and achieves a diversity equivalent to but not significantly exceeding all the Paleozoic Stenolaemata combined.

The morphologies of the zoaria of the orders of Stenolaemata are varied. All groups contain both encrusting and erect forms. The Trepostomata had long tubular zooecia and constructed massive zoaria. There were no interzooecial connections within the skeleton although the growing surface tissues were highly integrated across the colony surface. The Cryptostomata had much shorter zooecia in most instances but also had no skeletally internal interzooecial connections. The growth forms of the Cryptostomata include the elaborate frondescent and meshwork structures of the fenestellids. The internal communication between zooecia permitted by the porous wall of the Mesozoic and Cenozoic Cyclostomata permits living tissue to remain functioning in the interior of the colony, a feature not possible in Trepostomata, nor one for which the Cryptostomata are designed. However the basic tubular nature of the cyclostome zooecia and the growth habits of the zoaria are similar to the Paleozoic Stenolaemata and the Cyclostomata appear to occupy no different life habits than Paleozoic forms.

The Gymnolaemata contains two orders, the Ctenostomata and the Cheilostomata. The Ctenostomata are represented in the fossil record by from two to six genera in each Period from the Ordovician to the Neogene. They are the only Paleozoic Gymnolaemata and their low but relatively constant diversity forms the long narrow part of the Gymnolaemata clade in Figure 9. The Cheilostomata originate possibly in the Jurassic and diversify explosively in the Cretaceous (Figure 12). The Cheilostomata have numerous distinctive features. The zooid is reoriented so that the feeding polypide is extruded through an orifice in the side rather than at the end of the zooecium. This presents a wide range of new and different mechanisms of calcifying the open side of the zooid. Colony growth forms, shapes and life sites are of great variety. Another feature is the extreme variety of polymorphism that includes avicularia and vibracula among protective devices and a range of brood chambers, etc. While anatomically still obviously bryozoa, the Cheilostomata are quite different skeletally from both Paleozoic Gymnolaemata and all Stenolaemata. This new form of organization was exploited in the post-Paleozoic diversity expansion of the Bryozoa.

4. VARIOUS NON-SKELETAL PHYLA

The fossil record, imperfect at best, is obviously at its poorest with regard to soft-bodied unskeletonized organisms. More than half the class level taxa have such a poor fossil record that we can do little more than record their presence (Figure 2, bottom row). If it were not for the spectacular examples of unusual preservation such as the Burgess Shale (Middle Cambrian), Hunsruck Shale (Middle Devonian), Francis Creek Shale (Upper Carboniferous), Solenhofen Limestone (Late Jurassic) called lagerstätten, most of these groups would have no fossil record at all. Despite these severe problems there are some tentative observations to make.

Among the worm phyla that are most diverse today there is no record for free living Platyhelminthes and only a minuscule one for the Nematoda, but the Annelida does have a significant record. The Polychaeta, a dominant class in modern benthic environments, has a long history extending back to the Vendian (Figure 2, top row, fifth from the right). They diversified in the Cambrian and Ordovician and have maintained that high familial diversity throughout the rest of the Phanerozoic.

This contrasts with the fossil record of some of the other worm-like groups, which are not diverse today and which show no suggestion of any higher prior diversity. The fossil record of the Sipunculoidea, Nemertinea, Echiurida, and Chaetognatha is each that of a single family extending from Carboniferous to Recent. Although their records are sparse indeed, it is interesting to note that none of the post Paleozoic lagerstätten record a great diversity of these groups and they are not present in Early Paleozoic lagerstätten.

Several groups hint at higher prior diversity than they now possess. The Burgess Shale (Middle Cambrian) has a diverse assemblage of Priapulida, for example. The fossil record also reveals that several very distinct groups, now extinct, existed at times in the past and contributed to past levels of diversity. *Hallucigenia* (Conway Morris, 1977) and *Opabinia* (Whittington, 1975) from the Burgess Shale and *Tullimonstrum* from the Francis Creek Shale (Johnson and Richardson, 1969) are delightful bizarre examples.

The conclusion the record suggests is that the polychaete annelids have been the most diverse group of larger soft-bodied organisms throughout the Phanerozoic. There has been a significant diversity of other, quite varied forms as well. This diversity seems to have been maintained through the Phanerozoic by a Cambrian diversification of many groups that then declined and persisted (Pogonophora, Priapulida) or became extinct (*Hallucigenia*) as other groups originated and thereby maintained the former total diversity. Most groups which enter the record in the Middle and Late Paleozoic do not seem to have increased in diversity dramatically at any later time if the evidence from later lagerstätten is correct. We have no information about the Platyhelminthes and very little about the Nematoda. All we can do is assume they have had a similar pattern of relatively constant diversity. Because many of the members of those two phyla are parasitic we can further assume that the component of their diversity related to parasitic habits parallels the origins and diversification of the organisms they parasitize. It is most unlikely that their diversity was higher when the diversity of potential hosts was lower. The soft bodied groups have apparently contributed a relatively constant proportional amount of diversity to the biosphere through the entire Phanerozoic.

5. MOLLUSCA (Figure 13)

The Mollusca is one of the most diverse of all phyla and has been significant throughout the Phanerozoic. It contains several classes that diversified in the Cambrian and Early Ordovician and are unimportant or extinct today (Monoplacophora, Hyolitha, and Rostroconchia), three classes that have had high diversity since the Ordovician (Cephalopoda, Bivalvia, and Gastropoda) and two small classes that have persisted with little diversity change since the Ordovician (Scaphopoda and Polyplacophora). The Aplacophora lack a fossil record.

The classes that diversified in the Cambrian and early Ordovician each possess features that suggest a rather limited range of life habits. The living Monoplacophora are sluggish, apparently surface deposit feeding or grazing deep sea organisms. Early Paleozoic forms were similar in feeding habits but lived in a variety of environments. Because of their untorted anatomy they retain an anatomical bilateral symmetry that channels them only into the "limpet-like" style of life. The Hyolitha had long conical shells with opercula. It seems that they remained relatively immobile on the sea floor and were apparently suspension feeders or possibly passive surface deposit feeders. The Rostroconchia were infaunal or semi-infaunal. Calcified shell extended across the dorsal commissure and the animals were therefore immobile suspension feeders despite their superficial resemblance to the Bivalvia. The two classes that have persisted with low but nearly constant diversity since the Ordovician are also rather restricted in form and life habits. The Scaphopoda are sluggish burrowing infaunal deposit feeders with a slightly curved tapering tube-shaped shell. The Polyplacophora are creeping grazing organisms adapted primarily for life attached firmly to rocks. Their diversity history may not be

MOLLUSCA

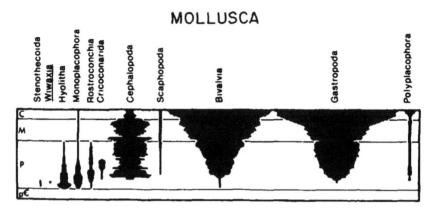

Figure 13. Family diversity in classes of Mollusca. Derived from Figure 2.

completely represented in the fossil record because the shell of eight plaques becomes disarticulated after death and the record of the Polyplacophora is probably poorer than any other molluscan class.

The three highly diverse molluscan classes each illustrate factors that are important in maintaining or increasing diversity. The Cephalopoda have fluctuated widely in diversity and display a succession of groups that have succeeded each other in diversity dominance. The Bivalvia and Gastropoda both contain groups that have maintained diversity with little change and have increased in diversity by developing new groups that have invaded new habitats.

A. Cephalopoda (Figure 14). The Cephalopoda display the greatest degree of diversity fluctuation of any class. They originated in the Cambrian and first diversified in the Ordovician when seven new orders were added to the single one extending up from the Cambrian. Figure 14 illustrates the diversity history of the subclasses of the Cephalopoda. The Nautiloidea are subdivided into primarily non-coiled forms (the orders Ellesmerocerida, Orthocerida, Ascocerida, Oncocerida, Discocerida, and Bactritoidea [which often are viewed as a separate subclass]) and the primarily coiled forms (the orders Tarphycerida, Barrandeocerida, and Nautilida). The subclasses Endocerida and Actinocerida along with the six orders of non-coiled Nautiloidea all diversified greatly in the Ordovician and dwindled in diversity in the later Paleozoic, and all were extinct by the end of the Triassic. The coiled Nautiloidea also became established in the Ordovician. They have fluctuated somewhat in diversity but not nearly as widely as the other shelled cephalopods. They have also persisted throughout the entire post-Cambrian interval. The Ammonoidea originated in

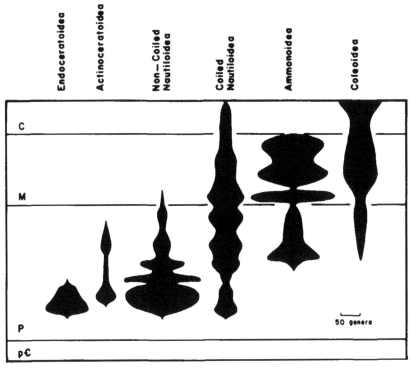

Figure 14. Generic diversity within subclasses of the Cephalopoda. Subclass Nautiloida is subdivided into non-coiled (orthoconic, cyrtoconic and brevionic) forms and fully coiled forms. Compiled from *Treatise on Invertebrate Paleontology* (Part K, 1964).

the Devonian and experienced several rapid expansions and collapses of diversity until becoming extinct at the end of the Cretaceous. The Coleoidea originated in the Middle Paleozoic and have expanded in diversity during the Mesozoic and Cenozoic. The precise diversity history of the Coleoidea is difficult to trace because many of them (the squids and octopuses) are non-skeletal.

The overall diversity history of the Cephalopoda is one of maintenance of the diversity established in the Ordovician but through a relay of replacement of one suite of morphologies by another. The Ordovician expansion of diversity in the Cephalopoda was by primarily non-coiled forms. Teichert (1967) has pointed out that most major features in cephalopod evolution seem to be responses to the need for buoyancy control in these nektonic and nekto-benthic organisms. The early diversified cephalopod groups used endosiphuncular deposits (Endoceratoidea, Actinoceratoidea, Orthoceratida, Oncoceratida, and Discoceratida), cameral deposits (Actinoceratoidea,

Orthoceratida) and various restrictions on shell growth to form brevicones (Ascocerida and Oncocerida) to aid in buoyancy control and shell orientation. The development of coiling placed the body chamber and the center of gravity of the organism underneath the buoyant chambers and relieved the need for elaborate counterweight structures. The secretion of fluids and their removal from the shell chambers by the siphuncle to control precise buoyancy response has continued, however (Ward et al., 1981 and references therein). The coiled Nautiloidea maintained a relatively stable diversity as the non-coiled forms declined and they remained fairly diverse long after the non-coiled forms were all extinct. With the development of the Ammonoidea, which also are usually fully coiled in form, the pattern of septal growth became complex. The folded septal form in ammonites had several consequences such as strengthening the shell. Ammonites diversified in the Devonian (Anarcestida, Clymenida, and Goniatitida) and achieved, with the coiled Nautiloidea, diversity similar to the variety of Ordovician Cephalopoda. The ammonites dwindled in the late Paleozoic and almost became extinct but a new group, the Ceratitida, diversified in the Triassic. The collapse and expansion pattern occurred twice more, with the collapse of the Ceratitida at the end of the Triassic and expansion of the Ammonitida in the Jurassic and the dwindling of the Ammonitida in the Cretaceous and expansion of the Lytocerida and Phyllocerida. Major changes in septal morphology characterize the major new ammonite groups and the Lytoceratida also included the heteromorph ammonites whereas most others had been planispirally coiled. The extinction of the Ammonoidea reduced cephalopod diversity (Figure 13) but the Coleoidea have apparently diversified during the Mesozoic and Cenozoic until there are 33 living families (Sepkoski, personal communication). This is equivalent to the average familial diversity of the shelled cephalopoda from the Ordovician to the Cretaceous. The Coleoidea are anatomically different from the shelled cephalopods, as represented by living *Nautilus*, in lacking a large external buoyant shell, in having a well developed eye with a lens rather than the simple, optically less precise "pin-hole" eye of *Nautilus*, in having suckers and hooks on the tentacles, and in having only two rather than four gills.

The diversity history of the cephalopods is one of diversity maintenance through replacement. The non-coiled shelled forms were replaced by coiled forms. These coiled forms may have expanded in depth range through shell strengthening features such as folded septa but essentially they remained nektonic or nekto-benthic predators as their predecessors had been. Today the diverse cephalopods are shell-less, efficient swimmers with advanced eyes and arm structures but they, too, are essentially nektonic and nekto-benthic predators. With all the changes in shell form, buoyancy control and anatomy the cephalopods have remained channeled into one general mode of life and have not invaded new ecospace. They have maintained but have not expanded the diversity they established in the Ordovician when they first developed the range of life habits they still command.

B. Bivalvia (Figures 15 and 16). The Bivalvia have had two episodes of diversification. This is masked when the class as a whole is considered (Figure 14). Figure 15 illustrates the diversity history of the subclasses of the Bivalvia. The Bivalvia originated in the Cambrian and all subclasses were established by the Middle to Late Ordovician. All but the Heterodonta diversified to near their present diversity by the Silurian and have maintained that level of diversity ever since. The superfamily level used in Figure 15 is a very high taxonomic level and has relatively low resolution but the same conclusion is reached with family level data as shown in Figure 16. The only groups of bivalves that contain siphonate forms are the Palaeotaxodonta and the Heterodonta. The Palaeotaxodonta with siphons are deposit feeding bivalves with a different mode of mantle fusion than occurs in the Heterodonta. They are the principal siphonate forms in the Middle and Late Paleozoic and remain at that level of diversity today. Therefore the increase in familial diversity in the Bivalvia in the Mesozoic and Cenozoic is a result of the diversification of the siphonate Heterodonta. The bivalves of the Middle Paleozoic had developed the full range of shallow infaunal, endo- and epibyssate, and epifaunal life habits that the living non-siphonate forms have today (Stanley, 1968, 1977; Bambach, 1971). These non-siphonate bivalves had become as diverse as they are now. The development of full mantle fusion and siphons in the Heterodonta enabled them to burrow more efficiently and permitted the invasion of ecospace previously closed to the bivalves. Because of the improved ability to maintain contact with the overlying water mass for respiratory and feeding purposes provided by siphons (Stanley, 1968, 1970, 1977), they could live in unstable substrates such as beach sands that had previously been unavailable as well as burrow to depths previously unobtainable. It is the siphonate heterodont bivalves that create the diversity expansion of the bivalves in the Mesozoic and Cenozoic. The other groups, including the non-siphonate heterodonts, simply maintain the level of diversity first achieved in the Ordovician and Silurian.

C. Gastropoda (Figure 17). The Gastropoda are another very diverse class which increased diversity dramatically in the Mesozoic and Cenozoic. The majority of marine gastropods are in the subclass Prosobranchia. Figure 17 illustrates the diversity history of the Prosobranchia. The Archaeogastropoda have a long Paleozoic history with little diversity fluctuation and they maintain that diversity through the Mesozoic and Cenozoic with only modest increase, primarily in the superfamily Trochacea which originates in the Mesozoic. By

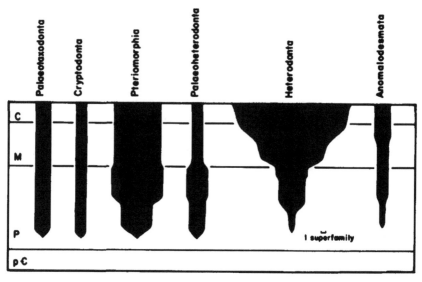

Figure 15. Superfamily diversity within subclasses of Bivalvia. Compiled from *Treatise on Invertebrate Paleontology* (Part N, 1969).

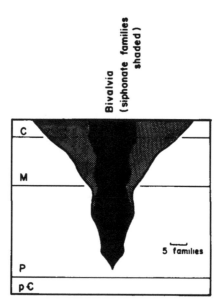

Figure 16. Family diversity in the Bivalvia contrasting nonsiphonate and siphonate forms. Modified from Stanley (1977).

far the largest share of Mesozoic and Cenozoic diversity increase occurs in the Caenogastropoda. The Caenogastropoda originate in the Ordovician but have a modest Paleozoic history. Only 3 or 4 genera are assigned to the Caenogastropoda during any one epoch of the Ordovician through Devonian and 20 genera or less for each period of the Late Paleozoic. In the Mesozoic the Canogastropoda begin to diversify and in the Cretaceous and Cenozoic they expand explosively.

The Archaeogastropoda have aspidobranch gills (with the filaments arranged on both sides of the axis) and complex, multi-toothed radulae. They are all grazers or raspers, primarily on firm substrata. The Caenogastropoda have pectinibranch gills (filaments on only one side of the axis) and radulae with fewer teeth. The altered gill structure permits a free left-right flow of water across the mantle cavity and the development of ciliated cleaning tracts. These features plus the development of siphons permit the Caenogastropoda to inhabit soft substrates and even burrow. The radular teeth, while fewer in number, are developed in various ways to permit predatory feeding, including boring as well as more conservative rasping functions. In some forms the teeth

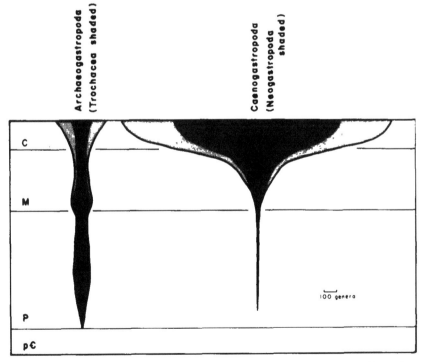

Figure 17. Generic diversity within the orders of the subclass Prosobranchiata (class Gastropoda). Compiled from Moore, Lalicker and Fischer (1952).

even become long and needle-like and are associated with poison glands. These features have permitted the Caenogastropoda to invade new ecospace, both in the sense of physical environments and by developing new trophic habits. The great Mesozoic-Cenozoic diversity increase in the gastropods is the result of the diversification of the Caenogastropoda as they underwent their adaptive radiation into ecospace previously not occupied by the Archaeogastropoda. Several aspects of this diversification are touched on by Vermeij (1977) in a discussion of the Mesozoic marine revolution and by Taylor et al. (1980) in a discussion of food specialization and the evolution of predatory prosobranch gastropods.

6. ARTHROPODA (Figure 18).

Arthropoda have been among the dominant organisms in marine faunas from the earliest Phanerozoic. Because of their multiple-part skeleton and because of the tendency in some groups to have little if any calcification, the quality of the fossil record of arthropods varies from class to class. The evidence from lagerstätten, however, and comparison of diversity of fossil groups with living forms suggest that in general the classes with high fossil diversity faithfully record the pattern of marine arthropod diversity. The only clear exception is the "Trilobitoidea" whose diversity in the Burgess Shale fauna (Middle Cambrian) is far greater than recorded through the rest of their range. Five classes: the Trilobita, Merostomata, Ostracoda, Malacostraca, and Cirripedia, contribute the most diversity to the marine arthropod record and are discussed separately below.

A. Trilobita (Figure 19). The Trilobita was the first metazoan class to achieve high diversity. The trilobites are overwhelmingly the most diverse Cambrian group. Concealed within the apparent general expansion and contraction of the class without much fluctuation (Figure 18) is a replacement sequence of orders and suborders with a variety of diversity histories (clade shapes) (Figure 19). All of the groups that originated in the Early Cambrian were extinct by the end of the Ordovician whereas those originating later all persisted at least through the Middle Devonian.

Three of the groups that had Early Cambrian origins were morphologically conservative. The Redlichiida had relatively large semicircular cephalons, commonly with genal spines, numerous thoracic segments, commonly with their ends produced as spines giving the outline of the thorax a serrated appearance, and small pygidia. Enrollment would not have produced a good opposition of cephalon and pygidium. The Corynexochiida retained the large, semicircular cephalon with genal spines, the thorax was restricted to 5 to 11 segments, but spinose terminations were still common whereas the pygidium

ARTHROPODA

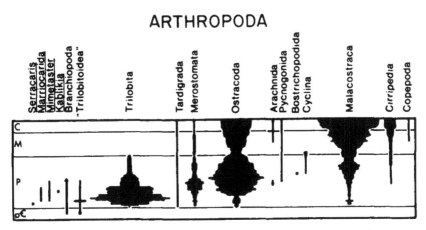

Figure 18. Family diversity in classes of Arthropoda. Derived from Figure 2.

Ptychopariida

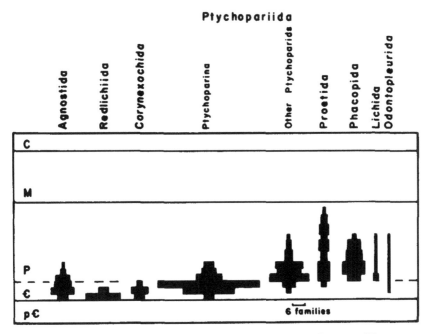

Figure 19. Family diversity within orders of the Trilobita. The Ptychopariida are separated into the suborder Ptychopariina and the other suborders. Compiled from the *Treatise on Invertebrate Paleontology* (Part O, 1959) and Fortey and Owens (1975).

was large and rounded. The Ptychopariina, a suborder of the Ptychopariida. generally retained the large semicircular cephalon, a relatively large thorax, often with spinose terminations to the segments, and a relatively small pygidium. The glabella was relatively simple and generalized in all three of these groups, also. The Redlichiida were the most diverse Early Cambrian group but became extinct by the Late Cambrian, whereas the other two groups expanded in diversity in the Middle Cambrian. In the Late Cambrian the Ptychopariina reached the greatest diversity of any trilobite group but dwindled in the Ordovician. The fourth trilobite group to originate in the Early Cambrian was the order Agnostida. They were small forms with symmetrical cephalon and pygidium and are thought to have been, at least in part, planktonic (or nektonic). The agnostids reached maximum diversity in the Middle Cambrian and declined rapidly in the Middle and Late Ordovician.

The other ptychoparid suborders besides the Ptychopariina are the Asaphina, Illaenina, Harpina, and Trinucleina. These groups plus the orders Proetida and Phacopida diversified in the Ordovician so that, taken together, they maintained the level of diversity achieved by the class in the Middle Cambrian and nearly that of the Late Cambrian maximum. These groups display a range of morphologies quite different from the dominant Cambrian forms. These include smooth, non-spinose thoracic segments, equal size cephalon and pygidium with precise opposition in enrollment, enlarged glabellas, schizochroal eyes, and enlarged fringes on the cephalon with various structures such as rows of pits. Two other orders, the Lichida and Odontopleurida persist through the Silurian and Devonian as well. They, too, have unusual morphologic features, bizarre spines being the most distinctive.

The trilobites had a diversity history in which generalized, conservative forms with small pygidia and numerous segments (Redlichiida) were replaced during the Cambrian by generalized forms with some development of the pygidium (Corynexochida) or reduction in thoracic segments (Ptychopariina). These forms were in turn replaced in the Ordovician by a variety of groups with widely varied morphologies. These groups persisted until the end of the Devonian and one survived until the Late Paleozoic. Interestingly, the longest surviving group, the Proetida, fluctuated rather little in diversity and never diversified as much as the shorter lived Redlichiida, Ptychopariina, other Ptychoparids or Phacopida. The Proetida simply maintained a rather constant modest diversity from the Late Cambrian to the Permian.

B. Merostomata (Figure 20). The diversity history of the Merostomata is composed of a succession of four different groups (Figure 20). The subclass Xiphosura contains the orders Aglaspida and Xiphosurida (subdivided into the suborders Synxiphosurina and Limulina). They are all forms with a relatively large prosoma with small chelicera and a series of relatively undifferentiated

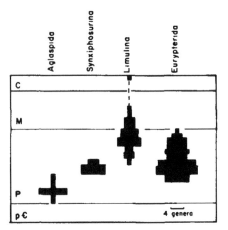

Figure 20. Generic diversity within orders of the Merostomata. Xiphosurida separated into suborders Synxiphosurina and Limulina. Compiled from *Treatise on Invertebrate Paleontology* (Part P, 1955).

walking legs, an opisthosoma with 9 to 12 segments and a spike-like long telson. The fourth group is the subclass Eurypterida, characterized by a moderate sized prosoma with chelicera (sometimes quite large), a series of walking legs of which the last pair is commonly transformed into swimming paddles, an elongate body of twelve segments commonly subdivided into a seven-segmented preabdomen and a five-segmented postabdomen, and a variable shaped telson. The Aglaspida, Synxiphosurina and Limulina form a sequence of diversity replacement without increase through the Phanerozoic. They appear to have all been primarily benthic and inhabited nearshore but primarily marine habitats. The Eurypterida are known from facies that represent brackish environments. They clearly developed nektonic as well as benthic habits. This group of Merostomata is the one that entered new ecospace and it is the source of the higher diversity of the Merostomata in the Middle Paleozoic.

C. Ostracoda (Figure 21). The diversity history of the ostracods resembles that of the Anthozoa and the Stenolaemata (Figure 2) and, as in those groups, it is produced by a diversity replacement of one suite by another group rather than a decline and rediversification within one or more of the constituent groups. The Paleozoic diversification of the ostracods includes representatives of all five orders, three of which become extinct by the end of the Permian. Although one Paleozoic suborder besides the Podocopina has persisted to the Recent and two new suborders have originated in the post-Paleozoic, the bulk of the Mesozoic-Cenozoic rediversification of the Ostracoda is confined to two superfamilies (the Cytheracea and the Cytheridacea) in the suborder Podocopina. Although these two superfamilies have some Paleozoic representatives there are only five genera of Cytheracea prior to the Late Carboniferous and only two pre-Late Carboniferous genera of Cytheridacea. Of the 896 genera of Ostracoda 421 belong to these two superfamilies. Those two superfamilies of the Podocopina create the post-Paleozoic diversity increase that returns Ostracoda diversity to its Paleozoic level.

D. Malacostraca (Figure 22). The diversity history of the Malacostraca appears to be one of almost continual diversification through the Phanerozoic with an increased rate of diversification in the Mesozoic and Cenozoic (Figure 18). More detailed analysis reveals that this pattern is produced by the successive addition of groups that diversify and then maintain a constant diversity (Figure 22). All arthropods are subject to disarticulation and disintegration after death. In Figure 22 the number of living genera in each category are listed for comparison with the number of fossil genera in the Late Cenozoic. In each case there are about four times as many living genera as there are Late Cenozoic fossil forms. The rather constant proportion of fossilized to extant genera suggests that the diversity pattern of the fossil record represents actual changes and ratios of diversity even though it only reveals a part of the total absolute diversity.

The Phyllocarida diversified in the Cambrian and then persisted with some fluctuation to the Recent. They disappear from the fossil record in the Early Mesozoic but four living genera are known. The low diversity of living Phyllocarids and the fact that they never achieve high diversity, even in lagerstätten, suggests that they never had very high diversity but their consistent presence

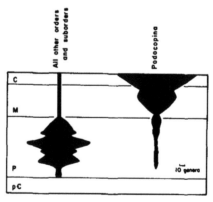

Figure 21. Generic diversity of the suborder Podocopina contrasted to the rest of the Ostracoda. Compiled from the *Treatise on Invertebrate Paleontology* (Part Q, 1961).

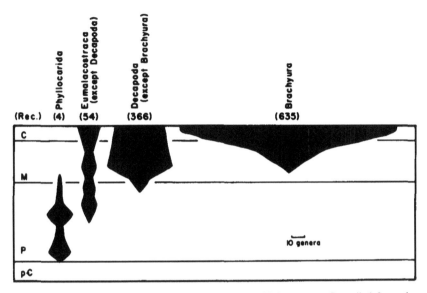

Figure 22. Generic diversity of groups within the Malacostraca. Compiled from the *Treatise on Invertebrate Paleontology* (Part R, 1966).

in the Paleozoic record suggests that they had somewhat higher diversity then than later.

The subclass Eumalacostraca originates in the Middle Devonian. It contains the superorders Eocarida (Middle Devonian-Permian), Syncarida (Early Carboniferous-Recent), Peracarida (Permian-Recent) and Eucarida (Permian-Recent). The diverse order Decapoda is in the Eucarida but will be considered separately. All Eumalacostraca excluding the Decapoda maintain a modest but nearly constant diversity from the Devonian to the Recent. The later Paleozoic diversity increase in the Malacostraca as a whole results from the addition of these groups to the continuing Phyllocarida.

The Decapoda first appeared in the Permian. In Figure 22 the Decapoda are treated in two groups: all Decapoda except the infraorder Brachyura, and the Brachyura (crabs). The Decapoda, excluding Brachyura, diversified in the first half of the Mesozoic, reached a maximum fossil diversity in the Jurassic, and have maintained a high but not increasing diversity since then. The Brachyura originate in the Early Jurassic and diversify greatly in the Cretaceous and Cenozoic. The diversity increase of the Malacostraca through the post-Paleozoic is created by the diversification of first the non-brachyuran Decapoda and then the Brachyura.

The Phyllocarida are Malacostraca with a bivalved carapace. They are primarily benthic detritus feeders although some are also nektonic. Eocarida were shrimp-like Malacostraca with a carapace that was not fused to the thoracic segments. They seem to have been primarily benthic scavengers and detritus feeders (they lacked chelae). The Syncarida lack a carapace. The living Syncarida are predominantly freshwater forms though some Paleozoic forms were marine. Small syncarids enter the interstitial habitat in coarser sediments. The Peracarida have the first thoracic segment fused to the carapace. They have a variety of life habits. Some are pelagic, most are benthic. Parasitic and carnivorous trophic habits are found in the Peracarida but most benthic forms are detritus feeders. The Decapoda have developed highly specialized appendages with three pairs of maxillipeds and five pairs of locomotory pereiopods (hence Decapoda) of which one or more pairs end in chelae. These features plus others permit the Decapoda to burrow deeply, swim and act as powerful predators on heavy shelled prey. In the Brachyura the whole form of the body has been altered with the abdomen shortened and turned under the carapace, which has become widened and shortened. Chelae are invariably present on the first pereiopods of the Brachyura.

In the development of high diversity in the Malacostraca it seems that the development of chelae and other modified appendages that permit both a variety of feeding activities and the development of varied life habits were of great importance. Paleozoic Malacostraca were primarily benthic shallow detritus feeders, scavengers and filter feeders. They developed a modest diversity in the Cambrian (the Phyllocarida) and added some to their overall diversity in the later Paleozoic as varied Eumalacostracan groups originated, but did not diversify much. These other groups also were primarily epifaunal and shallow infaunal deposit feeders and low level predators and filter feeders. They have persisted up to the Recent but their diversity remains modest in the living fauna. The origin of the Decapoda with their highly varied trophic and life habits caused the major addition to Malacostracan diversity. This occurred in two steps as the shrimp and lobster-like Decapoda invaded new ecospace and diversified in the Early and Middle Mesozoic and then the Brachyura, the crabs, with an entirely different body form, style of locomotion, and broad range of feeding activities diversified in the Cretaceous and Cenozoic This diversification was also associated with the diversification of the bivalves and gastropods, the prey animals of the crabs. The addition of diversity in the Malacostraca during the Mesozoic and Cenozoic was the result of both the invasion of new ecospace and the development of new opportunities for specialization. The first was a result of the development of the decapod limb sequence, the second was a co-evolutionary response to the concomitant molluscan diversification.

E. Cirripedia. The barnacles that diversify in the Mesozoic and Cenozoic

(Figure 18) are the non-pedunculate, fully cemented, skeletonized groups such as the Balanomorpha (Late Cretaceous to Recent) (Stanley and Newman, 1980; Newman and Stanley, 1981), all of which enter the fossil record only in the Mesozoic, and one family (also new in the Mesozoic) of the pedunculate barnacles. The Paleozoic barnacles are either boring forms, which have maintained a low diversity since the Carboniferous, or the pedunculate ("gooseneck") barnacles which enter the record in the Silurian. The scarcity of Paleozoic pedunculate barnacle fossils may be a preservational problem of disarticulation of the plates after death.

7. ECHINODERMATA (Figure 23).

The Echinodermata are exceptional for their Early Paleozoic class-level variety. Their later Phanerozoic history also illustrates a shift in diversity from less mobile to more mobile forms.

One of the distinctive features of the Early Paleozoic fauna is the great variety of small but distinctive classes of echinoderms. The classes that originated in the Cambrian (Helicoplacoidea, Edrioasteroidea, Eocrinoidea, Stylophora, Homostelea, Homiostelea, Ctenocystoidea, "Cycloidea," "Cymaoidea," and Holothuroidea) are all bizarre forms with irregular or simply circular (radial) growth patterns in comparison to the typical regular pentaradiate symmetry and regular plate arrangements of the diverse classes (Crinoidea, Blastoidea, Stelleroidea, and Echinoidea) of the later Phanerozoic. These Cambrian classes usually have a large number of thecal plates which are arranged irregularly, in radial symmetry or even in spiral form. Only five of the ten Cambrian groups extend into the Ordovician and all but the Holothuroidea were extinct by the Late Carboniferous. The worm-like Holothuroidea are known from the Burgess Shale (Middle Cambrian) and from dermal ossicles throughout the Phanerozoic but their precise diversity history is otherwise obscure, as is usual for primarily soft-bodied organisms. Their non-skeletonized nature sets them off from most of the other echinoderms.

Actually, a crinoid is also known from five specimens from the Burgess Shale of Middle Cambrian age (Sprinkle, 1973). It has an irregularly plated cup, and is attached by irregularly plated holdfasts rather than a stem. This organism has been placed in a separate subclass, the Echmatocrinea. This crinoid, although it has arms and a differentiation of calyx and tegmen, has many features that resemble the other Cambrian echinoderm groups. The diversification of the Crinoidea did not begin until the stem and regular plate arrangement had developed in the Ordovician.

In the Ordovician another wave of new classes appeared. These included both forms with radial or irregular symmetry and numerous plates reminiscent of the types originating in the Cambrian and the first forms of each of

ECHINODERMATA

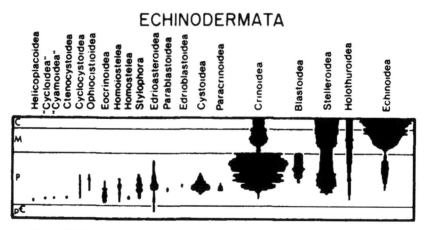

Figure 23. Family diversity in classes of the Echinodermata. Derived from Figure 2.

the highly symmetrical, typically pentaradiate, highly diverse classes with regular plate arrangements. The "archaic" groups without pentaradiate symmetry or with irregular or numerous plates include the Ophiocystoidea, Cyclocystoidea, Cystoidea, and Paracrinoidea. Of these classes only the Cystoidea became very diverse and all were extinct by the end of the Devonian.

In the Ordovician the stemmed Crinoidea diversified and the Stelleroidea and Echinoidea first appeared. Two small classes, the Edrioblastoidea and Parablastoidea, also put in an appearance. They resemble the Blastoidea, although true blastoids do not appear until the Silurian. These more diverse groups are all much more regularly organized in growth pattern than the Cambrian classes and usually display strong pentaradiate symmetry. The Crinoidea and Blastoidea were firmly attached, stemmed forms with arms or brachioles forming an efficient filtering system for feeding. The Stelleroidea and Echinoidea were mobile epifaunal or shallow infaunal predators, deposit feeders, grazers and filter feeders.

The diversity history of the echinoderms in the Early Paleozic was the replacement of a variety of small classes, of irregular or simple radial symmetry, by classes with more regular growth forms and more highly developed structures for locomotion or for attachment and filter feeding. As these various new groups established the structural features that permitted them to invade new ecospace they diversified, both recovering the diversity of the declining earlier groups and adding to the total diversity of the phylum.

A. Crinoidea (Figure 24). The Crinoidea diversified in the Ordovician, remained diverse through the rest of the Paleozoic, collapsed in diversity at the end of the Permian and recovered some diversity in the Mesozoic although they did not return to their former high diversity (Figure 23). The diversity history of groups within the class is shown in Figure 24. Three subclasses (the Camerata, Inadunata and Flexibilia) appeared in the Ordovician. Each had reached high diversity by the Silurian and each persisted at that level of diversity until the decline in the Permian. Figure 24 is a tabulation of genera whereas the clade in Figure 23 is for families. The Carboniferous expansion in the Inadunata is the result of an elaboration of genera within the same average number of families as were present in the Middle Paleozoic. This could be an artifact of taxonomic practice by specialists on the Carboniferous Inadunata but it is probably a real diversity expansion at the generic level, possibly related to a response by these attached benthic organisms to the disappearance of reef-building organisms at the end of the Devonian from the physical setting that both types of organisms inhabited.

All Mesozoic and Cenozoic crinoids are members of a subclass, the Articulata, that originated in the Triassic. This group diversified during the Mesozoic and has maintained that diversity through the Cenozoic. The Articulata have several features that distinguish them from Paleozoic crinoids, the most important of which is the plate and muscle articulation of the arms. This permits not only a variety of adjustments in displaying the filtration fan of arms and pinnules but has also permitted the Comatulida to abandon the use of a stem for attachment and to become mobile, using cirri and arms for locomotion. The Articulata have stalked forms, most of which are now restricted to bathyal depths. These were the most diverse Mesozoic fossil forms. It is interesting that the apparent decline in diversity of stalked Articulata may simply

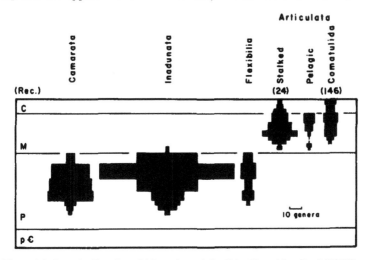

Figure 24. Generic diversity within orders of the Crinoidea. After Paul (1977).

reflect the shift of these forms to deeper water habitats whose facies are just poorly represented, especially in the youngest geologic record for which there has not been sufficient time for uplift to produce extensive exposures of deeper water facies. The diversity of living stalked Articulata (24 genera) is nearly that of the Late Jurassic (30 genera). Pelagic Articulata also developed in the Mesozoic diversification of the group but they became extinct at the end of the Cretaceous. The mobile, unattached Comatulida seem to have diversified in the Jurassic and then maintained their diversity at rather constant level through the Cretaceous and Cenozoic. Here again we should note that the proportional diversity history for a group may be correct but the absolute diversity may be poorly represented in the record. There are 146 living genera of Comatulida but no more than 10 genera are present in the fossil record at any one time. The Comatulida are notorious for their rapid disintegration after death (Liddell, 1975) and many live in habitats with relatively strong currents. Therefore their fossil record is scanty. The general pattern of crinoid diversification in the Mesozoic is one of moderate diversification but not to levels achieved in the Paleozoic if the fossil record is to be taken at face value. This is probably a correct proportional view because Paleozoic diversity is undoubtedly not fully represented either. However, the number of known living crinoid genera does match the number of Carboniferous crinoid genera. The diversity recovery of the Mesozoic-Cenozoic crinoids may have matched Paleozoic levels of diversity, but it certainly hasn't exceeded them. Modern crinoids are benthic suspension feeders, just as Paleozoic crinoids were.

B. Stelleroidea (Figure 25). The diversity of the Stelleroidea has been maintained without much fluctuation since they diversified in the Ordovician (Figure 23). This is borne out by an analysis of the diversity history of the orders within the class although this view also reveals that the diversity maintenance has been achieved in different ways by the two subclasses of the Stelleroidea (Figure 25). The Asteroidea have had continuous maintenance of all orders as the method of keeping diversity relatively constant whereas the Ophiuroidea have had a replacement of two diverse orders by a third. The replacement of diverse orders in the Ophiuroidea also incorporated a shift from slow-moving or shallow-burrowing suspension feeders (Stenurida and Oegophiurida) to groups with modified arms coiled upwards as a filtering net (Phyrnophiurida, the "basket-stars") or with arm articulation capable of providing rapid motion and mobility (the Ophiurida) so that the organisms could adopt predatory as well as suspension feeding habits. Although there are a few Paleozoic Ophiurida they had incompletely developed radial shield plates on the body disc and had not yet achieved the integrated armored structure that characterizes all post-Paleozoic Ophiurida.

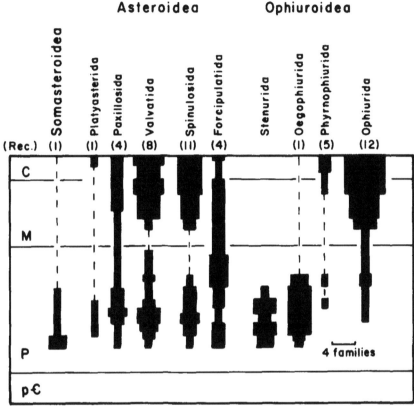

Figure 25. Family diversity within the orders of the Stelleroidea. Compiled from the *Treatise on Invertebrate Paleontology* (Part U, 1966).

C. Echinoidea (Figure 26). The Echinoidea are a clear case in which the development of particular features permitted diversification because new ecospace was opened up to them. The clade shape of the whole class has a long Paleozoic stem of low diversity and then a mushrooming of diversity in the Mesozoic and Cenozoic (Figure 23). The diversity histories of the orders and superorders of the Echinoidea reveal that the overall pattern of the class is a combination of diversity maintenance with replacement in one subclass (Perischoechinoidea) and diversity expansion in a second, new, subclass (Euechinoidea) (Figure 26).

All Paleozoic echinoids belong to the four orders of the Perischoechinoidea. These orders had different numbers of interambulacral plate columns than the two that characterize the Euechinoidea. Some had rigid and some had

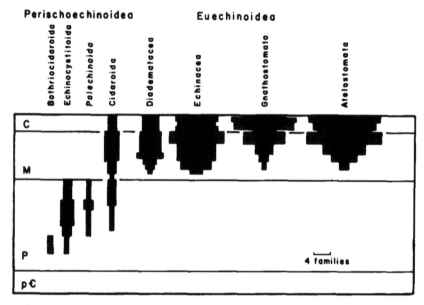

Figure 26. Family diversity of orders of the subclass Perischoechnoidea and the super-orders of the subclass Euechinoidea. Compiled from the *Treatise on Invertebrate Paleon-tology* (Part U, 1966).

flexible tests. All had some type of simple jaw (lantern) structure but only the Cidaroidea had a rudimentary beginning of a perignathic girdle. All were apparently epifaunal, sluggish to sedentary herbivores. Only the Cidaroida survived into the Mesozoic and Cenozoic. The total number of Paleozoic genera of Perischoechinoidea listed in the *Treatise on Invertebrate Paleontol-ogy* (Durham et al., 1966) is 43, of which only 8 are Cidaroida. There are 54 genera of Mesozoic and Cenozoic Cidaroida. The Perischoechinoidea have maintained a relatively constant low diversity with the Cidaroida expanding to recover, but not exceed, the diversity lost by the extinction of the exclu-sively Paleozoic orders.

The Euechinoidea have diversified widely in the Mesozoic and Cenozoic. Two of the superorders of Euchinoidea are comprised of regular sea urchins that retain pentaradial symmetry, the Diadematacea and the Echinacea. The Diadematacea are much more mobile than the Cidaroida, have a fully devel-oped perignathic girdle unlike the Cidaroida, but retain a similar jaw structure (aulodent lantern). They are primarily herbivorous epifaunal forms. They di-versified in the Triassic and Jurassic and have retained but not expanded their

Middle Mesozoic diversity. The Echinacea have developed several modifications to the lantern and have a well developed perignathic girdle. They have also developed solid spines. The Echinacea have carnivorous as well as herbivorous feeding habits. Some have developed the ability to bore into rocky substrates. Some of them cover themselves with available debris during daylight. These are habits that exploit more dimensions of the hypervolume of ecospace than either the Perischoechinoidea or Diadematacea exploited. The Echinacea also diversified in the Early Mesozoic and then maintained their Jurassic levels of diversity through the Cretaceous and Cenozoic. The family diversity of the Echinacea has been maintained at about twice that of the Diadematacea since the Jurassic and there are about four times as many genera of Echinacea (195) as Diadematacea (56).

The other two superorders of the Euechinoidea are the Gnathostomata and the Atelostomata, both irregular echinoid groups with bilateral rather than pentaradial symmetry. Both groups have exploited ecospace previously unavailable to echinoids. This has been possible because of the change of the position of the periproct (anal opening) to outside the oculogenital ring and modifications in the form of the test, spines, tube feet pattern, and the feeding system. The Gnathostomata retain the lantern but have highly modified test forms which include the "sand dollars." They include active burrowing in mobile sandy substrates and suspension and deposit feeding in their diverse modes of life. The Atelostomata have lost the jaw structure (lantern) and have developed a variety of modifications to the form of the test. This group includes the heart urchins. Many are deposit feeders, most are infaunal, and some are capable of burrowing and maintaining themselves up to five test lengths below the sediment surface. The Gnathostomata first appear in the Jurassic and they have continued to diversify through the Cretaceous and Cenozoic. The Atelostomata diversified during the Mesozoic and have maintained their Cretaceous diversity level through the Cenozoic.

The diversification of the Echinoidea in the Mesozoic clearly correlates with their invasion of new ecospace. The modest additional diversity of the more mobile Diadematecea with better jaw articulation over the Cidaroida was surpassed by the much greater diversification in the Echinacea that had developed new trophic abilities as well as the ability to live in different habitats (wave-swept rocky shorelines, for example). Even this diversification was matched by each of the two groups of irregular echinoids that invaded the infaunal habitat, one group emphasizing features to enhance active shallow burrowing in unstable shifting substrates, the other being channeled toward deposit feeding at a variety of depths within the substratum. All in all there are 407 genera of irregular echinoids listed in the *Treatise* and 251 genera of regular forms. The development of new groups with structures that permitted changes in feeding, mobility, and life site enabled the Echinoidea to add diversity in the Mesozoic and Cenozoic. The forms that remained essentially like the Paleozoic forms simply maintained similar levels of diversity to those achieved in the Paleozoic.

8. CHORDATA AND RELATED PHYLA (Figure 27).

The various invertebrate groups allied with the Chordata plus the classes of the Chordata form a sequence of replacement and increase in diversity that also reflects change in the utilization of ecospace. The earliest organisms with chordate affinities to diversify were the Conodontophorida, which were apparently nektonic or planktonic for the most part, and the colonial Graptolithina, also primarily with planktonic habits. The Conodontophorida were enigmatic organisms but the phosphatic composition and growth patterns of conodonts suggest chordate affinities. According to the *Treatise on Invertebrate Paleontology* (Part W, Supplement 2, 1981:W114), these organisms originated in the Late Precambrian, reached maximum diversity in the Ordovician, remained moderately diverse through the rest of the Paleozoic, and became extinct in the Triassic. Data in the newly published revision of the conodonts in the *Treatise on Invertebrate Paleontology* (Part W, Supplement 2, 1981) now make familial diversity in the Ordovician three times as great as at any time after the Silurian. The Graptolithina also originated in the Cambrian, diversified in the Ordovician and collapsed after the Silurian. Pterobranchia, Urochordata, and Cephalochordata are all soft bodied and have a poor fossil record. The first chordates were the Agnatha, the jawless fish. They are known from fragmentary bony materials in Ordovician rocks. The Agnatha diversified in the Silurian, then dropped to low levels of diversity after the Devonian but persist to the Recent. The Placodermi were the first

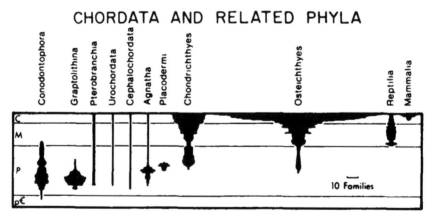

Figure 27. Family diversity in classes of the Chordata and related phyla. Derived from Figure 2.

jaw-bearing chordates. They included a diverse assortment of Devonian forms with heavy bony armor on the skull and part of the thorax. The Placodermi became extinct in the Early Carboniferous. The Chondrichthyes, the cartilaginous fishes, originated in the Devonian, diversified in the Carboniferous, contracted in diversity at the end of the Paleozoic and diversified again in the Mesozoic and Cenozoic. The Osteichthyes, the bony fishes, originated in the Late Silurian. They persisted without much diversification (although several lineages were present) until the Mesozoic when they began a diversification that increased enormously in the Cenozoic.

The general pattern of the diversity history of the chordates and allied groups in the Paleozoic is an Ordovician diversification of several primarily planktonic groups (Conodontophorida and Graptolithina), the decline of those groups, and the diversification of jawless and armored fish (Agnatha and Placodermi) in the Silurian and Devonian. These two groups then decline at the end of the Devonian and the Chondrichthyes, the shark-like cartilaginous fishes, diversify as the Osteichthyes persist at low levels of diversity. This is another case of diversity replacement but it is also clear that the ecospace occupied was also enlarging. All the groups were planktonic or nektonic, but they varied trophically. The feeding habits of conodonts are unknown. Graptolites were suspension feeders. Agnatha could have been both suspension feeders and deposit feeders, and they and their successors were obviously nektonic. Placodermi, Chondrichthyes, and Osteichthyes were all predators, although some may have been herbivorous.

The diversification of marine vertebrates beyond Paleozoic levels appears to have been related to the invasion of new ecospace and to co-evolution in response to the diversification of marine invertebrates. The Mesozoic-Cenozoic diversity history of the marine Chordata comprised a new diversification of the Chondrichthyes, a massive diversification of the Osteichthyes, a modest diversification of marine reptiles in the Mesozoic with a decline in the Cenozoic, and a diversification of marine mammals in the Cenozoic. The Chondrichthyes recovered and surpassed their Paleozoic level of diversity. The Osteichthyes have had the most dramatic Cenozoic diversification of any group. The Reptilia and Mammalia seem to represent a diversity replacement sequence.

The Mesozoic-Cenozoic diversification in the Chondrichthyes took place in groups that did not develop until the Mesozoic. The modern sharks are primarily members of the suborders Galeoidea and Squaloidea, both of which originated in the Mesozoic. These groups have a different jaw articulation and more varied teeth than the Paleozoic shark-like fishes. The skates and rays (order Batoidea) first appeared in the Jurassic. Whereas the new suborders of sharks have diversified to replace but not exceed Paleozoic diversity levels, the diversification of the skates and rays has added to the total diversity of the Chondrichthyes. The Batoidea are primarily bottom living types with a very flat body, dorsally directed eyes, ventrally placed gills and immense pectoral and pelvic fins. Although some Paleozoic sharks (Hybodontoidea) developed low, crushing or pavement teeth that effectively crushed shells of benthic invertebrates (Boyd and Newell, 1972), the skates and rays developed dental batteries highly specialized for this type of diet. It was the more complete exploitation of the nekto-benthic habitat with marine invertebrates as prey that permitted diversification of the Chondrichthyes beyond Paleozoic levels. The expansion of diversity by suitable food organisms, especially mollusks and crabs, was probably the factor that opened this dimension of ecospace for greater exploitation than before.

The huge diversification of the Osteichthyes is apparently related to two factors. For one, the Paleozoic history of the group is principally in freshwater facies. The Osteichthyes originated in the Devonian but few forms invaded marine environments until Mesozoic time. The principal diversification of the Osteichthyes in the Paleozoic was in non-marine settings. The second factor was the origin of the Teleostei. These fish seem to have originated in the marine environment (Romer, 1966) from some of the relatively rare marine Holostei. The Teleostei create the massive Cenozoic diversification of the Osteichthyes. The Teleostei have a different jaw articulation and placement of teeth than the other bony fish. Their diversification also reflects the wide range of ecospace they occupy. Teleosts are herbivores as well as scavengers and carnivores. They have an exceedingly wide range of body sizes and forms. It may be that the bony ray support for the thin fin tissues permits a rigid, functional fin at smaller sizes than can occur in the fleshy fins of Chondrichthyans. The fully ossified bones may also permit full swimming and other body support functions that are not as easily achieved by more flexible cartilaginous structures at small body sizes. The Osteichthyes have invaded a far greater range of ecospace than the Chondrichthyes and they have become far more diverse.

The Reptilia invaded the sea in the Triassic. A variety of forms, mosasaurs, ichthyosaurs, plesiosaurs and the like, are found in Mesozoic marine rocks. Most, except for the turtles, disappeared by the end of the Cretaceous. The marine mammals appear in the Early Cenozoic. The whales, porpoises, and seals have diversified in the Cenozoic and recovered the diversity that air-breathing forms had achieved in the Mesozoic.

DISCUSSION

INTERPRETATION OF DATA

The four basic clade shapes of classes mentioned above, (1) bottom-heavy, (2) irregular, or fluctuating, (3) relatively smooth even column, and (4) top-heavy, are also predominant at the suborder, order, and superorder levels. The bottom-heavy clades are those that diversify widely early in their history and then dwindle, either to extinction or with some members persisting and forming a relatively smooth column or tail post-dating the peak of diversity. Eighty percent of the bottom-heavy clades appear in the Lower Paleozoic (Cambrian through Devonian). Only the Lychniscosida (Hexactinellida), "true" (post-Paleozoic) Cyclostomata (Stenolaemata), stalked Articulata (Crinoidea), and marine Reptilia (Chordata) display bottom-heavy clade form in the Mesozoic and Cenozoic. The irregular or fluctuating clade shapes at the class level (Cephalopoda, Articulata [Brachiopoda], Crinoidea and the like) all represent replacement sequences of different included taxa. The replacement of the tabulate and rugose corals by the scleractinians and the replacement of non-coiled cephalopods by coiled nautiloids and ammonites that were in turn replaced by the dibranchiate cephalopods are examples. Such replacement sequences are found throughout the Phanerozoic from the turnover within the Trilobita in the Early Paleozoic to the turnover within the Rugosa in the Late Paleozoic to the turnover of the Ophiurida in the Mesozoic. The maintenance of diversity without much fluctuation creates the relatively smooth column clade shape. Clades of this type, with the groups persisting for over 200 million years without much change in diversity, are present in every phylum. The surprising thing about these clades is that many have rather low diversity. The long "tails" on many bottom-heavy clades (Inarticulata, Monoplacophora, Agnatha) also represent relatively unvarying diversity for those few representatives of these groups that survive the decline in early diversity of the bulk of the clade. The top-heavy clades are those which have diversified in relatively recent times, primarily the Mesozoic and Cenozoic. The clade shape itself has little importance in this case. It probably just signifies that not enough time has elapsed since the group has diversified to reveal the fate that awaits it. All diverse clades were top-heavy in appearance for fifty to one hundred and fifty million years after diversification of the group began.

The interesting facet of top-heavy class level clades (middle row of Figure 2) is that in every case they can be resolved into some included taxa that are simply maintaining diversity relatively unchanged and others that are contributing the increased diversity. Usually it is only one of these included taxa in a class that is diversifying. In all cases of large diversity increase in the Mesozoic and Cenozoic a newly developed structural feature or features is present in the diversifying taxon to permit it to expand its feeding or habitat variety (utilize more ecospace).

The reliability of clade shapes as representing real fluctuation in diversity can be questioned because the fossil record is incomplete at best. However, the complete suite of data supports the contention that these clade shapes represent true patterns of diversity change even though they do not give an absolute numerical accounting of diversity. The fossil record of well-skeletonized groups is apparently quite good even if the proportion of lightly skeletonized or soft-bodied forms in the fossil record is poor. Schopf (1975) reports on the proportion of living genera in several habitats occurring as identifiable remains in Recent sediment samples and as fossils. Remains of 75% of living genera that might be expected to produce fossils because they have heavily mineralized skeletons were found in recent sediment samples associated with living specimens and all of those robustly skeletonized genera present in the sediment samples actually have a fossil record. This contrasts with the living genera that were expected to yield few fossils. Remains of only 16% of these genera appeared in the sediment samples. Of these few genera, about two thirds had a fossil record but they represented only about 10% of the lightly skeletonized portion of the fauna. Approximately one third of all the genera studied were expected to yield no fossils and they did not. All of the bivalve and gastropod genera and two thirds of the brachyuran genera were represented in the fossil record. Also, the five polyplacophoran, three cirripedian, two echinoid and the single articulate brachiopod genera were also represented in the fossil record. The fossil record of the Bivalvia, Gastropoda, Brachiopoda, skeletonized Anthozoa, Stenolaemata, Gymnolaemata, Echinoidea and other easily preserved groups is clearly quite good. Their diversity histories are reasonably well represented by the fossil record. It is most unlikely that the different diversity patterns observed in these groups would be produced simply by preservational biases.

There are undeniable biases in preservation of less well-skeletonized forms, but the general pattern of diversity change actually seems to be represented in the record for most of those groups, too. The argument to support this is twofold. First, the presence of lagerstatten gives us a chance to census local diversity of less easily preserved taxa at various times. Surprisingly, although some groups do show conspicuous bulges at the times of lagerstätten formation (see Trilobitomorpha and Priapulida in Middle Cambrian on bottom row of Figure 2), most less well-preserved groups do not reveal large diversity fluctuations at such times. If these groups had been more diverse and common in the past, one would expect that lagerstätten would regularly record this variety in contrast to the normal geologic record. The second point is comparison of proportions of living forms in different taxa with their proportional representation in the fossil record. Such data can be compared in Figures 22, 24 and 25. In all cases the number of living taxa is considerably greater than the number recorded at any past time. However, in the case of the Stelleroidea and the Malacostraca, the proportional representation of the various groups is comparable in the fossil record and the Recent. The proportions of different Articulata (Crinoidea) are not accurately represented, however. Among the poorly-skeletonized groups it appears that many are represented reasonably well, with forms diverse now, such as Polychaetes, better documented than less diverse groups although some, such as the Platyhelminthes and Nematoda, are not (see discussion above). Therefore, although there are some distortions, a majority of even the poorly-preserved taxa seem to have a realistic pattern of their diversity history documented in the fossil record. If the various individual patterns of well-skeletonized taxa and of a majority of less well-preserved taxa are correct, then the sum of all is also going to be a correct reflection of total diversity change.

The proportional range of diversity change can be different at different taxonomic levels. Valentine (1969) has observed that diversity has not increased at or above the order level since the Early Paleozoic. Valentine (1969) also noted that in some groups, such as the brachiopods, the ratio of genera to families stayed about constant (diversity fluctuated by the same proportion at each taxonomic level) but that for some other groups the proportion of genera to families increased as they diversified in the Mesozoic and Cenozoic. In the data reported here some comparisons can be made, too. For corals (Anthozoa), brachiopods, bryozoa, cephalopods and ostracods the ratio of genera to families seems to be similar in the Mesozoic-Cenozoic and in the Paleozoic, whereas the ratio of genera to families in the gastropods and malacostracans appears greater in the Mesozoic-Cenozoic than in the Paleozoic. Although the data are not compiled here for all groups so no claim is made that a relationship has been established, it is interesting that the ratio of genera to families increases in the two groups that develop a broader range of trophic habits. Diversity increase (as different from diversity replacement) alone is not responsible for increase in the ratio of genera to families. This is demonstrated by the Bryozoa (Figure 12). The ratio of Gymnolaemata to Stenolaemata at the family and generic levels is about the same as are the ratios between Mesozoic-Cenozoic Stenolaemata and Paleozoic Stenolaemata at both generic and family levels.

There is no consistent taxonomic level that reveals most clearly the pattern of diversity change in a class. This is because diversity fluctuations seem to be associated with organisms with similar basic structural patterns rather than with any particular level of the taxonomic hierarchy. Taxonomic practice uses a variety of criteria, not just those factors associated with diversity and diversity change. Taxonomic hierarchies have been erected independently for each major taxon, too. Therefore we find that the Mesozoic-Cenozoic diversity replacement in the Ostracoda occurs primarily in one suborder (and predominantly in just two superfamilies), whereas the diversity expansion in the Echinoidea occurs in four superorders comprising an entire subclass. Several orders within a class may be grouped together to illustrate patterns of diversity change because they have similar diversity histories even though they are not taxonomically related in a recognized single taxonomic subdivision. Examples of this are the groupings used here in analyzing the Rugosa (Figure 8), articulate Brachiopoda (Figure 11), and Cephalopoda (Figure 14). This does not make the analysis either arbitrary or subjective, however. The various groupings used all have consistent features that were fundamental in determining how the organisms functioned and succeeded in the environment. The purpose of the study is to discover if such a relationship is consistently present in organisms with similar diversity histories. Because their previously developed taxonomic subdivisions were not created using such criteria (nor should taxonomy be based on features of primarily ecologic significance) it should not be expected that clades at any particular taxonomic level should reveal the basic patterning of diversity change demonstrated by the compilation used here.

THE ECOLOGIC RULES OF DIVERSIFICATION

The thesis of this study is that an ecologic relationship to diversification exists. The results suggest that increase in diversity (the addition of diversity beyond that previously attained) is achieved only by organisms that develop features that permit them to utilize more ecospace (the multidimensional hypervolume of ecologic resources such as food, mode of life, behavior, physiologic tolerance of ambient conditions, etc.). Several other factors associated with the diversity patterns recognized here also emerge. Diversity replacement sequences, in which new groups diversify and recover the diversity once attained by a precursor group, seem to be composed of groups in which new modifications arise that permit greater strength, more rapid growth or more precise or specific activity in each successively diversifying group. In these cases, however, the general range of ecospace utilized by the sequence of replacing taxa remains unchanged. The groups that maintain diversity without much fluctuation through long intervals of time, often at rather low levels of diversity, are apparently less subject to (or influenced by) the perturbations and vicissitudes that cause extinctions. It is tempting to use words such as "improvement" and "progress" in such a discussion but I shall try to avoid these and other philosophically difficult concepts. This discussion will attempt simply to relate the observations that link diversity change to ecospace utilization.

Diversity expansion first occurred in the Vendian-Cambrian (first five clades of top row and left half of bottom row of Figure 2). This diversification was related to the origins of the major metazoan phyla and the initial exploitation of ecospace by multicelled animals. At this time all new groups were invading previously unexploited ecospace. Valentine (1973, 1980, 1981), Stanley (1976), and Sepkoski (1978) have discussed this radiation and diversification extensively. Most classes had low diversity. The few that attained more than five families were epifaunal or very shallow burrowing infaunal forms (Figure 28). Only a few classes diversified in each of the epifaunal modes of life. Most pelagic and most infaunal ecospace was still vacant.

A second episode of diversity increase occurred in the Ordovician (most of top row and first five clades at left of middle row of Figure 2). Aspects of this pulse of diversification are discussed by Sepkoski (1979, 1981), Sepkoski and Sheehan (1983) and Sepkoski and Miller (this volume). This diversification featured full elaboration of the variety of epifaunal modes of life and some increase in the variety of pelagic and shallow infaunal modes of life (Figure 29). Colonial growth forms (Anthozoa, Bryozoa, and Graptolithina), varied morphologies accommodating or enhancing survival on different substrata (articulate Brachiopoda, Bivalvia, Gastropoda, and Ostracoda), erect forms extending up from the bottom with efficient filtration fans (Cryptostomata, Crinoidea), predators (Cephalopoda, Stelleroidea), and active nektonic forms (Cephalopoda, Chordata) first appear in abundance at this time. Organisms exploiting previously little utilized dimensions of ecospace contribute the increased diversity of the Middle and Late Paleozoic whereas the forms that diversified in the Cambrian either dwindle or simply maintain their Cambrian level of diversity.

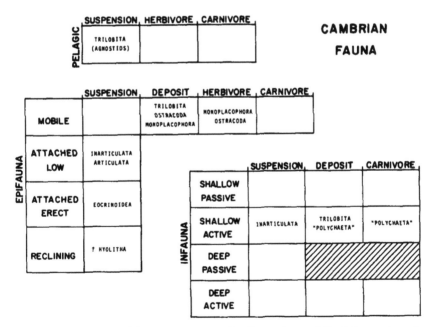

Figure 28. Modes of life common among the diverse classes of Cambrian marine animals. From Bambach (1983).

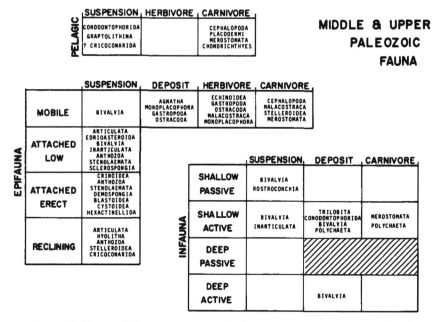

Figure 29. Modes of life common among the diverse classes of Ordovician to Permian marine mammals. From Bambach (1983).

The third episode of diversity expansion began in the Middle Mesozoic and has extended into the Cenozoic (middle row of Figure 2). This has brought about a marked increase in diversity at all levels from the biosphere down to the local community (Figure 1). Although a number of classes that were diverse in the Paleozoic and suffered severe diversity loss in the Permo-Triassic extinction recover some or all of their Paleozoic diversity in the Mesozoic and Cenozoic (Crinoidea, Ostracoda, Stenolaemata, and Anthozoa, for example), this re-diversification does not contribute to the increase in diversity of the last 150 million years. The same situation pertains for the lower level taxa that became diverse in the Paleozoic that are included within the classes that have diversified further in the Mesozoic and Cenozoic. The additional diversity of the Mesozoic and Cenozoic has been produced by newly developed groups that have occupied ecospace not utilized by Paleozoic representatives of their classes. Major expansion of infaunal habits (Thayer, 1983) and predatory feeding habits (Vermeij, 1977) occur with the diversification of the Mesozoic and Cenozoic (Figure 30).

Diversity expansion in the Mesozoic-Cenozoic occurred within several classes of the Porifera, the Gymnolaemata, Bivalvia, Gastropoda, Malacostraca,

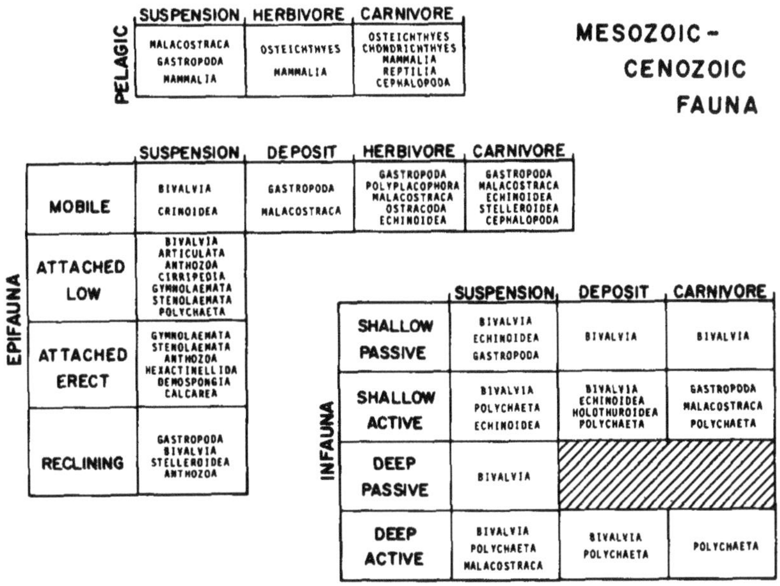

MESOZOIC - CENOZOIC FAUNA

PELAGIC	SUSPENSION	HERBIVORE	CARNIVORE
	MALACOSTRACA GASTROPODA MAMMALIA	OSTEICHTHYES MAMMALIA	OSTEICHTHYES CHONDRICHTHYES MAMMALIA REPTILIA CEPHALOPODA

EPIFAUNA	SUSPENSION	DEPOSIT	HERBIVORE	CARNIVORE
MOBILE	BIVALVIA CRINOIDEA	GASTROPODA MALACOSTRACA	GASTROPODA POLYPLACOPHORA MALACOSTRACA OSTRACODA ECHINOIDEA	GASTROPODA MALACOSTRACA ECHINOIDEA STELLEROIDEA CEPHALOPODA
ATTACHED LOW	BIVALVIA ARTICULATA ANTHOZOA CIRRIPEDIA GYMNOLAEMATA STENOLAEMATA POLYCHAETA			
ATTACHED ERECT	GYMNOLAEMATA STENOLAEMATA ANTHOZOA HEXACTINELLIDA DEMOSPONGIA CALCAREA			
RECLINING	GASTROPODA BIVALVIA STELLEROIDEA ANTHOZOA			

INFAUNA	SUSPENSION	DEPOSIT	CARNIVORE
SHALLOW PASSIVE	BIVALVIA ECHINOIDEA GASTROPODA	BIVALVIA	BIVALVIA
SHALLOW ACTIVE	BIVALVIA POLYCHAETA ECHINOIDEA	BIVALVIA ECHINOIDEA HOLOTHUROIDEA POLYCHAETA	GASTROPODA MALACOSTRACA POLYCHAETA
DEEP PASSIVE	BIVALVIA	(hatched)	(hatched)
DEEP ACTIVE	BIVALVIA POLYCHAETA MALACOSTRACA	BIVALVIA POLYCHAETA	POLYCHAETA

Figure 30. Modes of life common among the diverse classes of Mesozoic and Cenozoic marine animals. From Bambach (1983).

Echinoidea, Batoidea, Teleostei, and the Reptilia and Mammalia. Expansion of diversity within classes of the Porifera has occurred in orders that either developed rigid skeletal organization (Lithistida, Hexactinosida, Lychniscosida and Pharetronida) or achieved a level of spicular organization with different spicule types in different regions of the sponge (Poecilosclerida) or possessing tetraxons (Astrophorida, Spirophorida, and others). These varied sponges have been able to enter a variety of habitats not fully exploited by earlier differentiated sponge types. Expansion in diversity in the Bryozoa occurred entirely within the Cheilostomata with their radically different development of zooids, zooecia and zoaria permitting them to produce a seemingly infinite variety of growth forms and live abundantly in a broad spectrum of physical habitats. Within the Mollusca, diversity increase has been concentrated in the adaptive radiation of two groups, the siphonate heterodont bivalves with their ability to live as deep infauna and on or in unstable, shifting substrates, and the Caenogastropoda with their novel gill structure and modified radular organization which permit both life in a variety of habitats and the development of a wide range of feeding habits. It is these two groups of mollusks that have exploited new ecospace and diversified further while the other molluscan groups, including the other groups of the Bivalvia and Gastropoda, simply maintain prior levels of diversity (and also remain channeled into the same regions of ecospace they have always occupied). The Arthropoda have contributed to the Mesozoic-Cenozoic increase in marine diversity primarily by the expansion of the Decapoda with their development of specialized chelae, walking and swimming pereiopods that permitted them to successfully develop a wide variety of life habits. The Brachyura, with their abilities to prey upon the widely diversifying mollusks, have contributed the most to the increase in diversity by the Decapoda. All other groups of marine arthropods have simply maintained prior diversity except the Cirripedia in which cemented, fully calcified forms diversified as they invaded intertidal habitats, among others. The only class of the Echinodermata to add to marine diversity in the Mesozoic and Cenozoic is the Echinoidea. Diversity expansion in the Echinoidea is confined to the subclass Euechinoidea. The four superorders of the Euechinoidea each possess structural features that have permitted them to invade new ecospace and diversify. These features are added mobility in the Diadematacea, a more developed lantern providing a variety of new feeding habits in the Echinacea, and test shape alteration and feeding apparatus change that permitted both infaunal life styles in many habitats and new feeding habits in the Gnathostomata and Atelostomata. The Batoidea, with jaws, teeth and body form providing opportunities for predation on benthic shelled invertebrates, the Teleostei, with their highly varied fin, body and jaw forms permitting a vast spectrum of life styles, and the marine tetrapods (Reptilia and Mammalia) with their different physiology and limb and jaw forms which permit them to grow to enormous size, among other things, are the groups of the Chordata that participate in the Mesozoic-Cenozoic expansion of diversity. The increased utilization of ecospace by the groups that add to diversity in the Mesozoic and Cenozoic is shown by the presence of diverse classes in all infaunal and pelagic modes of life (Figure 30).

The sequences of diversity replacement or recovery all seem to be associated with the origin and diversification of groups that have more efficient or precise methods of adapting to the requirements for life in the ecospace also occupied by their precursor groups. The renewed success of a class as new groups with such features develop may be an expression of co-evolutionary response. The decline of precursor groups may have been for a variety of reasons but re-diversification does not seem to take place unless a new form develops that can be successful in the changed biosphere in which other new groups have also developed. Formerly successful (diverse) groups that have declined in diversity do not expand for a second time. The Scleractinia have a capacity for cementation, a lower ratio of mineralized skeleton to polyp mass and, in the hermatypic corals, a physiologic system not present in either the Tabulata or Rugosa. Within the Strophomenida, the Productidina possessed a greater range of various shell forms and spine arrays that aided in maintaining the life position in various environments than did the

Strophomenidina. Among pedunculate articulate brachiopods the Terebratulida, Spiriferida, and Rhynchonellida had altered pedicle insertion and pedicle foramena, as well as calcified supports for the lophophore, lacking in the Orthida and Pentamerida. The Cyclostomata have internal zooecial connections not present in any Paleozoic orders of the Stenolaemata. The Cephalopoda display a change from straight and curved, but not coiled, to coiled forms (with altered septa, too) to shell-less forms with well developed eyes. The Trilobita have much greater morphologic variety in Ordovician and later dominant groups than in Cambrian dominants. Arm mobility and other features are present in diverse Mesozoic and Cenozoic Crinoidea (Articulata) and Ophiuroidea (Ophiurida) that were not developed in Paleozoic representatives of these classes. The arm and pinnule arrangement in Paleozoic Crinoidea is a more precisely arranged and displayed filtration fan than the asymmetrical arms of Eocrinoidea, any of the carpoids, or the Cystoidea.

It would be overly simplistic to expect that change in ecospace always insures increase in diversity. The important point is that diversity does not increase without increase in utilization of ecospace. In the cases of diversity replacement sequences the replacing groups often do exploit some new aspect of ecospace. Intratentacular budding and more secure cementation have expanded the range of forms that scleractinians achieve compared to Paleozoic corals. Comatulid crinoids are free-living and mobile whereas all Paleozoic crinoids were attached forms. The reason these changes do not lead to increased diversity for the replacing groups is probably related to co-evolution of the entire fauna. Relatively small changes in effective ecospace utilization may not provide significant opportunity for diversification because of limitations imposed by the other members of the fauna. As faunas change some groups may contract in some ecospace dimensions while gaining in others. For example, it is likely that both expansion of bioturbation from new infaunal groups, creating greater instability in soft substrata, and increased predation have created new problems for organisms with epifaunal, sedentary modes of life. Solitary corals are now less diverse than in the Paleozoic and stratified assemblages of attached crinoids no longer inhabit shelf environments as they did in the Paleozoic. The success of any group will be related to the balance of interactions influencing the group within the context of the whole fauna.

Diversification is not a continuous process although evolution and evolutionary turnover in faunas operates constantly. Diversity within a group increases during what are traditionally called adaptive radiations, usually shortly after the development of new structures that permit invasion of new ecospace or the recovery of formerly inhabited ecospace. The rate of diversity increase for any group (whether for a group adding new diversity or for a group diversifying to replace prior diversity achieved within a class) is usually relatively rapid during an adaptive radiation compared to the length of time the group persists after diversifying. Examples of adaptive radiations in the marine biosphere include: the Jurassic expansion of various Porifera with rigid skeletons (Figure 5); the Ordovician appearance of all groups of Tabulata (Figure 8), the Early Carboniferous replacement within the Rugosa (Figure 8), and the Jurassic diversification of the Scleractinia (Figure 7); the Ordovician diversification of the Orthida, Strophomenidina, and Pentamerida and the Devonian diversification of the Spiriferida (Figure 10); the Ordovician expansion of all Stenolaemata and the Cretaceous diversification of the Cheilostomata (Figure 12); the Ordovician establishment of the Cephalopoda and the various later expansions of the Ammonidea (Figure 14); the Late Mesozoic expansion of the Caenogastropoda (Figure 17); the Cambrian expansion of the Trilobita and the establishment of replacing groups in the Ordovician (Figure 19); the Ordovician diversification of the Eurypterida (Figure 20); the Early Mesozoic expansion of the non-brachyuran Decapoda and the Cretaceous-Early Cenozoic expansion of the Brachyura (Figure 22); the Silurian diversification of the Camarata and the Jurassic diversification of the Articulata (Crinoidea) (Figure 24); the Ordovician diversification of the Stelleroidea (Figure 25); the Middle Mesozoic diversification of the Euechinoidea (Figure 26); and the Ordovician diversification of Conodontoforida and Graptolithina, Late Devonian diversification of the Chondrichthyes, Mesozoic and Cenozoic diversification of the Teleostei and the Mesozoic diversification of the marine Reptilia and Cenozoic diversification of the marine Mammalia (Figure 27).

After intervals of adaptive radiation a group either maintains the new diversity or declines in diversity. Often some members of a group that declines in diversity will persist, maintaining a low but relatively constant diversity for a long interval. During these long intervals of diversity maintenance (at high or low diversity) the group is in a dynamic balance in which extinctions and originations of constituent taxa are nearly equal. Apparently the evolution of new structures that alter the ecologic relationships of a group is the phenomenon that upsets the balance and permits diversification (originations exceed extinctions) until a new balance is reached. Although it is difficult to prove conclusively that such long term stability of diversity represents equilibrium or a more complex situation (Mark and Flessa, 1977), the data presented in this paper suggest that there is an ecologic influence on patterns of diversification and that diversification does indeed occur in pulses interspersed with long intervals with little diversity change. Sepkoski (1978, 1979, 1981), Carr and Kitchell (1980), Kitchell and Carr (this volume) and references therein discuss the mathematical aspects of diversity dynamics.

RELATIONSHIPS TO PHANEROZOIC DIVERSITY PATTERNS

As identified by Valentine (1970), Bambach (1977), Sepkoski (1981), and Sepkoski et al. (1981), marine diversity has increased in stepwise fashion through the Phanerozoic rather than continuously or gradually. This is because of the series of adaptive radiations that have occurred have been clustered at particular times. All classes present in the Early Cambrian were diversifying. Two-thirds of the clades of Ordovician age examined in this paper were diversifying in the Middle and Late Ordovician, whereas only ten percent expand their diversity from the Silurian through the Permian. Nearly half of the clades of Mesozoic age examined here were expanding in the Jurassic whereas, although diversification continued, only one-third of the clades in the Cenozoic continued to expand.

The intervals of diversity balance may be regarded as times of equilibrium when the world is filled to a carrying capacity. This does not mean that all possible modes of life are developed, however. It only means that the adaptive range of the existing fauna with its existing structural capabilities has reached a diversity balance related to the ecospace the fauna is able to utilize. Further diversification occurs when new structures arise that permit the utilization of more ecospace.

The utilization of ecospace has increased as diversity has increased (Figures 28, 29 and 30). Diversity is related to realized ecospace, not potential ecospace. The conquest of new ecospace since the Early Paleozoic has permitted increased diversity to develop, but this conquest has not required the origination of entirely new or novel body plans. Hence diversification has occurred by the acquisition of structural modifications that are reflected in taxonomy at levels below the class level. The data presented here support the arguments presented by Valentine (1980) that diversity increase occurs at lower taxonomic levels as time goes on.

The increased utilization of ecospace associated with increasing diversity also affects the species richness in communities. As observed by Bambach (1983), the increase in species richness in communities with time is paralleled by an increase in the variety of modes of life (guilds) present in communities. This change in community organization and structure through time is a direct product of the invasion of new ecospace by diversifying groups with their increased adaptive potential.

The structures that permit increases in adaptive variety, the invasion of new ecospace, and diversity increase are often called adaptations. Gould and Vrba (1982) note that many structures are actually co-opted for uses for which they did not initially develop. They suggest the term exaptation for such features and point out that adaptation should be used only for those structures initially shaped by natural selection for their current use. Many of the features that have contributed to the increase in diversity by permitting organisms to exploit more ecospace very well may be co-opted for their diverse functions rather than having been directly formed through selective processes for those purposes. Mantle fusion in heterodont bivalves permits siphon formation, but mantle fusion may not have developed to permit the formation of siphons. The shift of the periproct from inside to outside the oculogenital ring in the irregular echinoids may not have been produced by natural selection associated with infaunal life, but it certainly has enhanced the potential of irregular echinoids to radiate into infaunal habits. The prevalence of rapid diversification (adaptive radiations) might be interpreted as the product of the common conversion of pre-existing structures into new uses, rather than solely from the development of entirely newly organized structures. Adaptations and exaptations together are the basis for diversification.

Although the conclusions of this paper explain why each of the successive great marine faunas is more diverse than the previous fauna, it does not explain the timing of these expansions in diversity. The question still remains as to whether the three great evolutionary marine faunas are actually coherent assemblages with internally governed relationships that stimulated synchronous diversification or whether the apparent similarities in timing of diversification are simply coincidental. Considerable further analysis will be necessary before an answer is determined, but two aspects of the question deserve preliminary comment.

On the side of simple coincidence contributing what may be only an illusion of coherence to the timing of emergence of the three faunas are several historic constraints. The first (Cambrian) fauna must have developed almost synchronously simply because it marks the time of origin and initial diversification of the metazoa. No further coordinating mechanism is needed. Diversification of whatever groups that first originated would occur at the same time. The second (Middle and Late Paleozoic) fauna may be a composite of the independent waxing and waning of a variety of unconnected groups. For example, articulate brachiopod diversity increases whereas anthozoan diversity drops from Ordovician-Silurian to Carboniferous times. The scattering of timing of diversification of the groups typifying the third (Mesozoic-Cenozoic) fauna is smeared through the entire Mesozoic Era and into the Cenozoic and may still be in progress. The expansion of this fauna may not be coordinated by anything more than the influence of the Permian extinction resetting the evolutionary clock back to "go" by causing diversity to drop far below the established Paleozoic capacity of the marine ecosystem.

On the other hand, the coincidental timing mechanisms mentioned above do not explain why new ecospace exploitation develops in the second and third faunas nor what triggers that development. Also, it is certain that some coordination in diversification is caused by co-evolutionary linkages. For example, the diversification of the brachyurans depended on the diversification of suitable prey. Even if the initiation of an episode of diversification is simply fortuitous (because of the development of the ability to invade new ecospace in a group unrelated to established interactions in the existing fauna), the consequences of subsequent changes must reverberate through the whole structure of the ecosystem. In the Cambrian fauna we see the initial development of biotic interactions as the metazoan ecosystem becomes established. The development of the Middle and Late Paleozoic fauna includes the co-evolutionary development of tiered benthic communities. The Mesozoic-Cenozoic fauna elaborates on infaunal modes of life and their influence on the ecosystem and on the vast array of interactions in predator-prey systems.

The apparent coordination in the timing of the evolution of the three great faunas is probably a combination of coincidence and co-evolutionary linkages. As we develop a clearer picture of the timing of events in evolutionary time and as we discover cause-effect relationships for these events, a better understanding of the overall determinants of the patterns of evolution should emerge.

ACKNOWLEDGMENTS

My appreciation goes to Jim Valentine, Tom Schopf and Steve Gould for thoughtful reviews of the manuscript, and to Jack Sepkoski and other members of the NASA workshop on the evolution of complex life for stimulating my thinking on this subject. Many thanks to Llyn Sharp who did the bulk of the drafting and Donna Williams who typed the manuscript.

REFERENCES

Astrova, G. G., 1965. A new order of Paleozoic Bryoza. Internat. Geol. Rev. 7:1622-1628.

Ausich, W. I., and Bottjer, D. J., 1985. Phanerozoic tiering in suspension-feeding communities on soft substrata: Implications for diversity. In Valentine, J. W. (ed.), Phanerozoic diversity patterns: Profiles in macroevolution, Princeton, N.J.: Princeton Univ. Press and Amer. Assoc. Adv. Sci. (this volume).

Bambach, R. K., 1971. Diversity of life habits in middle paleozoic Bivalvia. Geol. Soc. Amer. Abstr. with programs 3:292.

Bambach, R. K., 1977. Species richness in marine benthic habitats through the Phanerozoic. Paleobiology 3:152-167.

Bambach, R. K., 1983. Ecospace utilization and guilds in marine communities through the Phanerozoic. In Tevesz, M. J. S., and McCall, P. M. (eds.),

Biotic interactions in recent and fossil benthic communities, New York: Plenum, 719-746.

Bergquist, P. R., 1978. Sponges. Univ. Calif. Press, Berkeley, Calif.

Boyd, D. W. and Newell, N. D., 1972. Taphonomy and diagenesis of a Permian fossil assemblage from Wyoming. J. Paleontol. 46:1-14.

Carr, T. R. and Kitchell, J. A., 1980. Dynamics of taxonomic diversity. Paleobiology 6:427-443.

Conway Morris, S., 1977. A new metazoan from the Cambrian Burgess Shale of British Columbia. Paleontology 20:623-640.

Copper, P., 1966. Ecological distribution of Devonian atrypid brachiopods. Palaeogeog., Palaeoclimat., Palaeoecol. 2:245-266.

Copper, P., 1967. Adaptations and life habits of Devonian atrypid brachiopods. Palaeogeog., Palaeoclimat., Palaeoecol. 3:363-379.

Crimes, T. P., 1974. Colonization of the early ocean floor. Nature 248:328-330.

Finks, R. M., 1967. The structure of *Saccospongia laxata* Bassler (Ordovician) and the phylogency of the Demospongea. J. Paleontol. 41:1137-1149.

Fortey, R. A. and Owens, R. M., 1975. Proetida–a new order of trilobites. In Martinsson, A. (ed.), Evolution and morphology of the Trilobita, Trilobitoidea and Merostomata, Oslo: Universitetsforlaget, Fossils and strata (4): 227-240.

Gould, S. J., Raup, D. M., Sepkoski, J. J., Jr., Schopf, T. J. M., and Simberloff, D. S., 1977. The shape of evolution: A comparison of real and random clades. Paleobiology 3:23-40.

Gould, S. J., and Vrba, E. S., 1982. Exaptation–A missing term in the science of form. Paleobiology 8:4-15.

Hartman, W. D., 1980. Systematics of the Porifera. In Hartman, W. D.,Wendt, J. W., and Wiedenmayer, F., Living and Fossil Sponges, Sedimenta VIII, Miami, Fla.: Comparative Sedimentology Laboratory, Univ. of Miami, 24-51.

Heckel, P. H., 1974. Carbonate buildups in the geologic record: A review. In Laporte, L. F. (ed.), Reefs in time and space, Tulsa, Okla.: Soc. Econ. Paleontologists and Mineralogists Spec. Publ. (18):90-154.

Johnson, R. G., and Richardson, E. S., 1969. Pennsylvanian invertebrates of the Mazon Creek area, Illinois: The morphology and affinities of *Tullimonstrum*. Fieldiana 12:119-149.

Kitchell, J. A., and Carr, T. R., 1985. Non-equilibrium model of diversification: Faunal turnover dynamics. In Valentine, J. W. (ed.), Phanerozoic diversity patterns: Profiles in macroevolution, Princeton, N.J.: Princeton Univ. Press and Amer. Assoc. Adv. Sci. (this volume).

LaBarbera, M., 1977. Brachiopod orientation to water movement. I. Theory, laboratory behavior and field orientations. Paleobiology 3:270-287.

Liddell, W. D., 1975. Recent crinoid biostratinomy. Geol. Soc. of Amer. Abstr. with Programs 7:1169.

Mark, G. A. and Flessa, K. W., 1977. A test for evolutionary equilibria: Phanerozoic brachiopods and Cenozoic mammals. Paleobiology 3:17-22.

Moore, R. C., Lalicker, C. G., and Fischer, A. G., 1952. Invertebrate fossils. New York: McGraw-Hill.

Newman, W. A., and Stanley, S. M., 1981. Competition wins out overall: Reply to Paine. Paleobiology 7:561-569.

Paul, C. R. C., 1977. Evolution of primitive echinoderms. In Hallam, A. (ed.), Patterns of evolution, New York: Elsevier, 123-158.

Raup, D. M., 1976. Species diversity in the Phanerozoic: A tabulation. Paleobiology 2:279-288.

Raup, D. M., 1977. Stochastic models in evolutionary paleontology. In Hallam, A. (ed.), Patterns of evolution, New York: Elsevier, 59-78.

Raup, D. M., 1978. Cohort analysis of generic survivorship. Paleobiology 4: 1-15.

Raup, D. M., Gould, S. J., Schopf, T. J. M., and Simberloff, D. S., 1973. Stochastic models of phylogeny and the evolution of diversity. J. Geol. 81: 525-542.

Richardson, J. R., 1981. Brachiopods and pedicles. Paleobiology 7:87-95.

Romer, A. S., 1966. Vertebrate Paleontology. 3rd ed. Chicago, Ill.: Univ. of Chicago Press, 468.

Ryland, J. S., 1970. Bryozoans. London: Hutchinson.

Schopf, T. J. M., 1977. Patterns and themes of evolution among the Bryozoa. Chapter 6. In Hallam, A. (ed.), Patterns of evolution, New York: Elsevier, 159-207.

Schopf, T. J. M., 1978. Fossilization potential of an intertidal fauna: Friday Harbor, Washington. Paleobiology 4:261-270.

Schopf, T. J. M., 1979. Evolving paleontological views on deterministic and stochastic approaches. Paleobiology 5:337-352.

Seilacher, A. , 1974. Flysch trace fossils: Evolution of behavioural diversity in the deep sea. Neues Jahrbuch fur Geologie und Palaontologie. Monatshefte (4):233-245.

Sepkoski, J. J., Jr., 1978. A kinetic model of Phanerozoic taxonomic diversity. I. Analysis of marine orders. Paleobiology 4:223-251.

Sepkoski, J. J., Jr., 1979. A kinetic model of Phanerozoic taxonomic diversity. II. Early Phanerozoic families and multiple equilibria. Paleobiology 5:222-251.

Sepkoski, J. J., Jr., 1981. A factor analytic description of the Phanerozoic marine fossil record. Paleobiology 7:36-53.

Sepkoski, J. J., Jr., Bambach, R. K., Raup, D. M., and Valentine, J. W., 1981. Phanerozoic marine diversity and the fossil record. Nature 293:435-437.

Sepkoski, J. J., Jr., and Miller, A. I., 1985. Evolutionary faunas and the distribution of Paleozoic benthic communities in space and time. In Valentine, J. W. (ed.), Phanerozoic diversity patterns: Profiles in macroevolution, Princeton, N.J.: Princeton Univ. Press and Amer. Assoc. Adv. Sci. (this volume).

Sepkoski, J. J., Jr., and Sheehan, P. M., 1983. Diversification, faunal change, and community replacement during the Ordovician radiations. In Tevesz,

M. J. S., and McCall, P. M. (eds.), Biotic interactions in recent and fossil benthic communities, New York: Plenum, 673-717.

Sprinkle, J., 1973. Morphology and Evolution of blastozoan echinoderms. Harvard Univ. Mus. Compar. Zool., Spec. Publ., 283.

Stanley, S. M., 1968. Post-Paleozoic adaptive radiation of infaunal bivalve mollusks—A consequence of mantle fusion. J. Paleontol. 42:214-229.

Stanley, S. M., 1970. Relation of shell form to life habits of the Bivalvia. Geol. Soc. Amer. Mem. 125; 296.

Stanley, S. M., 1976. Fossil data and the Precambrian-Cambrian evolutionary transition. Amer. J. of Sci. 276:56-76.

Stanley, S. M., 1977. Trends, rates and patterns of evolution in the Bivalvia. Chapter 7. In Hallam, A. (ed.), Patterns of evolution, New York: Elsevier, 209-250.

Stanley, S. M., and Newman, W. A., 1980. Competitive exclusion in evolutionary time: The case of the acorn barnacles. Paleobiology 6:173-183.

Stanley, S. M., Signor, P. W., III, Lidgard, S. L., and Kerr, A. F., 1981. Natural clades differ from "random" clades: Simulation and analyses. Paleobiology 7:115-127.

Taylor, J. D., Morris, N. J., and Taylor, C. N., 1980. Food specialization and the evolution of predatory prosobranch gastropods. Paleontology 23: 375-409.

Teichert, C., 1967. Major features of cephalopod evolution. In Teichert, C. and Yochelson, E. L. (eds.), Eassays in paleontology and stratigraphy, Lawrence, Kan.: Univ. Kansas Press, 163-210.

Thayer, C. W., 1975. Strength of pedicle attachment in articulate brachiopods: Ecologic and paleoecologic significance. Paleobiology 1:388-399.

Thayer, C. W., 1983. Sediment-mediated biological disturbance and the evolution of the marine benthos. In Tevesz, M. J. S., and McCall, P. M. (eds.), Biotic interactions in recent and fossil benthic communities, New York: Plenum, 474-669.

Thomas, W. R., and Foin, T. C., 1982. Neutral hypotheses and patterns of species diversity: Fact or artifact? Paleobiology 8:45-55.

Treatise on invertebrate paleontology, Parts E through W, 1953-1981, Moore, R. C., et al. (eds.), Lawrence, Kan.: Univ. of Kansas Press.

Valentine, J. W., 1969. Patterns of taxonomic and ecological structure of the shelf benthos during Phanerozoic time. Palaeontology 12:684-709.

Valentine, J. W., 1970. How many marine invertebrate fossil species? A new approximation. J. Paleontol. 44:410-415.

Valentine, J. W., 1973. Evolutionary paleoecology of the marine biosphere. Englewood Cliffs, N.J.: Prentice Hall.

Valentine, J. W., 1980. Determinants of diversity in higher taxonomic categories. Paleobiology 6:444-450.

Valentine, J. W., 1981. Emergence and radiation of multicellular organisms. In Billingham, J. (ed.), Life in the universe, Washington, D.C.: U.S. Govt. Printing Office, NASA Conf. Publ. (2156):229-257.

Vermeij, G. J., 1977. The Mesozoic marine revolution: Evidence from snails, predators and grazers. Paleobiology 3:245-258.

Walker, T. D., 1985. Diversification functions and the rate of taxonomic evolution. In Valentine, J. W. (ed.), Phanerozoic diversity patterns: Profiles in macroevolution, Princeton, N.J.: Princeton Univ. Press and Amer. Assoc. Adv. Sci. (this volume).

Ward, P. D., 1979. Cameral liquid in *Nautilus* and ammonites. Paleobiology 5: 40-49.

Ward, P., Greenwald, L., and Magnier, Y., 1981. The chamber formation cycle in *Nautilus macromphalus*. Paleobiology 7:481-493.

Wendt, J., 1980. Affinities of calcareous sponges to extinct groups. In Hartman, W. D., Wendt, J. W., and Wiedenmayer, F., Living and fossil sponges, Miami, Fla.: Univ. of Miami, Comp. Sedimentol. Lab., Sedimenta VIII: 215-225.

Whittington, H. B., 1975. The enigmatic animal *Opabinia regalis*, Middle Cambrian, Burgess Shale, British Columbia. Philos. Trans. Royal Soc. of London, Ser. B 271:1-43.

Williams, A., and Hurst, J. M., 1977. Brachiopod evolution. Chapter 4. In Hallam, A. (ed.), Patterns of evolution, New York: Elsevier, 79-121.

Williams, A., and Rowell, A. J., 1965. Classification. In Moore, R. C. (ed.), Treatise on invertebrate paleontology, Lawrence, Kan.: Univ. Kansas Press, Part H. Brachiopoda, H214-H237.

Chapter 7

PHANEROZOIC TIERING IN SUSPENSION-FEEDING COMMUNITIES ON SOFT SUBSTRATA: IMPLICATIONS FOR DIVERSITY

WILLIAM I. AUSICH

Department of Geology & Mineralogy, The Ohio State University, Columbus

DAVID J. BOTTJER

Department of Geological Sciences, University of Southern California, Los Angeles

INTRODUCTION

Tiering is the establishment of a vertical community structure with organisms distributed at different levels (in ecological studies the term stratification is commonly used [Odum, 1971]). Tiering has been documented in a wide array of settings including temperate forest communities (MacArthur and MacArthur, 1961), benthic deposit-feeding communities (Levinton and Bambach, 1975), benthic epifaunal suspension-feeding communities (Lane, 1963, 1973; Ausich, 1980), and benthic infaunal suspension-feeding communities (Stump, 1975; Hoffman, 1977; Peterson, 1977). From site-specific studies, Ausich and Bottjer (1982) developed a comprehensive history for tiering of soft-substrata suspension-feeding communities throughout the Phanerozoic (Figure 1).

This Phanerozoic tiering model was developed from the fossil record of all organisms that have been consistently interpreted as benthic suspension feeders, as well as from trace fossils of suspension feeders. Only soft-substrata

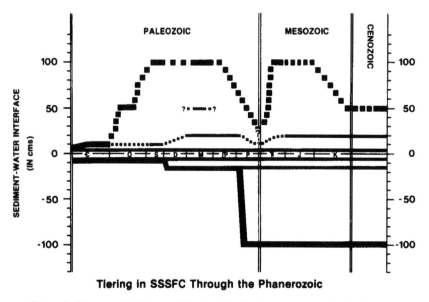

Tiering in SSSFC Through the Phanerozoic

Figure 1. Phanerozoic history of tiering of soft-substrata suspension-feeding communities. The heaviest lines represent maximum tiering levels with other lines tier subdivisions. Solid lines represent data; dotted lines are inferred. (From Ausich and Bottjer, 1982). (Copyright 1982 by the American Association for the Advancement of Science; used with permission.)

settings from non-reef habitats in either subtidal shelf or epicontinental sea shallow-water environments were considered. This specific range of habitats was chosen in order to eliminate some environmental variability which affects community development. In addition, soft-substrata suspension-feeding communities were chosen because soft-substrata habitats constitute the majority of marine shallow-water settings and because both infaunal and epifaunal tiering may be developed in soft substrata.

The tiering model presented in Figure 1 implies a regimented structure for marine suspension-feeding benthos on soft substrata. Despite this implication, we do not presume that the tier levels were rigidly subscribed to by organisms. Rather, the tiering model represents logical subdivisions of community structure, as presently understood.

Bias may be introduced by non-preservation and by consideration of information from a single range of habitats. Information loss by non-preservation cannot be resolved, but data from a single range of habitats may not be a restrictive problem. We would argue that temporal changes in soft-substrata habitats would be mirrored, albeit in perhaps a more complex manner, in other contemporaneous habitats, such as hardgrounds or reefs. Consequently, the Phanerozoic tiering model (Figure 1) that reflects community structure in soft-substrata suspension-feeding communities may be an accurate measure of community structural changes among benthic marine suspension feeders in many other habitats. This model is insensitive to other trophic groups.

TIERING HISTORY

The tiering history proposed by Ausich and Bottjer (1982) (Figure 1), as presently understood, shows that marked temporal changes in tiering have occurred during the Phanerozoic. The Paleozoic was characterized by epifaunal tiering, whereas the Mesozoic had both epifaunal and infaunal tiering, although infaunal organisms were much more abundant. The Cenozoic was typically dominated by infaunal tiering.

Maximum epifaunal tier subdivision developed during the Early Mississippian in diverse crinoid communities where four subdivisions have been recognized (Lane, 1963, 1973; Ausich, 1980). At other times a three-tier subdivision was more common. This included a tier established by sediment-water interface organisms at 5 cm above the seafloor (+5 cm). During the Paleozoic fenestrate bryozoans, pelmatozoan echinoderms, graptolites, sponges and others and in the post-Paleozoic corals, sponges, alcyonarians and others maintained a tier which varied from +10 cm by at least the Silurian to +20 cm from at least the Devonian on (except in the late Paleozoic). The uppermost tier level was maintained by pelmatozoan echinoderms, primarily crinoids, from the Cambrian to the end of the Jurassic, while alcyonarians and sponges have maintained this upper level from the Cretaceous to the present.

Cambrian pelmatozoan echinoderms were elevated up to about 10 cm above the sediment-water interface. By at least the Middle Silurian, crinoids attained heights of +100 cm and maintained this level through the Jurassic, except for the Permian-Triassic extinctions. After the Jurassic a maximum level of about +50 cm is proposed for organisms in soft-substrata settings.

Development of infaunal tiering has been independent of trends above the sediment-water interface. Primarily inarticulate brachiopods, bivalves, gastropods and arthropods have been responsible for infaunal suspension-feeding tiering. A three-tier infaunal structure is also recognized with subdivisions at -6, -12, and -100 cm. From a previous low of -6 cm, maximum tiering was increased to -12 cm by lucinid bivalves during the Early Devonian. By at least the Early Permian, evidence from anomalodesmatan bivalves and possibly suspension-feeding arthropod burrows indicates that the maximum depth had increased to -100 cm. This maximum depth was reached, as well, by siphonate heterodont bivalves by at least the Cenozoic. A more thorough discussion of this history is given in Ausich and Bottjer (1982).

The tiering model has been presented with the hope that it will be tested. Although a few minor adjustments to the temporal trends should be expected as more data becomes available, we believe that the basic trends in Figure 1 will remain valid.

IMPLICATIONS FOR DIVERSITY

TIERING AND ECOLOGICAL DIVERSITY

If tiering is a real biological aspect of the ecological structure of soft-substrata suspension-feeding communities, changes in tiering should be manifested by changes in other ecological parameters. The effects of tiering on diversity in an ecological time perspective will be examined here.

Diversity throughout this paper will refer to numeric diversity which is real and easily interpreted, unlike some diversity indices. Many biological and physical factors influence the development of diversity. Thomas and Foin (1982) have recently summarized limiting factors that affect diversity within an ecological time framework. Thomas and Foin make a useful distinction between biological factors and neutral factors (non-biological, physical-chemical). Among others, biological factors include competition (e.g., MacArthur and Wilson, 1967; Grant and Abbott, 1980), predation (e.g., Paine, 1966), species packing (Valentine, 1973), and productivity (e.g., Connell and Orias, 1964; Valentine, 1971). Neutral factors include stability-time hypothesis (Sanders, 1968, 1969), spatial heterogeneity (e.g., Richerson et al., 1970), space (e.g., MacArthur and Wilson, 1967), plate tectonics (Valentine, 1971), and stochastic processes (Strong et al., 1979). These and other factors have been identified to exert an influence on diversity, although considerable debate has ensued over many of these factors. Does the stability-time hypothesis really work; does competition increase or decrease diversity, etc.?

From the vast array of ideas and often contradictory results, it is perhaps prudent to assume an intermediate position and acknowledge that many, perhaps all, factors mentioned above and elsewhere do have an effect on diversity under certain conditions. Likewise, under differing conditions the same factor may have opposite effects.

We offer tiering, an aspect of community structure, as another factor which should influence diversity. If tiering is important, it should be positively correlated with ecological diversity. The implications of this postulate are straightforward. As tiering increases within a given habitat, the number of niches should increase either by utilization of new space or by compression of existing niches (species packing, Valentine, 1973). This should result in higher diversity.

To test the prediction that tiering is positively correlated with ecological diversity, examples among contemporaneous habitats or localities and contiguous stratigraphic samples will be considered. Five examples are offered: Late Ordovician of the central Appalachians (Bretsky, 1970), Early Mississippian of Indiana and Kentucky (Ausich et al., 1979), Late Cretaceous of the Gulf Coast region (Jablonski, 1979; Bottjer, 1981; Jablonski and Bottjer, 1983), Pliocene of California (Stanton and Dodd, 1976), and Holocene of the San Francisco Bay area (Stanton and Dodd, 1976). The analysis of ecological diversity will be both qualitative and quantitative, as published data allow.

In a study of the fossil communities and facies relationships of the Reedsville Formation (Late Ordovician: central Appalachians), Bretsky (1970) identified three communities which were generally aligned along an onshore-offshore gradient. These include the generally onshore *Orthorhynchula-Ambonychia* community, the *Sowerbyella-Onniella* community which generally occupied an intermediate position, and the generally offshore *Zygospira-Hebertella* community (Table 1). The onshore-offshore community relationships were influenced by clastic input from the east. Environmental stability probably increased farther offshore. As used throughout this report, increased environmental stability-predictability refers to a decrease in the effect of physical-chemical limiting factors on the organic production of suspension feeders. As depicted in Table 1, the onshore, relatively poorly tiered *Orthorhynchula-Ambonychia* community had a relatively low diversity (the lowest diversities occurred at the shoreline), and the offshore, relatively well-tiered *Zygospira-Herbertella* community had a high diversity. The *Sowerbyella-Onniella* community, with intermediate tiering, had intermediate diversities.

Different community types have been identified by Ausich et al. (1979) on the different physiographic areas of the Early Mississippian Borden deltaic complex of Indiana and Kentucky. In general the delta platform and prodeltaic settings would have offered more stable habitats than would have the delta slope, which would have been subjected to episodic turbidite deposition (Figure 2). The skeletal communities of the delta platform and the communities in well-oxygenated settings in the prodelta (Kammer, 1982) were well-tiered crinoid, blastoid, bryozoan and brachiopod communities; whereas the delta slope communities were generally untiered. Diversity varied by an order of magnitude between tiered and untiered communities on the Borden delta. Untiered delta slope community diversities ranged from 2-23 species, and tiered platform and prodeltaic communities displayed diversities ranging from 40-155 species (Table 2).

Diversity and tiering can also be considered by examining samples from within a single facies. Two data sets are considered, each from a different community of the Borden delta platform. These include (1) interbedded siltstones of the skeletal carbonate bank facies with each siltstone separated by a

Fossil Communities of the Borden Delta
Early Mississippian: Indiana & Kentucky

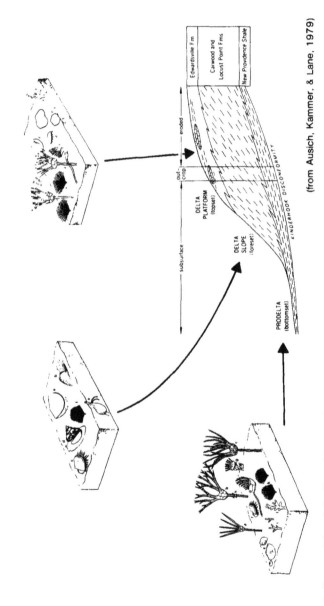

(from Ausich, Kammer, & Lane, 1979)

Figure 2. Communities of the Early Mississippian Borden delta. Characteristic communities are shown from each physiographic part of the delta (from Ausich et al., 1979).

Table 1. Relative tiering, diversity, and onshore-offshore relationships
of Late Ordovician communities of the central Appalachians
(from Bretsky, 1970).

Community	Relative Tiering	Relative Diversity	Relative Onshore-Offshore Position
Orthorhynchula-Ambonychia	poor	intermediate	onshore
Sowerbyella-Onniella	intermediate	intermediate-high	intermediate
Zygospira-Hebertella	good	high	offshore

limestone (Allens Creek Bank, Indiana University locality IU 15114, see Ausich and Lane, 1980), and (2) interdistributary mudstone facies where contiguous stratigraphic sampling was possible (Boy Scout Camp, IU 15109). In each sample set, the combined number of intermediate and higher level organisms are plotted against the number of organisms in the lower level epifaunal tier (Figure 3). The r^2 for linear regression lines fitted to these plots is relatively high (Figure 3). If r^2 is high, the slope of the line can be considered as a rough index of the degree of tiering development. If the slope is extremely low or high certain levels became preferentially more diverse as overall diversity increased. Intermediate values of m for both cases (Figure 3) indicate that with increased overall diversity higher level tiers increased in diversity at nearly the same rate as the lower level tier. Consequently, as tiering increased within a single habitat, overall diversity increased.

Fossil assemblages which are representative of a Gulf Coast Late Cretaceous onshore-offshore transect have been examined by Jablonski (1979), Bottjer (1981), and Jablonski and Bottjer (1983). Although the extent of the data set varies, it is evident that faunas from the Coon Creek Formation (Tennessee) and the Owl Creek Formation (Mississippi), which have intermediate tiering, are more diverse than those from the Saratoga and Annona Formations (Arkansas), which are less well tiered (Table 3). For this example degree of tiering decreased in an offshore direction. Poor levels of tiering in Late Cretaceous Gulf Coast middle to outer shelf environments are attributed to their fine-grained carbonate (now chalk) soft substrata, which limited the development of both infaunal and epifaunal suspension-feeders (Bottjer, 1981; Jablonski and Bottjer, 1983). This phenomenon made the substratum of these environments less stable than those nearer shore.

Stanton and Dodd (1976) compared the trophic structure and environmental

**Interbedded Siltstones of
Skeletal Carbonate Bank Facies
(Allens Creek Bank)**

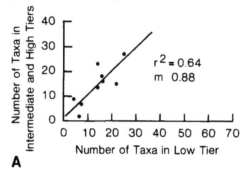

A

**Interdistributary Mudstone Facies
(Boy Scout Camp)**

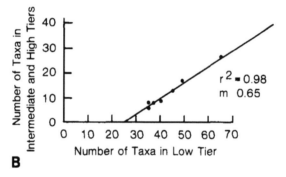

B

Figure 3. Correlation of specific diversity between low tier with intermediate and high tiers in stratigraphic samples from single localities on the Early Mississippian Borden delta platform of Indiana. Examples are from the interbedded siltstones of the skeletal carbonate bank facies (N=9) and interdistributary mudstone facies (N=7); (m is the slope).

limiting factors of communities of the *Pecten* zone, San Joaquin Formation, Kettlemen Hills, California (Pliocene) and the San Francisco Bay area (Holocene). From their extensive data set, tiering, again, can be examined along an onshore-offshore transect. Total genera in the -12 and -100 cm tiers were considered as an index of tiering. In both the Pliocene and Holocene examples total tiering peaked at an intermediate point along the transect (Figure 4). In addition, with increased overall diversity, combined -12 and -100 cm tiers increased in diversity at almost the same rate as combined low level epifaunal and -6 cm tiers (Figure 5). Consequently, as tiering increased with these environments, overall diversity increased.

Table 2. Tiering and species diversity relationships on the Early
Mississippian Borden deltaic complex (from Ausich
et al., 1979; Kammer, 1982).

Physiographic Areas on Delta	Relative Tiering	Diversity Range
Delta Platform	good	49 - 155
Delta Slope	poor	2 - 23
Prodelta	good	40 - 128

Therefore we can conclude that a positive correlation does appear to exist between tiering and ecological diversity. In addition available data indicate that tiering and ecological diversity increase in environmental settings that are more favorable for organic production of suspension feeders.

TIERING AND PHANEROZOIC DIVERSITY TRENDS

Work on long-duration temporal diversity trends has been of two major types: (1) studies that numerically document diversity trends, and (2) studies that attempt to explain the causes for diversity fluctuations.

Phanerozoic marine diversity trends have been compiled by a variety of means both parochial and worldwide, and a general consensus has been reached among many of the investigators involved (Sepkoski et al., 1981). After an initial diversity increase during the Cambrian and Ordovician, diversity levels reached an equilibrium threshold throughout the remainder of the Paleozoic until the Permian-Triassic extinctions. After the Permian-Triassic extinction low, diversity has increased until the present to a level much higher

Table 3. Relative tiering, diversity, and onshore-offshore relationships of Late Cretaceous assemblages of the Gulf Coast region (from Jablonski, 1979; Bottjer, 1981; Jablonski and Bottjer, 1983).

Assemblage	Relative Tiering	Generic Diversity	Relative Onshore-Offshore Position
Coon Creek	intermediate	55 (1 sample)	nearshore
Owl Creek	intermediate	63 (1 sample)	nearshore
Saratoga	poor-intermediate	2-14 (range for 30 samples)	middle shelf
Annona	poor	2-6 (range for 19 samples)	outer shelf

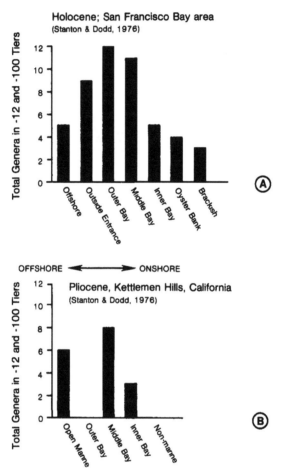

Figure 4. Onshore-offshore variations in infaunal tiering (genera in -12 and -100 cm tiers). A, Holocene, San Francisco Bay area; B, Pliocene, *Pecten* zone, San Joaquin Formation, Kettlemen Hills, California. (Data in part from Stanton and Dodd, 1976).

than the Paleozoic threshold, yet with no apparent stabilization. The most recent and comprehensive data set which illustrates this diversity history was presented by Sepkoski (1981) (Figure 6). Sepkoski's compilation considered numeric abundance of families of marine organisms throughout the Phanerozoic and will be used as a basis for comparison to tiering.

Phanerozoic diversity trends have been attributed to both biological and non-biological causes. Non-biological causes, among others, include (1): biogeographic provincialism (Boucot, 1975; Valentine et al., 1979; Schopf,

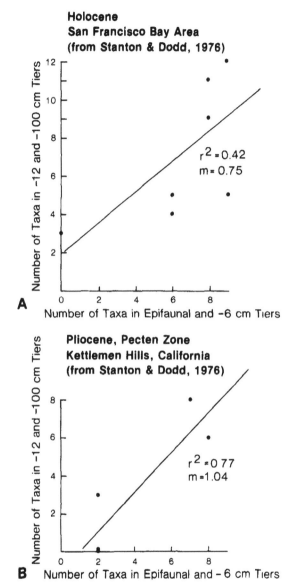

Figure 5. Correlation of generic diversity between the epifaunal and -6 cm tier and -12 and -100 cm tiers in infaunally tiered communities. A, Holocene, San Francisco Bay area (N=7); B, Pliocene, *Pecten* zone, San Joaquin Formation, Kettlemen Hills, California (N=5). (Data, in part, from Stanton and Dodd, 1976).

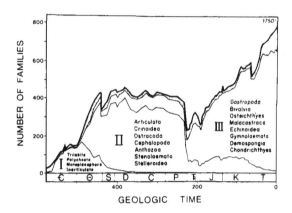

GEOLOGIC TIME

Figure 6. Familial diversity through the Phanerozoic with three major faunal components delineated (from Sepkoski, 1981). (Used with permission of Paleontological Society and J. J. Sepkoski).

1979); (2) changes in habitat space (Schopf, 1974); (3) combinations of (1) and (2) (Valentine, 1971; Valentine and Moores, 1972), and (4) principles of growth during an adaptive radiation (Sepkoski, 1977, 1979). Biological influences on temporal diversity, among others, include species packing (Valentine, 1973).

We offer tiering as another factor that shows correlation with Phanerozoic diversity trends. If ecological diversity is correlated with tiering, then tiering potential through time should correlate with diversity trends. This prediction will be examined here. At this point we will not speculate on the cause or effect of tiering on diversity, but will document their relationship.

In Figure 7 the tiering curves generated by Ausich and Bottjer (1982) are superimposed on familial diversity trends of Sepkoski (1981). If familial diversities are plotted against total additive tiering for the entire Phanerozoic, no significant correlation exists (r^2 = 0.29; see Figure 8). The same data of Figure 8 are displayed on Figure 9 with Paleozoic, Mesozoic and Cenozoic points partitioned. A general correlation is apparent for the Paleozoic, whereas post-Paleozoic data contribute significant scatter. If Paleozoic epifaunal tiering alone is considered in relationship to familial diversities (Figure 10), a strong correlation exists with r^2 equal to 0.89.

The strong correlation between tiering and diversity was present during the Paleozoic, which had a fossil record that was composed of epifaunal suspension-feeding benthos from shallow water marine settings. The diversity-tiering correlation decoupled after the Paleozoic. Mesozoic diversities were generally

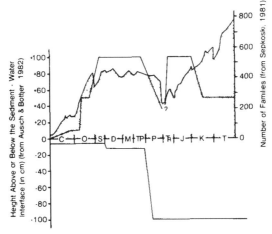

Figure 7. Phanerozoic tiering curves (Figure 1) superimposed on Phanerozoic diversity trend (Figure 6). (From Ausich and Bottjer, 1982, and Sepkoski, 1981, respectively).

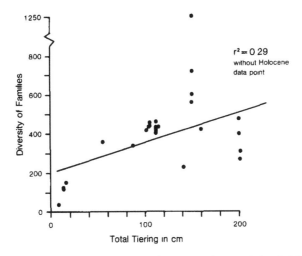

Figure 8. Correlation between familial diversity and total tiering (additive infaunal and epifaunal) throughout the Phanerozoic. Data points from the start, midpoint, and end of each period of the Paleozoic and Mesozoic and from the start, midpoint and present end of the Cenozoic [N=23]). Holocene data point is that with a diversity of 1250. (Data from Ausich and Bottjer, 1982; Sepkoski, 1981).

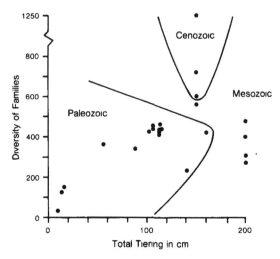

Figure 9. Paleozoic, Mesozoic and Cenozoic contributions to comparison of Phanerozoic familial diversity and total tiering, as presented in Figure 8. The Permian-Triassic point is included with the Paleozoic and the Cretaceous-Cenozoic point is included with the Cenozoic.

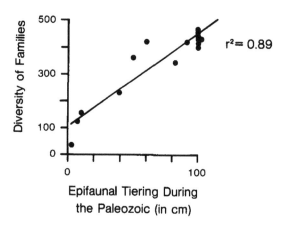

Figure 10. Correlation between familial diversity and epifaunal tiering through the Paleozoic (N=14). (Data from Ausich and Bottjer, 1982; Sepkoski, 1981).

lower than expected for the tiering structure (as compared with the Paleozoic trend), whereas Cenozoic diversities were generally higher.

Several possibilities are offered for the decoupling of tiering and diversity after the Paleozoic. First, tiering may have made no long-term contributions to familial diversity trends during this time, although this seems unlikely because ecological diversity and tiering were apparently correlated throughout the Phanerozoic. Deterministic factors which have been attributed to post-Paleozoic changes include all the potential causes listed above, which have been attributed to have influenced Phanerozoic diversity trends. Of these factors, changes in biogeographic provincialism (Valentine, 1971, 1973; Schopf, 1979) coincides most readily with changes in the tiering-diversity relationship. The high diversities of the Cenozoic may be a result of increased provincialty during that time. However, the relatively low diversities of the early and middle Mesozoic seem not to have been a result of only provinciality changes.

Other less important factors which may have produced scatter away from a correlation between tiering and diversity are (1) much post-Paleozoic diversity data (Sepkoski, 1981) are not that of benthic marine suspension feeders, and (2) the unknown contribution of tiered benthos on marine sea grasses which proliferated after the Paleozoic (Brasier, 1975).

DISCUSSION

A correlation exists between tiering and diversity, both within an ecological framework and for long-duration trends for at least the Paleozoic. The data presented here imply that the correlation of tiering with ecological diversity has been due to a large extent on the partitioning of available niche space. Maximum tiering and maximum diversity tend to be realized in environmental settings where the roles of physical-chemical factors are minimized, and where biological limiting factors have a controlling influence. The role of tiering, as an aspect of community structure, is to increase the number of niches by utilization of new space or by niche compression of existing niches, thereby increasing the potential diversity threshold.

Any given tiering development will have a diversity threshold which may be approached in environmentally stable settings. An increase in tiering and a concomitant increased diversity threshold should not occur without reason. An increase in maximum diversity potential should not occur without an increase in tiering, with other factors equal. A diversity threshold, maintained by tiering complexity, could be increased by an increased need for tiering caused by a tendency for increased diversity. Increases in tiering and a diversity threshold should accompany one another and be driven together by an independent cause (e.g., increased spatial heterogeneity, exponential growth curve, etc.). Once established, tiering reductions should be caused either by extinctions or by migration out of the habitat under study (e.g., Permian-Triassic extinctions and migration of crinoids at the end of the Jurassic, respectively). Either mechanism would result in a concomitant decrease in diversity.

Tiering and diversity potential appear to be inexorably linked. Neither tiering nor diversity has a direct causal relationship with the other. Rather, other biological and non-biological factors should cause parallel changes in tiering and diversity.

It is unrealistic to assume that ecological processes that operate to influence ecological diversity could have had a direct affect upon long-duration diversity trends. Nevertheless, diversity and tiering trends are correlated for at least the Paleozoic, and we believe that an indirect relationship exists between ecological processes and long-duration trends. Tiering levels at individual times throughout the Phanerozoic were established in concert with ecological factors. At each point in time, the resultant tiering would have been correlated to a diversity threshold. The addition of an infinite number of times throughout the Phanerozoic, when tiering and diversity were adjusted together, resulted in correlated long-term trends. Consequently the long-duration trends of tiering are not directly the result of factors operating in ecological time but are the cumulative result of an infinite number of diversity-tiering adjustments which occurred in ecological time.

The correlation between epifaunal tiering and diversity (Figure 10) during the Paleozoic was probably the consequence of the accumulation of tiering-diversity adjustments during that time. During the Paleozoic biogeographic provinciality was generally low (Valentine, 1971, 1973). The cumulative effect of ecological factors dominated the tiering-diversity relationship as the marine biosphere was populated and ecologically filled during the early Paleozoic, and as the middle to late Paleozoic diversity threshold was reached and maintained.

After the Paleozoic the tiering-diversity relationship decoupled, as provinciality increased. Increased provinciality should have resulted in higher diversities than expected for the tiering structure present. In general, higher than expected diversities existed for the Cenozoic, however lower than expected diversities were present for much of the Mesozoic.

The Triassic and Jurassic had the highest potential tiering of any time throughout the Phanerozoic, yet diversities were low. Not only are they well below an expected diversity threshold for such potentially highly tiered communities, but during much of this time, diversities fell below the middle to late Paleozoic threshold. Apparently Triassic and Jurassic provinciality was low enough not to affect diversity as it did during the Cretaceous and throughout the Cenozoic.

One possible reason for the combination of low diversities and high potential tiering is that much of the epifaunal suspension-feeding component of Mesozoic shallow subtidal shelf and epicontinental sea communities probably represented a holdover of various body plans that had been successful in these environments during the Paleozoic. The diversity at higher taxonomic levels for these epifaunal benthos was greatly diminished from the Paleozoic (three Paleozoic subclasses of crinoids, one post-Paleozoic crinoid subclass; five Paleozoic bryozoan orders, three post-Paleozoic bryozoan orders; etc.). Although the Triassic and Jurassic forms must have flourished to be represented in the record, perhaps the potential for epifaunal diversity was so diminished that possible epifaunal niches were "unfilled." This contention of "unfilled" niche space by epifaunal Paleozoic holdovers during the Mesozoic is supported by modification of the data presented in Figures 8-10. As stated above, r^2 for diversity and total tiering throughout the Phanerozoic is 0.29 (Figure 8). If the epifaunal aspect of Triassic and Jurassic communities did not exert a major influence on the tiering-diversity correlation, then removal of the epifaunal tiering should improve the correlation (epifaunal diversity is not removed from data and may be a source of error). Epifaunal Paleozoic tiering-diversity data (Figure 10) combined with unmodified Triassic-Jurassic tiering data have an r^2 of 0.27, whereas Paleozoic data and Triassic and Jurassic data with the epifaunal tiering removed have an r^2 of 0.74 (Table 4). This improved r^2 implies a diminished effect of epifaunal tiering on diversity during the Triassic and Jurassic.

During the Cretaceous and throughout the Cenozoic, diversities have been higher to much higher than expected for the tiering present. From the Triassic and Jurassic to the Cretaceous and Cenozoic, maximum tiering potential

Table 4. Correlation of tiering and diversity for select times both unmodified and modified (data from Sepkoski, 1981; Ausich and Bottjer, 1982).

Data	r^2	N
Phanerozoic	0.29	22
Paleozoic, epifaunal tiering only	0.89	15
Paleozoic, epifaunal tiering only; Triassic and Jurassic	0.27	19
Paleozoic, epifaunal tiering only; Triassic and Jurassic, infaunal tiering only	0.74	19

decreased and diversity increased. This trend is not consistent with ecologically based arguments presented to this point. Rather, this marked increase in diversity relative to tiering corresponds well with the marked increase in biogeographic provincialism that accompanied continental breakup and plate divergence that accelerated during the Cretaceous and has been maintained to the present (Valentine, 1971, 1973; Valentine and Moores, 1972; Schopf, 1979).

In summary, the Paleozoic tiering-diversity correlation is believed to represent the filling and maintenance of ecological space in a biosphere with relatively low and constant provinciality. It is suggested the relatively lower than expected Triassic and Jurassic diversities are the result of niches "unfilled" by Paleozoic "body-plan holdovers" in the epifaunal component of these communities. Relatively high diversities from the Cretaceous to the present are thought to correspond to post-Jurassic high levels of biogeographic provinciality.

ACKNOWLEDGMENTS

James W. Valentine and C. Bryan Gregor offered thoughtful comments to an earlier draft of this paper. WIA and DJB each were supported by separate grants from the American Chemical Society during the tenure of this project. Acknowledgment is made to the Donors of the Petroleum Research Fund, administered by the American Chemical Society, for support of this research.

REFERENCES

Ausich, W. I., 1980. A model for niche differentiation in Lower Mississippian crinoid communities. J. Paleontol. 54:273-288.

Ausich, W. I., and Bottjer, D. J., 1982. Tiering in suspension-feeding communities on soft substrata throughout the Phanerozoic. Science 216:173-174.

Ausich, W. I., and Lane, N. G., 1980. Platform communities and rocks of the Borden Siltstone delta (Mississippian) along the south shore of Monroe Reservoir, Monroe County, Indiana. In Shaver, R. H. (ed.), Field trips 1980 from the Indiana University Campus, Bloomington, Indiana: Dept. Geology, Indiana Univ. and Indiana Geol. Survey, 36-67.

Ausich, W. I., Kammer, T. W., and Lane, N. G., 1979. Fossil communities of the Borden (Mississippian) delta in Indiana and northern Kentucky. J. Paleontol. 53:1182-1196.

Bottjer, D. J., 1981. Structure of Upper Cretaceous chalk benthic communities, southwestern Arkansas. Palaeogeog., Palaeoclimat., Palaeoecol. 34: 225-256.

Boucot, A. J., 1975. Standing diversity of fossil groups in successive intervals of geologic time viewed in the light of changing levels of provincialism. J. Paleontol. 49:1105-1111.

Brasier, M. D., 1975. An outline history of seagrass communities. Palaeontology 18:681-702.

Bretsky, P. W., Jr., 1970. Upper Ordovician ecology of the central Appalachians. New Haven, Conn.: Yale Univ., Peabody Mus. Nat. Hist. Bull. (34), 150.

Connell, J. H., and Orias, E., 1964. The ecological regulation of species diversity. Amer. Naturalist 98:339-414.

Grant, P. R., and Abbott, I., 1980. Interspecific competition, island biogeography and null hypotheses. Evolution 34:332-341.

Hoffman, A., 1977. Synecology of macrobenthic assemblages of the Korytnica Clays (Middle Miocene; Holy Cross Mountains, Poland). Acta Geologica Polonica 27:227-280.

Jablonski, D., 1979. Paleoecology, paleobiogeography, and evolutionary patterns of Late Cretaceous Gulf and Atlantic Coastal Plain mollusks. Unpubl. Ph.D. dissertation, Yale Univ., New Haven, Conn., 604.

Jablonski, D., and Bottjer, D. J., 1983. Soft-bottom epifaunal suspension-feeding assemblages in the Late Cretaceous: Implications for the evolution of benthic paleocommunities. In Tevesz, M. J. S., and McCall, P. L. (eds.), Biotic interactions in recent and fossil benthic communities. New York: Plenum, 747-812.

Kammer, T. W., 1982. Fossil communities of the prodeltaic New Providence Shale Member of the Borden Formation (Mississippian), north-central Kentucky and southern Indiana. Unpubl. Ph.D. dissertation, Indiana University, Bloomington, Indiana, 303.

Lane, N. G., 1963. The Berkeley crinoid collection from Crawfordsville, Indiana. J. Paleontol. 37:1001-1008.

Lane, N. G., 1973. Paleontology and paleoecology of the Crawfordsville fossil site (Upper Osagian: Indiana). Univ. Calif. Publ. Geol. Sci. (99), 147.

Levinton, J. S., and Bambach, R. K., 1975. A comparative study of Silurian and Recent deposit-feeding bivalve communities. Paleobiology 1:97-124.

MacArthur, R. H., and MacArthur, J., 1961. On bird species diversity. Ecology 42:594-598.

MacArthur, R. H., and Wilson, E. O., 1967. The theory of island biogeography. Princeton, N.J.: Princeton Univ. Press, 203.

Odum, E. P., 1971. Fundamentals of ecology, 3rd ed. Philadelphia: W. B. Saunders Co., 524.

Paine, R. T., 1966. Food web complexity and species diversity. Amer. Naturalist 100:65-75.

Peterson, C. H., 1977. Competitive organization of the soft-bottom macrobenthic communities of southern California lagoons. Marine Biol. 43: 343-359.

Richerson, P., Armstrong, R., and Goldman, C. R., 1970. Contemporaneous disequilibrium, a new hypothesis to explain the "Paradox of the Plankton." Natl. Acad. Sci. (U.S.A.) Proc. 67:1710-1714.

Sanders, H. L., 1968. Marine benthic diversity: A comparative study. Amer. Naturalist 102:243-292.

Sanders, H. L., 1969. Benthic marine diversity and the stability-time hypothesis. In Diversity and stability in ecological systems, Brookhaven Symposium in Biol. (22):71-81.

Schopf, T. J. M., 1974. Permo-Triassic extinctions: Relation to sea-floor spreading. J. Geol. 82:129-143.

Schopf, T. J. M., 1979. The role of biogeographic provinces in regulating marine faunal diversity through geologic time. In Gray, J., and Boucot, A. J. (eds.), Historical biogeography, plate tectonics, and the changing environment, Corvallis, Oregon: Oregon State Univ. Press, 449-457.

Sepkoski, J. J., Jr., 1978. A kinetic model of Phanerozoic taxonomic diversity. I. Analysis of marine orders. Paleobiology 4:223-251.

Sepkoski, J. J., Jr., 1979. A kinetic model of Phanerozoic taxonomic diversity. II. Early Phanerozoic family and multiple equilibria. Paleobiology 5: 222-251.

Sepkoski, J. J., Jr., 1981. A factor analytical description of the Phanerozoic marine fossil record. Paleobiology 7:36-53.

Sepkoski, J. J., Jr., Bambach, R. K., Raup, D. M., and Valentine, J. W., 1981. Phanerozoic marine diversity and the fossil record. Nature 293:435-437.

Stanton, R. J., Jr., and Dodd, J. R., 1976. The application of trophic structure of fossil communities in paleoenvironmental reconstruction. Lethaia 9:327-342.

Strong, D. R., Jr., Szyska, L. A., and Simberloff, D. S., 1979. Tests of community-wide character displacement against null hypotheses. Evolution 33:897-913.

Stump, T. E., 1975. Pleistocene molluscan paleoecology and community structure of the Puerto Libertad region, Sonora, Mexico. Palaeogeog., Palaeoclimat., Palaeoecol. 17:177-226.

Thomas, W. R., and Foin, T. C., 1982. Neutral hypotheses and patterns of species diversity: Fact or artifact? Paleobiology 8:45-55.

Valentine, J. W., 1971. Plate tectonics and shallow marine diversity and endemism, an actualistic model. Systematic Zool. 20:253-264.

Valentine, J. W., 1973. Evolutionary paleoecology of the marine biosphere. Englewood Cliffs, N.J.: Prentice-Hall, Inc., 511.

Valentine, J. W., and Moores, E. M., 1972. Global tectonics and the fossil record. J. Geol. 80:167-184.

Valentine, J. W., Foin, T. C., and Peart, D., 1978. A provincial model of Phanerozoic marine diversity. Paleobiology 4:55-66.

Part III

ANALYTIC STUDIES OF MAJOR FAUNAL
PATTERNS AND EVENTS

Within the last decade, theoretical aspects of ecology and evolution have come to play major roles in the progress of paleontology. Although the contributions in previous sections have been by no means free of theory, the papers in this section have been designed specifically to develop or to test explanations of some of the major biotic patterns and events associated with Phanerozoic diversity.

Kitchell and Carr present a pioneering effort to construct a mathematical description for the pattern of diversity and replacement of the Phanerozoic faunas. Among other things they show that relatively simple non-equilibrium models can generate a complexity of behavior comparable to that displayed by the faunal patterns. Replacement patterns need not require mass extinction or other special events. Walker examines the analogy which seems always to be made between population growth models and clade diversifications. He shows that, under the most prevalent assumptions of ecology, the implied identity is erroneous. This leads to useful discussions as to why and when exponential models may provide reasonable estimates of diversifications and when alternate models should be employed.

Jablonski asks how higher taxa with distribution patterns similar to those of today would weather a mass extinction associated with the extirpation of the entire fauna of continental shelves. The answer (taxa would survive very well indeed at the family level) is surprising at first but seems so plausible upon reflection that one wonders how it has been overlooked. This finding brings into question many popular explanations for mass extinctions, particularly those based on marine regression and on species-area effects. Flessa and Thomas ask how the present size-frequency distribution of species ranges, which is a hollow curve, may have come about, and what the consequences may be for the success of lineages on different parts of the curve. Again, the conclusions are novel but upon reflection seem as if they should have been inferred long ago.

Chapter 8

NONEQUILIBRIUM MODEL OF DIVERSIFICATION: FAUNAL TURNOVER DYNAMICS

JENNIFER A. KITCHELL

Museum of Paleontology, University of Michigan, Ann Arbor

TIMOTHY R. CARR

Geological Research, ARCO Resources Technology, Dallas

> "Seek simplicity but distrust it."
> —Lagrange, in Hutchinson (1978)

INTRODUCTION

What can patterns of diversity reveal about processes? Our purpose in this paper is to assess quantitatively the complexity of patterns that can result from the most general mathematical model of diversification. The central question we address is: Does the observed pattern of diversification of global marine taxa significantly deviate from predictions of a deterministic model whose potential behaviors range from simple to chaotic or apparently random? Such an analysis provides a measure of subtraction for the range of diversification patterns that can arise from simple assumptions regarding nonlinearity or diversity-dependent feedback and nonequilibrium.

The topics we cover fall into six general categories. Posed as questions, these categories are: (1) can a very simple model adequately recreate the empirical diversification pattern of Phanerozoic marine taxa; (2) can the model be treated as a null hypothesis; (3) is there a hierarchical cohesiveness of evolutionary groups; (4) what is the importance of evolutionary innovation to nonequilibrium diversification; (5) what is the causal role of perturbations to faunal turnover; and (6) under what conditions can determinism mimic randomness?

Evolutionary paleobiologists have been concerned about the high degree of apparent nonrandomness in pattern that can be generated by purely random processes (e.g., Raup et al., 1973; Raup and Gould, 1974; Anderson and Anderson, 1975; Raup, 1977a,b; Gould et al., 1977). We stress the alternate pitfall—that completely nonrandom systems, if nonlinear and iterated (possessing feedback), can generate apparently random patterns (see Carr and Kitchell, 1980). To illustrate that completely nonrandom systems possess apparently random behaviors is not to endorse a deterministic philosophy (cf. Schopf, 1979) any more than previous discussions of the potential of random systems to produce apparently deterministic patterns endorsed a neutralist philosophy of the evolutionary process (Gould et al., 1977). Both demonstrations serve as warnings against an intuitive bias that determinism implies order and randomness implies chaos. Determinism may yield chaos, and randomness may yield order. Consequently, a familiarization of the dynamical properties of both nonlinear deterministic systems and stochastic systems is needed to replace intuition.

Secondly, as a criterion of subtraction, it is as important to establish how much of an observed pattern can be replicated with a deterministic model (particularly one with apparently random behaviors), as with a random model. Such an analysis delineates the limits of pattern as data to infer causality or to choose between explanations. The consequence of such an analysis is to query if one could determine—by pattern alone—whether the generating process is random, pseudorandom or nonrandom. This question has philosophical as well as practical significance (Wimsatt, 1980). Its answer restricts our ability to understand processes *when our only evidence of process is historical pattern*.

The approach of this paper, however, is stochastic as defined by Schopf (1979) and discussed by Hoffman (1981): the model analyzes general patterns of the history of life, treats groups of taxa as comprised of like particles at one hierarchical level of analysis, and assumes an equilibrium solution to be possible. We show, however, that although the model has an equilibrium solution, it is not an equilibrium model. As should be obvious, a stochastic approach to paleontology may utilize either (mathematically) deterministic or stochastic models; conversely, a deterministic approach may utilize either stochastic or deterministic models.

In our earlier study of taxonomic diversification (Carr and Kitchell, 1980), we focused attention on changes in total diversity in response to a given perturbation. Since then, Sepkoski (1981) has shown that the Phanerozoic marine fossil record at the familial level is statistically defined by three great evolutionary faunas (see also Sepkoski, 1980, ms). We now test the hypothesis that the simultaneous patterns of diversification of these three evolutionary faunas can be closely mimicked by a very simple model of diversification formulated in discrete time mathematics.

Although the model can produce a broad range of diversification patterns, a meaningful test is provided by deriving parameter values from empirical data. A conclusion that follows is that mass extinctions represent extrinsic perturbations and not manifestations of instability. These perturbations, including the "pivotal event" at the close of the Permian, are shown to alter the timing of faunal turnover but not the outcome. Finally, we show that simultaneously diversifying faunas, perturbations, and evolutionary innovations may indefinitely maintain the diversification process at nonequilibrium.

MODELS OF DIVERSIFICATION

THE LOGISTIC GROWTH EQUATION

Unbridled growth in an infinite or unbounded environment exhibits an exponential growth pattern, the well-known Malthusian parameter. A more realistic equation of growth in a finite or bounded environment is the familiar logistic growth equation, first introduced by Verhulst in 1845. The historical development of the logic involved in the progression to the logistic growth equation is nicely summarized by Hutchinson (1978), where, for our purposes, diversity (D) is analogous to population size (N), origination is analogous to birth, and extinction is analogous to death.

The general model of diversification is based on the fundamental assumption that the rate of change of diversity over time depends "in some way" on existing diversity. The simplest expression of our model is $dD/dt = f(D)$, which says that the rate of change of diversity is some function of existing diversity. If the function (f) is specified as the term a, then $dD/dt = aD$, and a ($\geqslant 0$) represents the unlimited rate of diversification, or origination rate minus extinction rate. By expanding this equation to the second term in a Taylor expansion series, one obtains the logistic equation, $dD/dt = aD + bD^2$, where again (aD) represents infinitely increasing diversity, but (bD^2) represents negative feedback with increasing diversity ($b = -a/\hat{D}$ where $\hat{D} = dD/dt = 0$). The negative term, $-bD^2$, causes rate of diversification to decline with increasing diversity: although faunas may have the capacity to diversify exponentially, diversity-dependent feedback limits this capacity. In summary, the logistic growth model represents the second step in a Taylor expansion series, and the simplest mathematical formulation that meets these two basic assumptions: all taxa arise from previously existing taxa; global diversification is finite rather than infinite.

TAXONOMIC DIVERSIFICATION MODELS:
DIFFERENTIAL vs. DISCRETE FORMULATION

The logistic growth equation was used by Sepkoski (1978, 1979) to examine patterns of Phanerozoic taxonomic diversity. We subsequently relied on the logistic growth model to examine the dynamics of diversification to perturbations, but formulated it in discrete time mathematics, such that $dD/dt = aD - bD^2$ is replaced by $D_{t+1} = \alpha D_t - \beta D_t^2 + D_t$ (see Carr and Kitchell, 1980). We pointed out a general asymmetry in these dynamics: the response of diversity from oversaturation is rapid compared to the response of diversity from undersaturation. Hence, an abrupt diversity decline is the predicted response to a perturbation, whereas a relatively slow recovery of diversity is the predicted response to cessation of the perturbation. We showed that an asymmetry in diversity decline and recovery, as observed during the Permo-Triassic crisis, is the predicted response of diversity-dependent diversification to even a symmetrical perturbation. Sepkoski (1979) had previously omitted the Permo-Triassic crisis from his analysis of diversity change, suggesting that this event must represent diversity-independent diversification.

We originally formulated the model in discrete time mathematics because we feel it is unrealistic to model diversification as an instantaneously responding process. Our justification for using difference equations rather than differential equations, as used by Sepkoski, was based on the noninstantaneous nature of the response of origination and extinction to changing abiotic and biotic conditions. We considered the process of evolutionary diversification, particularly at higher taxonomic levels, to be a discrete process rather than a continuous one. Hoffman (1981) has more recently expressed a similar opinion: ". . . the evolution of species diversity, and more so of suprageneric diversity, should be treated in terms of difference rather than differential equations because—contrary to a 'billiard ball' behaviour of physical-chemical particles—a lag time can be expected in each evolutionary response." Difference equations automatically incorporate a lag time—of one analytical period, for first-order difference equations. The length of the analytical period in absolute time can be variable.

Secondly, we argued that measurements of geological time represent geologic intervals rather than points of time. All geological data of global diversity change over time are discrete rather than continuous. Even high-resolution intervals of geological time represent large intervals of absolute time. Stage-level data, for example, represent high-resolution stratigraphy on a global scale, but the average length of a Phanerozoic stage in absolute time is 7.4 Myr (see Raup and Sepkoski, 1982).

The discrete time formulation of evolutionary diversification is biologically appropriate because it assumes that diversification is not continuous and instantaneous, but may be jerky and delayed. However, as Roughgarden (1979) has pointed out for discrete time equations of population growth "the price for this added realism . . . is complication in the stability of the equilibrium." Whereas the continuous-time formulation of the logistic equation can display only stable monotonic behaviors, the discrete time formulation (or the continuous time equation with an added delay term; see Appendix, Carr and Kitchell, 1980) can display a wide suite of behaviors, because feedback is not instantaneous. (For discussion of stochastic models of logistic growth in discrete time, see Bartlett, 1960; Barnett, 1962; Pielou, 1977; Smith and Mead, 1980.)

DYNAMICAL BEHAVIORS OF DISCRETE MODELS

Most simple nonlinear difference equations, as well as differential equations with nonlinear feedback or delay, possess a nonintuitive and remarkable array of dynamical behaviors, ranging ". . .from stable points, through cascades of stable cycles, to a regime in which the behaviour (although fully deterministic) is in many respects 'chaotic,' or indistinguishable from the sample function of a random process" (May, 1976; see also May, 1974, 1975, 1979; May and Oster, 1976; Li and Yorke, 1975; Mackey and Glass, 1977; Guckenheimer et al., 1977; Feigenbaum, 1980; Sparrow, 1980; also Pounder and Rogers, 1980). The complex dynamics of these deterministic models result from the combination of nonlinearity and feedback. Feedback in the discrete time model of diversification is provided by the old diversity value (previous D_{t+1}) becoming the new input diversity value (D_t) for each successive iteration. The two parameters of the model, (α) and (β), determine the strength of the nonlinearity. For all (β) values greater than zero, the model describes a parabolic curve in which the steepness of the hump of the parabola is determined by the (α) parameter (Figure 1 A–D; also see Hofstadter, 1981 for a particularly readable account of the dynamical behavior of nonlinear, parabolic functions). A high value for this parameter, corresponding to an initially rapid rate of diversification, i.e., a rapid rate of divergence between rate of origination and rate of extinction, results in a steeply rising parabola. The steepness or severity of the nonlinearity determines the dynamical behavior of this model (see Carr and Kitchell, 1980).

We previously showed by graphical stability analysis that the dynamics of the general model of diversification are varied, including not only monotonic approaches to equilibrium diversity as illustrated by Sepkoski (1978, 1979; Sepkoski and Sheehan, 1983), but also approaches to equilibrium involving overshoots (Figure 1A, C), unstable divergences away from equilibrium (Figure 1D), and apparent chaos (see Figure 2). We outlined all possible behaviors of the model, as a function of its two parameters (see Figure 4, Carr and

Kitchell, 1980). It is important to note that all oscillations in diversity are internally generated.

Another important feature of the discrete model is that not all equilibrium points have the same dynamical properties. For a parabolic function, there are two possible equilibrium points: the trivial equilibrium point at zero, and the nontrivial equilibrium point where $D_t = D_{t+1}$. Graphically, these points are equivalent to the points at which $f(D_t)$ intersects the 45° line in the phase plane plot of D_t vs D_{t+1}. The stability of this latter equilibrium point can be determined graphically by the slope of the phase line at its point of intersection with the 45° line in phase space. (The stability of the equilibrium point can also be analytically determined by its eigenvalue [see Vandermeer, 1981].) For the equilibrium point to be stable and to attract iterates of the

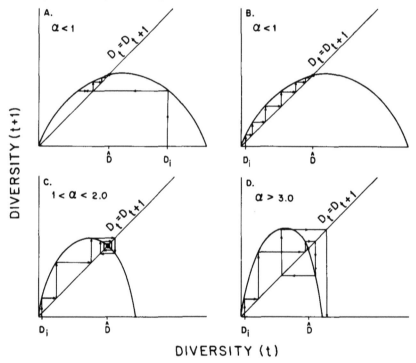

Figure 1. Phase diagrams illustrating the varied behaviors of the equation for diversification, $D_{t+1} = D_t + \alpha D_t - \beta D_t^2$; α varies from <1 to >3.0. Arrows depict trajectory of diversity over time, from initial diversity (D_i) to equilibrium diversity ($\hat{D}$). A. $\alpha < 1$. Decrease in diversity overshoots equilibrium diversity, followed by damped exponential increase to equilibrium diversity. B. $\alpha < 1$. Logistic behavior of diversity increase over time. C. $1 < \alpha < 2.0$. Convergent or damped oscillatory behavior of diversity change. D. $\alpha > 3.0$. Divergent oscillatory behavior of change in diversity, and eventual extinction (modified from Carr and Kitchell, 1980).

diversity function, the slope must be less than 45°. If the equilibrium point acts as a stable attractor, it results in the system approaching equilibrium diversity from all small displacements away from equilibrium.

At slopes in excess of 45°, the equilibrium point is unstable and acts as a repeller (Figure 1; refs. cited in Carr and Kitchell, 1980; Hofstadter, 1981). If the equilibrium point is a repeller, it causes the system to diverge increasingly away from equilibrium, and to ultimately display behavior that is indistinguishable from chaotic (termed "ergodic"). Hence, some rate functions cause trajectories of diversity to converge toward the equilibrium point. Such equilibrium points are stable. Other rate functions cause trajectories of diversity to diverge away from the equilibrium point. Such equilibrium points are unstable. In the former case, the equilibrium point is attracting, and equilibrium diversity is possible, whereas in the latter case, the "equilibrium" point itself is repelling and the system is unstable at every point away from the equilibrium value. (Even an ordinary differential equation that assumes continuous time and instantaneous response, when coupled with other first-order ordinary differential equations, displays dynamical behaviors that are similar to first-order difference equations [May, 1976].)

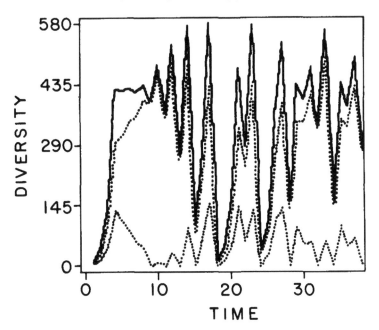

Figure 2. Simulated pattern of chaotic or apparently random diversification, resulting from high values of the α (i.e., rate of initial diversification) parameter. Solid line depicts total diversity; upper dotted line, α = 2.17, β = .005, of hypothetical evolutionary fauna; lower dotted line, α = 1.00, β = .0025, of second hypothetical evolutionary fauna.

In biological terms of population fluctuations, species that display unstable equilibrium points are those with high values of r, the intrinsic rate of natural increase (e.g., Mueller and Ayala, 1981). In evolutionary terms of diversity fluctuations, taxa that display unstable behaviors are those with high values of α, the net rate of diversification. We can speculate that α is a function of both a tendency to geographically isolate, and a function of how open or responsive the genetic system is to change (e.g., Dover, 1982).

NEW THEORY

EVOLUTIONARY INNOVATIONS

A problem inherent in the logistic growth equation as a model of diversification is that there is a specified upper bound or absolute limit on diversity. The existence of an absolute upper bound on diversity is disquieting in the sense that it assumes some a priori limit on "environment" when in reality organisms and environment are mutually interacting and an "environment" cannot be defined without reference to an organism (Levins, 1979). We would prefer a model of diversification that is not unlimited but that is indefinite. We would like our model to assume that although diversification occurs within a limited system of spatial and energy resources, the limits to diversity are not predetermined but are dependent on the historical events of evolution. Diversification is ultimately limited, but its limits are not foreordained but historical.

To achieve such a model of diversification, we need only add to the existing formulation of the model the potential for evolutionary innovation, which may continually relax the upper bounds on diversity. Consequently, we assume that evolutionary innovations (new taxa) can enter the system at any time during its diversification history, and that these taxa interact with existing taxa. Newly evolved, interacting taxa or faunas constitute "structural fluctuations" in the terminology of Prigogine et al. (1972, 1977; see also Nicolis and Prigogine, 1977; Jantsch, 1980). The interactions between faunas are deterministic, but the appearance of an evolutionary innovation provides stochastic structural fluctuation.

Displacement dynamics may now continue indefinitely. Taxa differ in rates of origination, extinction, and diversity-dependence. By analyzing the structural stability of the equations that comprise the model, it is easy to show that only those newly evolved taxa (evolutionary innovations) whose ratio of net rate of diversification : diversity-dependent feedback (the $\alpha{:}\beta$ ratio) is higher than existing taxa can successfully invade the system and displace other diversifying taxa (see Allen, 1976 for a similar discussion of population dynamics and the introduction of mutant alleles). Hence, the potential equilibrium level of diversity ($\hat{D}$) rises with each successful faunal displacement. An absolute upper bound on diversity can now be reached only in the absence of innovations. Because the model is open to both evolutionary innovations and diversity-dependent perturbations, nonequilibrium conditions are more likely to be maintained than equilibrium conditions.

EQUILIBRIUM MODELS vs NONEQUILIBRIUM MODELS

The model we develop in this paper represents a nonequilibrium approach to diversification. The major distinction between equilibrium and nonequilibrium models is not whether an equilibrium solution is possible, but to what extent a system stays in the neighborhood of equilibrium (see Caswell, 1978). Equilibrium theories and models are concerned only with the behavior of systems at equilibrium; nonequilibrium theories and models are interested in the temporal dynamics of systems away from equilibrium. Hence, although an equilibrium solution may be possible within a nonequilibrium model, it is not generally attained. (The term "disequilibrium" seems a more apt descriptor than "nonequilibrium" because of the identity of its prefix in terms such as displacement and disturbance, but nonequilibrium is the conventional term used to describe the dynamical behavior of systems not at equilibrium.) Steady-state, by contrast, simply means no change over time, a condition which could be maintained either at equilibrium or at nonequilibrium.

Any model of diversification must have at least one stable solution, in the absence of continued or large perturbations. If all solutions are unstable, then the process of diversification would not be persistent over evolutionary time. Hence, the mathematical formulation of the model assumes the existence of an equilibrium point, i.e. that rates of origination and extinction are convergent. As we pointed out previously (Carr and Kitchell, 1980), the only *mathematical* options whereby an equilibrium solution would be impossible are to portray origination and extinction rates as constantly diverging or exactly parallel. Either possibility, however, seems extremely unlikely. The first condition would result in a runaway system of infinitely increasing (or decreasing) diversity at an accelerating rate. The latter condition would require perfectly balanced functions over all of geological time.

The existence of an equilibrium solution, however, does not mean that an equilibrium diversity is ever reached nor that an equilibrium diversity is maintained. Equilibrium diversity may even be impossible to reach, despite the existence of an equilibrium solution (e.g., when the rate functions define a repelling equilibrium point and the system is unstable).

If a system of diversification remained at equilibrium over evolutionary time, it would display a static point in phase space, the space defined by the axes D_{t+1} vs D_t. By contrast, the empirical or observed pattern of marine metazoan diversification exhibits movement in phase space. Such a pattern represents a nonequilibrium system whose dynamics are known only to the present. Nonequilibrium systems display what Jantsch (1980) has termed "self-organization dynamics." In our model the system is maintained away from its equilibrium by the internal, deterministic mechanism of simultaneously and interactively diversifying faunas, as well as by the external and stochastic mechanisms of perturbations (e.g., mass extinctions) and the evolution of new faunas (evolutionary innovations).

Such a system is both open and closed: open in the sense that evolutionary innovations are possible, but closed in the sense that extinctions are irreversible. Maximum diversity represents the maximum potential of existing faunas to diversify, limited by the historical juxtaposition of simultaneously diversifying faunas. Thermodynamic or spatial limitations to diversity are not directly considered.

For the discrete model of diversification, the length of time the system remains at nonequilibrium is dependent on a) the relative parameter values for each fauna, b) the number of interacting faunas, c) the timing and frequency of invasion of the system by new faunas (evolutionary innovations), and d) the frequency, amplitude and differential effect of perturbations. Total diversity may indefinitely increase under nonequilibrium conditions if appropriate evolutionary innovations are sufficiently frequent, or diversity may attain either a temporary or permanent equilibrium solution in the absence of perturbations and innovations, provided there has been sufficient time for completion of all faunal turnover events.

SUFFICIENT PARAMETERS AND FAUNAL COHESIVENESS

α and β are what are termed sufficient parameters. A sufficient parameter is defined as "a many-to-one transformation of lower-level phenomena" (Levins, 1966). Sufficient parameters are invariant over changes at lower levels, giving sufficient parameters the property of robustness (Wimsatt, 1980). Hence, there is a wide variety of values for lower-level parameters whose *net sums* yield the same values for higher-level sufficient parameters. Hence, although sufficient parameters are time-constant, lower-level parameters may infinitely vary with time, within the bounds of higher-order net constancy.

The discrete model

$$D_{t+1} = D_t + \alpha D_t - \beta D_t^2 \qquad (1)$$

can also be written in terms of lower-level rates of origination and extinction.

Total origination rate (O_t) and total extinction rate (E_t) are defined by the equations

$$O_t = aD_t - bD_t^2 \quad , \tag{2}$$

$$E_t = cD_t + dD_t^2 \tag{3}$$

(see Carr and Kitchell, 1980 for discussion; see also Sepkoski, 1978, 1979). By definition,

$$D_{t+1} = D_t + O_t - E_t \quad ; \qquad \qquad \cdot \tag{4}$$

diversification is Markovian: diversity in any time period is equal to diversity in the preceding time period, plus the change in diversity resulting from net gains (or losses) of total successful originations (O_t) minus total extinctions (E_t). Substitution of equations (2) and (3) into equation (4) yields

$$D_{t+1} = D_t + (a-c)D_t - (b+d)D_t^2 \quad . \tag{5}$$

By defining $\alpha = (a-c)$ and $\beta = (b+d)$, one returns to equation (1), the first-order difference equation representing the logistic growth model. Consequently, there is a nearly infinite number of values of lower-level parameters (a and c) and (b and d) that will yield equivalent values of the higher-level parameters, α and β. Hence, the constancy of sufficient parameters may be giving statistical pattern to the great evolutionary faunas of the Phanerozoic, comprised of numerous lower-level taxa with variable rates.

We postulate that the evolutionary cohesion displayed by each of the three great evolutionary faunas represents similarity of its sufficient parameters, but any number of dissimilar values of the lower level rates of origination and extinction of its constituent taxa. Hence, evolutionary cohesion at the level of the three great evolutionary faunas is not evidence of, nor need imply, similar lower-level rates of origination and extinction. These rates may widely differ, provided their net rates are similar. Similarly, any ecological or phylogenetic relationship among members of an evolutionary fauna may be incidental or epiphenomenal. The evolutionary coherency of each of the three great evolutionary faunas may simply represent grouping by similar sufficient parameter values. This explanation is hierarchical: evolutionary cohesiveness is defined only at the higher level of sufficient parameters. Taxon-specific rates of origination and extinction may offer little insight into the temporal diversification pattern of an evolutionary fauna. A wide variety of taxon-specific rates is possible within the limits of constant sufficient parameters, i.e. within the net rate of initial diversification of an evolutionary fauna (α) and its rate of negative-feedback (β).

TESTING THE MODEL

THE THREE GREAT EVOLUTIONARY FAUNAS

The proposed hypothesis is that the pattern of global marine diversification can be described by a discrete time model. The pattern of global marine diversification we will use as the observed pattern is that of the three great evolutionary faunas of the Phanerozoic marine fossil record, statistically defined at the familial level by Sepkoski (1981). The three great evolutionary faunas are the Vendian-Cambrian fauna (Fauna I), the post Cambrian-Paleozoic fauna (Fauna II), and the Mesozoic-Cenozoic fauna (Fauna III). The statistical coherency of these faunas was identified using factor analysis. Factor analysis identifies taxa whose temporal patterns of diversification are similar, and confirms what is visually apparent.

The bewildering array of clade diagrams of marine taxa (see Figure 1, Sepkoski, 1981), if simply rearranged, appears less bewildering and more like three groups with distinctive histories of diversification (cf. Figure 3). Many of the taxa put into each of the three great evolutionary faunas by factor analysis, however, are minor, and some groups begin to appear in more than one fauna (see Table 1, Sepkoski, 1981). Consequently, we will consider only those taxa that dominate each of the three great evolutionary faunas (Figure 4). Fauna I (actually Factor 1) is comprised of the Trilobita (77%) and "Polychaeta," which, cumulatively, contribute 83% to this fauna. Fauna II is dominated by 5 groups: the Articulata, Crinoidea, Ostracoda, Cephalopoda, and Anthozoa, which cumulatively contribute 86% to this fauna. Fauna III is similarly dominated by 5 groups: the Gastropoda, Bivalvia, Osteichthyes, Malacostraca, and Echinoidea, which account for 83% of this fauna.

In compiling the temporal patterns of diversification of these three great evolutionary faunas, we have plotted a comparison compilation of only the well-preserved or shelly taxa. We refer to this latter compilation as the modified fauna. The difference between the two is the presence or absence of fossils with low fossilization potential: the modified faunas include only groups with good fossilization potential (cf. Raup and Sepkoski, 1982). In Fauna I (modified), the Polychaeta are removed. In Fauna III (modified), the Osteichthyes and Malacostraca are omitted. For Fauna II there is no difference

---

Figure 3 (facing page). Clade diagrams of marine families within metazoan classes. Individual diagrams are from a chart published by Sepkoski (1980). The clade diagrams are now arranged into three evolutionary faunas, in order of their factor scores for each fauna (see Table 1, Sepkoski, 1981). Classes that have loadings on more than one factor are placed in the fauna for which they have the highest factor score.

THE MARINE FOSSIL RECORD

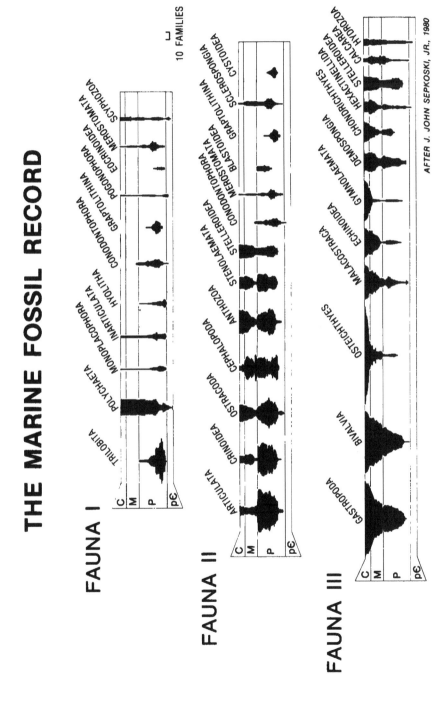

FAUNA I

FAUNA II

FAUNA III

AFTER J. JOHN SEPKOSKI, JR., 1980

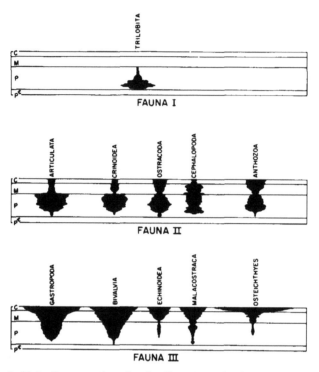

Figure 4. Clade diagrams of marine families within dominant metazoan classes of each evolutionary fauna. These groups represent 77% to 86% of each fauna.

between the total compilation and the modified compilation. The known history of the unmodified and modified evolutionary faunas is shown in Figures 5 and 6, respectively.

MODEL TESTING

We now want to test the model by quantifying the appropriate parameter values for each fauna, using the fossil record. The critical parameter α is estimated from the diversity data for each evolutionary fauna only during its brief initial period of diversification (see following section). Do these actual parameter values produce a pattern that closely replicates the known, observed pattern? There are three possible outcomes to testing the model: (1) the model may be rejected; the fit between simulated and observed patterns of diversification is not good. (The model obviously could be modified and retested as well). Or one of two alternative hypotheses, based on the model, may be rejected: (2) the observed pattern represents internal fluctuations of

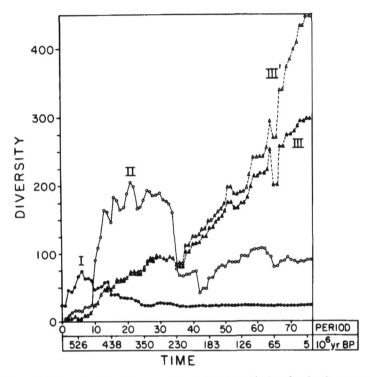

Figure 5. Change in diversity or number of families with time for the three great evolutionary faunas, labeled I, II, and III. Data are plotted for 76 stratigraphic stages, labeled analytical period. Corresponding time in Myr as a function of stratigraphic stages is plotted on a nonlinear scale for comparison. III[1] and III represent evolutionary Fauna III with and without Osteichthyes, respectively.

an unstable oscillatory system; or (3) the observed pattern represents the diversification of a stable system, experiencing external perturbations.

To test between these alternative predictions of the same model, one cannot work to fit the model to the data. Such a method assumes that the model is correct, and uses the model to solve for the unknowns: the parameter values. To test whether or not this model adequately describes the fossil record requires instead that estimates of parameter values be independently derived from data. The test lies in examining the fit of the simulated runs of the model to the observed fossil record, using these restricted values of the critical parameters. If, for example, the resultant behavior of the model is chaotic using the independently determined parameter values, but the observed data are not chaotic in pattern, then the model is inadequate and must be rejected. Alternatively, if the simulated data are stable at these parameter values, but the observed data display unstable fluctuations, then the model is again inadequate, and must be rejected (see Wimsatt, 1980).

Secondly, there are alternative patterns within each of these three categories of behavior. If several faunas are initially present and there are no subsequent innovations, the faunas may display either a) a fixed pattern of constant co-existence, b) a pattern of continuous single fauna dominance and eventual extinction of all other faunas, or c) a pattern of changing dominance and faunal displacement.

The major shortcoming to date of proposed fits of observed diversification patterns to simulated diversification patterns is that parameter values have not been specified (e.g., Sepkoski, 1978, 1979; Sepkoski and Sheehan, 1983). Hoffman (1981) also pointed out this deficiency when stating that "...any curves describing the course of diversity evolution can be produced by manipulation of the total origination and extinction rates." This is a valid criticism: any set of data can be fitted by an endless variety of models, and possibly by a single model if one is free to choose parameter values at will. Such a procedure is best termed curve-fitting, to be distinguished from model-testing. One cannot test a model by running simulations until one arrives at a good fit. The simple exercise of simulating with no attention paid

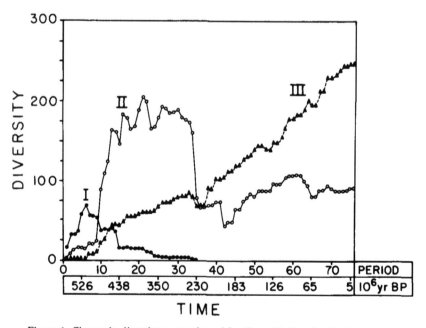

Figure 6. Change in diversity or number of families with time for the three great evolutionary faunas (modified). Trilobita comprise Fauna I. See text for further discussion.

to calculation of parameter values appropriate to the system being examined shows us only a picture book of possibilities, given a different "run" in evolutionary history. The resultant "fit" represents the possible and not the actual, to borrow a phrase from Jacob (1982).

Sepkoski and Sheehan (1983) do reveal that the parameter values of r and $\hat{D}$ (equivalent to α and the ratio of $\alpha:\beta$) for the three great evolutionary faunas have "the qualitative properties that $r_1 > r_2 > r_3$ and $\hat{D}_1 < \hat{D}_2 < \hat{D}_3$." These qualitative statements, however, do not set bounds either on the model's behavior or its numerical solution. The stability behavior of the model, and hence the simulated patterns, are dependent not only on qualitative relationships between parameter values but directly on the quantitative values of the critical parameters. Most parameter values lead to simulated patterns of diversification that have no statistical relationship to the pattern or to the scale of pattern (e.g., see Stanley et al., 1981) observed in the fossil record. In particular, the dynamical behavior of diversification using the discrete model can dramatically vary depending on the quantitative values of its critical parameter, α.

In this study, the rate parameters, α and β, are assumed to be time-invariant. Although it would be a simple matter to model the rates as time-variant, such an analysis could collapse the modeling effort from testing, given prescribed assumptions, to simple curve fitting. The model thereby serves as a strict criterion of subtraction: how much of the observed pattern can be mimicked by pattern produced by the stringent assumptions of constant rates of net initial diversification and constant rates of diversity-dependence for each fauna.

We will quantitatively estimate α for each of the three great evolutionary faunas, by using their initial rates of diversification. We will first compare the simulated results of the entire Phanerozoic pattern with the observed fossil record for the case where there are no perturbations. We will then compare simulated vs. observed for the same parameter values but with the periods of mass extinction included as perturbations. Do these calculated parameter values result in dynamic patterns of diversifications that closely mimic the observed patterns of the three great evolutionary faunas? If the answer is no, the model must be rejected. Goodness-of-fit for case 1 provides an estimate of the significance of inherent or characteristic properties of the faunas represented solely by the model's two parameters. Improvement in the goodness-of-fit between case 1 and case 2 serves as a quantitative estimate of the significance of perturbation to the observed pattern of Phanerozoic diversification.

CALCULATING CRITICAL PARAMETER VALUES

The critical parameter α is estimated from the diversity data for each evolutionary fauna only during its brief initial period of diversification, or continuously increasing diversity. Familial diversity data of marine metazoans from lower Tommotian to Pleistocene time comprise the diversity data for this study and were generously provided by Sepkoski (pers. comm., 1981). Time resolution of these data is predominantly at the stage level; there are 76 intervals of time (see Raup and Sepkoski, 1982). The method of calculating α assumes it to be an exponential growth rate, according to the equation

$$\hat{\alpha} = \frac{\log_e (D_{t_x}) - \log_e (D_{t_y})}{t_x - t_y} . \tag{6}$$

D represents diversity (number of families, in this case), and t_x and t_y represent different intervals of time. This method of calculating α attempts to best measure α before the effects of interaction between simultaneously diversifying faunas are felt, and before periods of mass extinction have ensued. Because taxa are not diversifying in an ecological and evolutionary vacuum, α values will be underestimated. Hence, (t_y) for Fauna I occurs at one-half maximum diversity, the maximum duration of time for exponential growth. T_y for Fauna II extends to the first plateau in its diversity rise, and t_y for Fauna III extends to the subsequent plateau in diversity rise.

To show that the model is robust to the absolute time used for the analytical interval of discrete time, we have calculated α for two measures of time: time in stage intervals (the finest temporal resolution of the data), and time in Myr intervals. The estimated $\hat{\alpha}$ values for stage-level intervals of time are 0.852 for Fauna I (or 1.73 if only trilobites are considered for Fauna I), 0.331 for Fauna II, and 0.115 for Fauna III. If time is in Myr intervals, then the estimates of $\hat{\alpha}$ are 0.131, 0.037, and 0.014 for Faunas I, II, and III, respectively. Because the length of the time step in absolute time can vary for any given set of simulations, provided there is equivalency in scale for calculating α, we will first run the model using a stage-level time step. We will then run the model with a 1 Myr time step.

If faunas were not interacting, then knowledge of α and maximum diversity would allow one to solve directly for β: $\beta = \alpha/\hat{D}$. But faunas are interacting in the coupled model (see below). β values were consequently estimated as a proportion of β if calculated using maximum diversity and decoupled or noninteractive diversification. The estimates of β that best achieve the maximum diversity of these simultaneously interacting faunas are 0.6×10^{-2} (Fauna I), 0.1×10^{-2} (Fauna II), and 0.335×10^{-3} (Fauna III). Initial D equals one for each fauna.

COUPLED vs. DECOUPLED MODE

The discrete time model can be implemented as either a coupled or a

INDEPENDENT DIVERSIFICATION

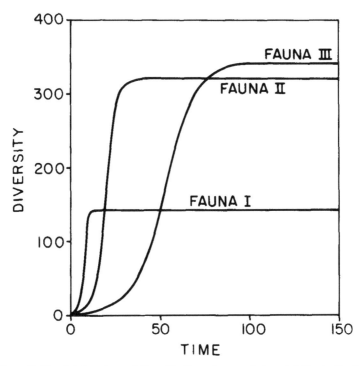

Figure 7. Simulated pattern of diversification of uncoupled model, illustrating lack of correspondence between observed diversification pattern and a model of independent or non-interactive diversification. α and β values are the same as for Figure 8.

decoupled model of diversification: faunas can be either diversity-dependent and interactive, or diversity-independent and noninteractive. We have been assuming a coupled model. We can, however, test which mode of the model is more appropriate. The simplest implementation of the model assumes no interactions between faunas. Diversity-independent diversification treats each fauna separately by specifying that the negative feedback parameter (βD_t^2) is intrafaunal rather than inter- as well as intra-faunal. The output of the de-coupled simulation, inputting the same parameter values for the three great evolutionary faunas as calculated for the subsequent coupled simulation, is shown in Figure 7. As is obvious, the simulated pattern does not resemble the actual pattern. The decoupled or diversity-independent form of the model of diversification can be rejected. We now construct the model specifying

interactive diversity-dependent diversification by allowing all faunas to interact through the D_t^2 term (see Sepkoski, 1978, 1979), and test the appropriate-
ness of this coupled discrete model for replicating the observed pattern of
global marine diversification.

GENERAL BEHAVIOR OF THE COUPLED MODEL

The coupled discrete model of diversification describes the process of diversification as resulting from differential rates of origination-extinction, and differential diversity-dependent feedback with simultaneously diversifying taxa.
Decline of a fauna does not allow expansion of a subsequent fauna (faunal replacement) so much as expansion of a fauna forces the decline of a previously
diverse fauna (faunal displacement).

Because the interaction term is incorporated in the diversity-dependent negative feedback term (βD_t^2) where D_t represents the combined diversity of all
taxa, the greater the quantitative value of β for a given group the more rapid
will be that group's displacement. Linear stability analysis (e.g., May, 1974)
shows that displacement is predictable: a taxon or fauna with a higher $\alpha:\beta$ ratio will displace faunas with lower ratio values. Faunas may indefinitely coex-
ist only if all ratios of $\alpha:\beta$ are identical. The persistence of a new fauna is dependent on its relative $\alpha:\beta$ ratio. The more closely this ratio matches that of
the most diverse fauna, the longer the persistence time before displacement.

Such a model of diversification acts only in an analogous manner with the
classic Lotka-Volterra equations of competitive interaction between species,
in that the ratio of $\alpha:\beta$ for each fauna affects the potential maximum diversity for all other faunas. In a similar manner, the competition coefficients are
constants describing the negative effect of each species on the potential maximum population size of other species.

If two taxa have equal α values (equal initial rates of divergence between
origination and extinction), then that taxon with the smaller β value will displace the other fauna, resulting in faunal turnover. If two taxa have equal β
values, then it follows that the taxon with the greater α value will be the displacer taxon during faunal turnover. If a fauna has both the highest α and the
lowest β value, this fauna will persistently dominate and there will be no long-
term establishment of other faunas. Only the trade-off between α and β values
between faunas allows for faunal turnover.

To what extent does rate of faunal turnover depend on the rate of diversification of a subsequent fauna? Both time to faunal turnover and time to faunal extinction depend on the absolute difference between α and β values for
each fauna. If β values are equal, then both time to faunal turnover and time
to faunal extinction increase with decreasing differences in α between two
faunas. If α values are constant among interacting faunas, then time to faunal

turnover and time to faunal extinction decrease with decreasing β differences. A decrease in α of the displacing fauna will increase the time to faunal turnover and extinction time of the displaced fauna, whereas a decrease in the β value of the displacing fauna decreases time to turnover and extinction.

EFFECT OF MULTIPLE FAUNAS

How are the dynamics of multiple faunas different from the dynamics of single faunas? In our earlier study (Carr and Kitchell, 1980), we were concerned only with the dynamics of total diversity. We are now concerned with the dynamics of individual faunas diversifying interactively. How does the behavior of the model change when multiple faunas diversify simultaneously? We will show that the simple addition of faunas, by increasing the quantitative sum of α, drives the system closer to instability. When taxa diversify interactively, the α values are additive. Consequently, the conditions necessary to generate unstable behavior are more easily attained in multiple fauna mathematical systems.

The coupling of multiple faunas can cause the model to move through the progression of behaviors from stable to unstable to chaotic, due not to an increasing α value for any single fauna, but by summation of total α for the system. An example is provided in Figure 8, where each fauna has the same time-constant α value of 1.0. As the system proceeds from a single fauna to adding a second fauna, to incorporating a third, and finally a fourth fauna, the behavior of *each* fauna dramatically changes, from simple logistic increase in diversity over time (Figure 8A), to damping oscillations (Figure 8B), to apparent disorder (Figure 8C), and rapid extinction (Figure 8D), as an oscillation drives through zero diversity.

If α values are additive, then the coupling of faunas places more severe limits on rates of diversification per fauna if the system is not to cross into stability fields of unstable behavior. Because we are dealing with three evolutionary faunas, we would predict a priori that the system may manifest unstable behavior. Even if each fauna would display stable behavior if independently diversifying, it is increasingly likely to display less stable behavior as additional faunas are coupled.

TEST OF THE MODEL: NO PERTURBATIONS

By inputting the estimated values of α, and iterating over stages of evolutionary time, the simulated output mimics the essential features of the observed pattern at fixed β values (Figure 9, upper left panel). Fauna I is initially successful, achieving highest diversity at low total diversities, and is then displaced by Fauna II, which dominates at intermediate total diversities. In the

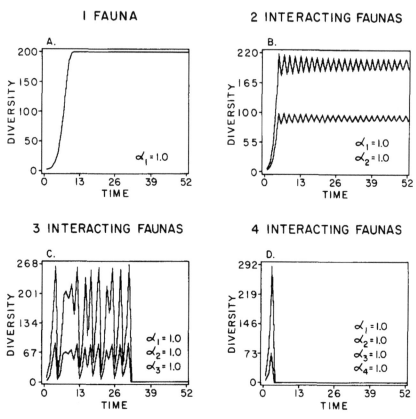

Figure 8. Change in behavior of diversification as faunas diversify interactively, due to increased sum of α. A, stable diversification, nonoscillatory logistic growth. B, stable diversification, oscillatory convergence. C, unstable diversification, chaos. D, unstable diversification, rapid extinction. Upper pattern is total diversity; lower pattern is diversification of a single fauna, superimposed for cases B, C, and D due to identical α and β parameter values.

Figure 9 (facing page). Effects of perturbations on diversification of three great evolutionary faunas, I, II, and III. Upper left, model simulation with no perturbations. Upper middle, model simulation with six mass extinction events. Upper right, model simulation without Guadalupian and Dzhulfian extinction events. Lower left, model simulation with same timing of mass extinction perturbations, but diversity decline per perturbation per fauna is equal. Lower middle, model simulation with sequence of mass extinction events in reverse order within both the Paleozoic and Mesozoic-Cenozoic. Lower right, model simulation with differential effect of perturbations on Faunas II and III reversed. α and β values are the same for all simulations.

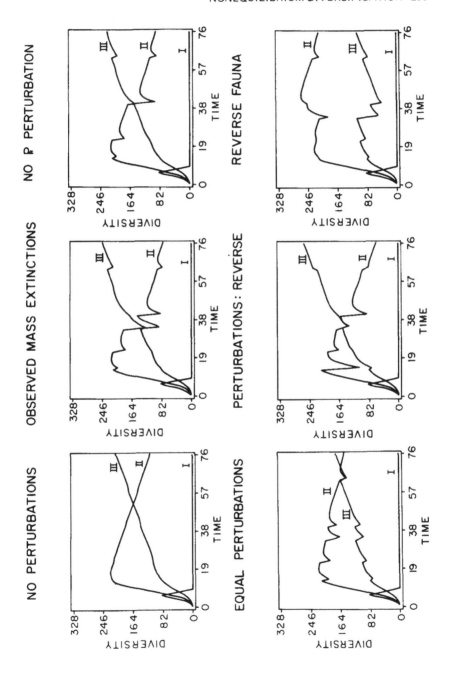

absence of any perturbations, Fauna II is subsequently displaced by Fauna III, which will achieve maximum diversity at high total diversities.

The entire behavior of the system is stable. The single exception is when only trilobites comprise Fauna I. The relatively high α value of trilobites now produces an overshoot of maximum diversity, resulting in a short-lived pattern of convergent oscillation.

With no perturbations, the correlation coefficient between the observed pattern of diversification (unmodified) and the simulated pattern is 0.72 for Fauna I, 0.79 for Fauna II, and 0.87 for Fauna III. Major differences between simulated and observed are that the modelled initial diversification of Faunas II and III rises too rapidly and, at faunal turnover between Faunas II and III, modelled Fauna III has a higher diversity.

TEST OF MODEL: WITH PERTURBATIONS

We first want quantitatively to assess the effect of major perturbations on diversification patterns. We will show that addition of the six mass extinction episodes as perturbations produces a simulated pattern that remarkably replicates the observed pattern. More importantly, we will show that faunal turnover is not dependent on perturbations, whose major role has been to maintain the system at nonequilibrium throughout the entire time corresponding to the Phanerozoic.

Background extinction operates continuously throughout all iterations of the model. It is represented by the two lower-level parameters of the total extinction rate equation, $(c+d)$, that comprise part of both the higher-level parameters, α and β. Mass extinction episodes represent brief periods of time when the frequency of extinction is suddenly accelerated. Raup and Sepkoski (1982) identified four mass extinction periods—the Ashgillian, Guadalupian, Dzhulfian, and Maastrichtian—as statistically distinct from background extinction at the 99% level. The Norian extinction event is significantly distinct at the 95% level. The Late Devonian (Frasnian) event is also unusually high, and will be included as a mass extinction event. The action of any perturbation in these simulations is limited to one time iteration or a stage duration of absolute time, because each of the mass extinctions occurred within a single stratigraphic stage (Raup and Sepkoski, 1982). The degree of perturbation associated with each period of mass extinction per fauna as a % decline in diversity during a single stage interval of time is given in Table 1. The perturbation algorithm for each evolutionary fauna is simply $D_{t+1} = D_t (1 - \text{fractional faunal decline})$.

These simulations even more satisfactorily replicate the observed pattern of diversity change over Phanerozoic time (Figure 9, upper middle panel). The correlation coefficient or goodness-of-fit (using least-squares linear regression)

Table 1. Relative diversity declines for each evolutionary fauna
(unmodified) and each stage of a mass extinction event.

Extinction Stage	Fauna I	Fauna II	Fauna III
Ashgillian	-34%	-9%	-2%
Frasnian (Givetian-Famennian)	-19%	-16%	–
Guadalupian	-8%	-52%	-11%
Dzhulfian	-4%	-14%	–
Norian	–	-44%	–
Maastrichtian	–	-14%	-8%

between the simulated numerical solution of the model and the observed pattern is 0.92 for both Fauna II and Fauna III (Figure 10). The major times of poor correlation for Fauna III are not the times when perturbations are active, but the first six and the last six time intervals (pull of the Recent; Raup, 1972) when predicted diversity exceeds and falls short of the observed diversity, respectively. Addition of the perturbations to the simulation explains 20% more of the pattern of Fauna II in comparison with the simulation with no perturbations. The fit is unchanged for Fauna I (0.72) because its simulated history chiefly precedes the timing of the first perturbation. The two most significant effects of perturbations are that (1) the timing of faunal turnover between Faunas II and III is accelerated to the stage interval corresponding to the Dzhulfian, and (2) the overall pattern of Fauna II during the time interval corresponding to the Paleozoic is flattened.

Fauna II displays a relatively level diversity curve during its period of dominance during the Paleozoic. This flatness has previously been interpreted as evidence of an equilibrium diversity (e.g., Sepkoski, 1978, 1979, and numerous references to this study, e.g., Gould, 1980). The simulations, with vs. without perturbations, suggest that an apparent equilibrium may be related to perturbations. Both the rate of rise and the rate of descent of the diversity curve may be slowed down by perturbations, giving the curve a more pronounced appearance of steady-state diversity. The apparent stability in diversity during the Paleozoic may represent a flattening of diversity resulting from closely spaced perturbations and the coincident timing of the perturbations across the humped portion of Fauna II's diversity curve. Hence, the apparent Paleozoic equilibrium in diversity may more properly represent a prolonged nonequilibrium diversity. Ironically, if small perturbations are frequent enough, their effect is to so slow down the rate of faunal displacement that the resultant pattern can be mistaken for stability: prolonged nonequilibrium can resemble equilibrium.

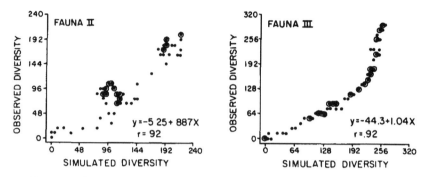

Figure 10. Plot of observed diversity from fossil record versus simulated diversity with mass extinction perturbations, for Faunas II and III.

To show that the analysis is robust with respect to time interval, we also tested the numerical solution of the model when α is calculated as a rate per Myr, and the time step of the iteration is 1 Myr. The observed pattern of diversification of the three evolutionary faunas and the simulated pattern, plotted over the same time scale of 570 Myr, closely agree. The goodness-of-fit, determined by linear correlation between observed and predicted including the mass extinction perturbations, is now 0.79 for Fauna II and 0.83 for Fauna III, or 0.91 for Fauna III, modified. Simulated Fauna I, however, is now too rapidly displaced, and the correlation coefficient is only 0.45.

PERTURBATIONS AND FAUNAL TURNOVER

We now want to test the hypothesis that perturbations have played a significant causal role in faunal turnover. It has been suggested that perturbations (abrupt declines in diversity) during major mass extinction episodes account for the taxonomic turnover observed during these extinction events. Examples include the replacement of the great evolutionary fauna II by III during the terminal Permian event (see Sepkoski and Sheehan, 1983), and the replacement of brachiopods by clams during the terminal Permian event (see Gould and Calloway, 1980). An alternative possibility is that the timing of the terminal Permian extinction event may have coincided with the timing of a faunal turnover event. To assess the potential role of perturbations in faunal turnover events, a series of simultations were run in which the timing, magnitude, and differential effect of perturbations were varied.

The Permian event has been called "the pivotal event" that "destroyed the Paleozoic stability and ushered in the Mesozoic-Cenozoic diversification" (Sepkoski and Sheehan, 1983). We will test this prediction by asking, What if the Permian event is completely eliminated from the diversification process? The results obtained after eliminating both the Dzhulfian and Guadalupian extinctions are shown in Figure 9 (upper right panel). The overall effect is surprisingly minor, given the magnitude of this extinction episode. Faunal turnover still occurs: Fauna III displaces Fauna II. Only the timing of the turnover event is delayed, and even the delay is moderate (9% of total time). Hence, deletion of even the terminal Permian crisis, quantitatively the most severe perturbation, has a nonsignificant effect on faunal turnover.

What would happen to the patterns of diversification of the three evolutionary faunas if all perturbations were equal in magnitude for each period of mass extinction and for each fauna? The mean decline in diversity for all mass extinction events on all faunas is 12%. Using this value for each perturbation, the major simulated patterns are generally unchanged. The dominant effect of equal perturbations is a delay in the timing of faunal turnover between Faunas II and III (Figure 9, lower left panel). Time to displacement is now substantially prolonged, from iteration (i.e., stage) 34 to 64.

What is the effect on diversification patterns if the mass extinction perturbations are run in reverse order? The terminal Permian event now happens first within the Paleozoic sequence of mass extinctions, and the terminal Cretaceous event now occurs first within the Mesozoic-Cenozoic sequence. The simulated pattern is obviously changed in detail, but again the major pattern persists (Figure 9, lower middle panel). Moreover, faunal turnover between Faunas II and III is unchanged: the time of turnover is at the same time as when the perturbations are in correct sequence.

Finally, the differential effects of the observed perturbations on each of the three evolutionary faunas was switched so that Fauna II experienced the relatively minor diversity declines actually experienced by Fauna III, whereas Fauna III experienced the more substantial diversity declines actually experienced by Fauna II. Time to faunal turnover is now significantly delayed (Figure 9, lower right panel). However, it is highly improbable that two of the above possibilities—reversal of the differential effect of perturbations or equal diversity responses among faunas—would occur. The phase plane plots of Faunas I, II, and III illustrate that an equal perturbation (e.g., from D_p to $\hat{D}$) will affect each fauna differentially (Figure 11). Fauna I has the most steeply humped phase curve due to its high α parameter value. Fauna III has the least steeply humped phase curve. The arrows outlining the time paths of diversity response to a fixed perturbation illustrate that the effect will be largest for Fauna I, intermediate for Fauna II, and least severe for Fauna III. This prediction matches the observed: the same mass extinction event produces differential diversity responses in this order among the three evolutionary faunas.

In summary, the timing, magnitude, and differential effect of perturbations generally prolong the time to faunal turnover. In all cases, however, the

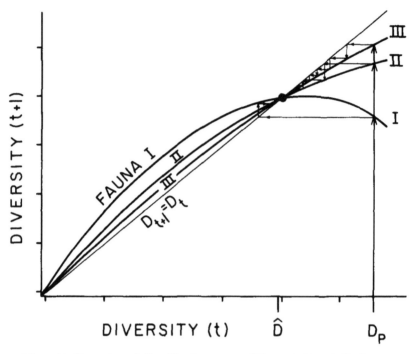

Figure 11. Phase plots of diversification curves of Faunas I, II, and III, showing differences in steepness. Intersection of phase curves with 45° line labeled $D_{t+1} = D_t$ represents the equilibrium point. A given perturbation, from D_p to $\hat{D}$, results in different diversity trajectories for each fauna, as illustrated by arrows. Fauna I experiences most severe and rapid diversity decline.

eventual outcome was not altered: Fauna II displaces Fauna I, and is in turn displaced by Fauna III. By prolonging the evolutionary persistence of a subsequently displaced fauna, perturbations thereby maintain the interactions of diversifying faunas at nonequilibrium. Perturbations rarely change the basic pattern of diversification (see also Sepkoski, 1980). Timing, magnitude, and differential effect of perturbations determine the detailed nonequilibrium pattern of diversification, but have no effect on its final solution.

DISCUSSION

Our analysis of Phanerozoic marine diversification patterns indicates that diversity may have remained in continual nonequilibrium rather than achieved early equilibrium. Nonequilibrium has been prolonged by faunal displacement dynamics and disruption. Future evolutionary innovations (a fourth evolutionary fauna?) or continued perturbations are shown to be adequate to maintain the system at continual nonequilibrium. Mass extinction perturbations are shown to represent extrinsic perturbations to a stable system at nonequilibrium, and not oscillations of an unstable or chaotic system. A similar result has been found for growth rates of natural populations (e.g., Hassel et al., 1976; Mueller and Ayala, 1981; it has been suggested that instability due to high rates of population growth may represent a selective force that keeps parameter values low and such systems stable [Thomas et al., 1980]).

The model of diversification quantitatively describes the time trajectory of each evolutionary fauna, a complex product of characteristic parameter values, interaction with simultaneously diversifying faunas, and differential responses to perturbations. Because perturbations have differential rather than equivalent effects on each fauna, perturbations do not simply "reset" conditions but change conditions. Perturbations act principally on the rate of faunal displacement. Although Sepkoski and Sheehan (1983) have postulated that "incorporation of time-specific perturbations into the model *accounts for.* . .the taxonomic turnover associated with major extinction events," this study demonstrates that mass extinction perturbations do not cause (i.e., "account for") taxonomic turnover. Taxonomic turnover is generally delayed by perturbations, rather than precipitated by them.

Our analysis emphasizes cohesive evolutionary (i.e., speciation and extinction rate) characteristics, rather than cohesive ecological characteristics. Fauna I can be typified as a "hedonist" evolutionary fauna: it has the highest rate of initial diversification and the highest degree of diversity-dependent negative feedback. Fauna I is also most likely to display oscillatory overshoots. It is most sensitive to increasing levels of diversity (highest β value) and to perturbations (most steeply humped phase curve, or highest α value). "Hedonist" characteristics may be related to a high tendency to isolate.

By contrast, Fauna III can be characterized as evolutionarily "stoic." It has the lowest rate of net diversification (least humped phase curve) and lowest rate of negative feedback. Consequently, it is least sensitive to increasing levels of diversity and to perturbations. "Stoic" characteristics may be associated, for some taxa, with a low tendency to isolate.

Displacement, of course, does not infer adaptive "superiority" of the displacing fauna with respect to the displaced fauna. Instead, the displacer is superior only in the sense that its realized diversity—during a particular period of evolutionary time—is higher. Since total diversity changes over time, displacement dynamics ensue.

The consequence of Fauna II displacing Fauna I, and Fauna III subsequently displacing Fauna II, is that both α and β values over evolutionary time have proportionately decreased. Hence, overall stability of the system to perturbations is predicted to have increased. Such a conclusion is consistent with the empirically observed decrease in extinction rate over Phanerozoic time (Raup and Sepkoski, 1982; Van Valen, 1984; Kitchell and Pena, 1984). In modeling the three evolutionary faunas, both α and β have proportionately decreased, meaning that both the total origination rate and total extinction rate curves have decreased in slope, or that only the total extinction rate curve has decreased. Either case results in a less steep phase curve and the intersection of these rate curves (where $D_t = D_{t+1} = \hat{D}$) at progressively higher total diversity values. The overall result is increased diversity and increasing stability.

SUMMARY

We have formulated and tested a deductive model of diversification, using discrete mathematics, and satisfactorily recreated the Phanerozoic pattern of marine diversification. The model's assumptions regarding nonlinearity, diversity-dependent feedback, and nonequilibrium are quite simple. The analysis demonstrates the potential of interactive diversification to mimic the known record of faunal turnover. Moreover, perturbations combined with evolutionary innovations and turnover maintained the system at nonequilibrium throughout the entire Phanerozoic.

Although the model is admittedly simplified, its complex dynamics, which include apparent randomness, cause one to wonder at what complexity of behaviors would be manifested by more realistic models, models that include environmental change or evolving rather than constant rate sufficient parameters (see Wimsatt, 1980). Whereas other paleobiologists have warned against the tendency to see determinism in pattern when there may be none, we also have shown that one must be on guard against the tendency to assume that deterministic processes generate only orderly patterns. We have provided a measure of subtraction for simple, complex, and chaotic patterns of diversification that can result from a single model of diversification, but different values for the critical parameter, α. However, to show that the Phanerozoic pattern of diversification can be fit by a deterministic (logistic) model is not to say that the causal mechanisms of diversification are determined by these simple rules. Such an analysis simply says that the pattern does not require an interpretation based on an assumption of randomness. The model, moreover, does not address which species within a fauna persist and which go extinct. This question requires a different level of analysis, associated with the palimpsest of patterns at different taxonomic levels. At the species level, patterns may be quite different, and may require a different model.

ACKNOWLEDGMENTS

We are indebted to Jack Sepkoski for generously providing unpublished data. We thank Antoni Hoffman and Jack Sepkoski for their reviews of this manuscript. We express our gratitude to James Valentine for organizing the symposium (June 1982) and this symposium volume. We thank Stephen Hewett for writing a Pascal program to implement our model. J.A.K. was supported by NSF grant DEB-8109914. Manuscript submitted August 9, 1982.

REFERENCES

Allen, P. M., 1976. Evolution population dynamics and stability. Natl. Acad. Sci. USA. 73:665-668.

Anderson, S., and Anderson, C. S., 1975. Three Monte Carlo models of faunal evolution. Am. Mus. Novit. (2563), 6.

Barnett, V. D., 1962. The Monte Carlo solution of a competing species problem. Biometrics 18:76-103.

Bartlett, M. S., 1960. Stochastic population models. London: Methuen & Co.

Carr, T. R., and Kitchell, J. A., 1980. Dynamics of taxonomic diversity. Paleobiology 6:427-443.

Caswell, H., 1978. Predator-mediated coexistence: A nonequilibrium model. Am. Nat. 112:127-154.

Dover, G., 1982. Molecular drive: A cohesive mode of species evolution. Nature 299:111-117.

Feigenbaum, J., 1980. Universal behavior in non-linear systems. Los Alamos Science/Summer 4-27.

Gould, S. J., 1980. G. G. Simpson, paleontology, and the modern synthesis. In Mayr, E., and Provine, W. B. (eds.), The evolutionary synthesis, perspectives on the unification of biology, Cambridge, Mass.: Harvard Univ. Press 153-172.

Gould, S. J., and Calloway, C. B., 1980. Clams and brachiopods—ships that pass in the night. Paleobiology 6:383-396.

Gould, S. J., Raup, D. M., Sepkoski, J. J., Jr., Schopf, T. J. M., and Simberloff, D. S., 1977. The shape of evolution: A comparison of real and random clades. Paleobiology 3:23-40.

Guckenheimer, J., Oster, G. R., and Ipaktchi, A., 1977. The dynamics of density-dependent population models. Math. Biol. J. 4:101-147.

Hassell, M., Lawton, J., and May, R. M., 1976. Pattern of dynamical behavior in single-species populations. Anim. Ecol. J. 45:471-486.

Hoffman, A., 1981. Stochastic versus deterministic approach to paleontology: The question of scaling or metaphysics? N. Jb. Geol. Palaont. Abh. 162: 80-96.

Hofstadter, D. R., 1981. Metamagical themas. Sci. Amer. 245 (5):22-43.

Hutchinson, G. E., 1978. An introduction to population ecology. New Haven, Conn.: Yale Univ. Press, 260.

Jacob, F., 1982. The possible and the actual. Seattle, Wash.: Univ. of Washington Press, 96.

Jantsch, E., 1980. The self-organizing universe. New York: Pergamon Press.

Kitchell, J. A., and Pena, D., 1984. Periodicity of extinctions in the geologic past: Deterministic versus stochastic explanations. Science 226:689-692.

Levins, R., 1966. The strategy of model building in population biology. Am. Sci. 54:421-431.

Levins, R., 1979. Coexistence in a variable environment. Am. Nat. 114:765-783.

Li, T.-Y., and Yorke, J. A., 1975. Period three implies chaos. Am. Math. Monthly 82:985-992.

Mackey, M. C., and Glass, L., 1977. Oscillation and chaos in physiological control systems. Science 197:287-289.

May, R. M., 1974. Biological populations with nonoverlapping generations: stable points, stable cycles, and chaos. Science 186:645-647.

May, R. M. 1975. Biological populations obeying difference equations: stable points, stable cycles, and chaos. Theor. Biol. J. 51:511-524.

May, R. M., 1976. Simple mathematical models with very complicated dynamics. Nature 261:459-467.

May, R. M., 1979. The structure and dynamics of ecological communities. In Anderson, R. M., Turner, B. D., and Taylor, L. R. (eds.), Population dynamics, Oxford: Blackwell Scientific Publications, 385-407.

May, R. M., and Oster, G. F., 1976. Bifurcations and dynamic complexity in simple ecological models. Am. Nat. 110:573-599.

Mueller, L. D., and Ayala, F. J., 1981. Dynamics of a single-species population growth: Stability or chaos? Ecology 62:1148-1154.

Nicolis, G., and Prigogine, I., 1977. Self-organization in nonequilibrium systems: From dissipative structures to order through fluctuations. New York: John Wiley & Sons.

Pielou, E. C., 1977. Mathematical ecology. New York: John Wiley & Sons.

Pounder, J. R., and Rogers, T. D., 1980. The geometry of chaos: Dynamics of a nonlinear second-order difference equation. Math. Biol. Bull. 42: 551-597.

Prigogine, I., Allen, P. M., and Herman, R., 1977. Long term trends and the evolution of complexity. In Laszlo, E., and Bierman, J. (eds.), Goals in a global community, New York: Pergamon Press, 1-63.

Prigogine, I., Nicolis, G., and Babloyantz, A., 1972. Thermodynamics of evolution. I, II. Phys. Today 25(11):23-28; (12):38-44.

Raup, D. M., 1972. Taxonomic diversity during the Phanerozoic. Science 177:1065-1071.

Raup, D. M., 1977a. Probabilistic models in evolutionary paleobiology. Am. Sci. 65:50-57.

Raup, D. M., 1977b. Stochastic models in evolutionary paleontology. In Hallam, A. (ed.), Patterns of evolution, New York: Elsevier, 59-78.

Raup, D. M., 1981. Extinction: Bad genes or bad luck? Acta Geol. Hispanica 16:25-33.

Raup, D. M., and Gould, S. J., 1974. Stochastic simulation and evolution of morphology—towards a nomothetic paleontology. Syst. Zool. 23:305-322.

Raup, D. M., Gould, S. J., Schopf, T. J. M., and Simberloff, D. S., 1973. Stochastic models of phylogeny and the evolution of diversity. J. Geol. 81: 525-542.

Raup, D. M., and Sepkoski, J. J., Jr., 1982. Mass extinctions in the marine fossil record. Science 215:1501-1502.

Roughgarden, J., 1979. Theory of population genetics and evolutionary ecology: An introduction. New York: MacMillan Publ. Co., 634.

Schopf, T. J. M., 1979. Evolving paleontological views on deterministic and stochastic approaches. Paleobiology 5:337-352.

Sepkoski, J. J., Jr., 1978. A kinetic model of Phanerozoic taxonomic diversity. I. Analysis of marine orders. Paleobiology 5:337-352.

Sepkoski, J. J., Jr., 1979. A kinetic model of Phanerozoic taxonomic diversity. II. Early Phanerozoic families and multiple equilibria. Paleobiology 5:222-251.

Sepkoski, J. J., Jr., 1980. Evolution of taxonomic diversity in the oceans. Unpubl. ms.

Sepkoski, J. J., Jr., 1980. The three great evolutionary faunas of the Phanerozoic marine fossil record. Geol. Soc. Amer. Ann. Meeting:520 (abst.).

Sepkoski, J. J., Jr., 1981. A factor analytic description of the Phanerozoic marine fossil record. Paleobiology 7:36-53.

Sepkoski, J. J., Jr., and Sheehan, P. M., 1983. Diversification, faunal change, and community replacement during the Ordovician radiations. In Tevesz, M. J. S., and McCall, P. L. (eds.), Biotic interactions in recent and fossil benthic communities, New York: Plenum Press (in press).

Smith, R. H., and Mead, R., 1980. The dynamics of discrete-time stochastic models of population growth. Theor. Biol. J. 86:607-627.

Sparrow, C., 1980. Bifurcations and chaotic behavior in simple feedback systems. Theor. Biol. J. 83:93-105.

Stanley, S. M., Signor, P. W., III, Lidgard, S., and Karr, A. F., 1981. Natural clades differ from "random" clades: Simulations and analyses. Paleobiology 7:115-127.

Thomas, W. R., Pomerantz, M. J., and Gilpin, M. E., 1980. Chaos, asymmetric growth, and group selection for dynamical stability. Ecology 61:1312-1320.

Vandermeer, J., 1981. Elementary mathematical ecology. New York: John Wiley and Sons, 294.

Van Valen, L., 1984. A resetting of Phanerozoic community evolution. Nature 307:50-52.

Wimsatt, C., 1980. Randomness and perceived-randomness in evolutionary biology. Synthese 43:287-329.

Chapter 9

DIVERSIFICATION FUNCTIONS
AND THE RATE OF TAXONOMIC EVOLUTION

TIMOTHY DANE WALKER

Department of Geological Sciences, University of California, Santa Barbara

INTRODUCTION

Certain data of taxonomic turnover are analogous to data of population biology; for example, first and last occurrences of taxa correspond to births and deaths, respectively, of individuals in a population. This has proven to be an important analogy because there is a well-developed mathematical theory of population ecology (see Pielou, 1977; May, 1981), which includes models of the growth of populations and their interactions with populations of other species. From a formal viewpoint, the analogy indicates that many of the analytical strategies, models and methods which have already been developed for population biology are at least potentially applicable to the data of taxonomic turnover. But from a biological point of view, the analogy between populations composed of individuals and ecosystems composed of species is very imperfect. This necessitates an increasing conceptual independence and sophistication in paleobiological models; and indeed this is what we have witnessed in the last few years.

Two elementary population growth models—the exponential and logistic equations—have been proposed as models of taxonomic diversification in adaptive radiations. Following a critical discussion of these two models and their uses by previous workers, I shall give an outline of some new mathematical models of diversification in adaptive radiations. One of the conclusions to

©1985 by Princeton University Press
Phanerozoic Diversity Patterns:
Profiles in Macroevolution

be drawn from this analysis is that if diversification is modeled as a process *sui generis*, the predicted patterns do not conform to those predicted by analogy to simple population growth models. Not only is this exactly what we expect, given the imperfection of the analogy, but it also allows us to envision a test of alternative hypotheses.

Every distinct diversification model entails a different method of calculating the intrinsic rate of diversification. Therefore, the discussion of diversification models must often lead us to consider one of the perennial themes of paleobiology: the rate of taxonomic evolution. There are several ways of measuring this rate (not all of which will be mentioned here), and it is a matter of great theoretical interest to know whether any of these is meaningful in terms of evolutionary processes. The analysis of diversification models can help to elucidate these processes, and therefore to see how the rate of taxonomic evolution should be measured.

If patterns predicted by a diversification model are reflected in the fossil record, this would suggest a correspondence between the model process and the natural process which created that record. This kind of inference is valid only if the models are realistic; therefore such models should be based on defensible propositions about the natural process itself. The models developed here can further be characterized as general and strategic, as opposed to precise and tactical (for a discussion of these distinctions see May, 1974). They incorporate only those factors which affect adaptive radiations in general, rather than going into the details of a particular taxonomic group or a particular span of geological time. Although the models generate quantitative predictions, their range of applicability is restricted by many simplifying assumptions which are often not independently verifiable. Attempts to find empirical support for such models can therefore yield only suggestive results; but this is a degree of testability appropriate to general models.

PREVIOUSLY PROPOSED MODELS

A diversification function is a mathematical statement of the trajectory of taxonomic diversification, expressed as the number of taxa as a function of time, or $N(t)$. The concept of a diversification function was introduced to paleontology by Zeuner (1946), who originated the device of drawing straight lines between data points on a graph of numbers of taxa (y-axis) versus time (x-axis). This kind of graph, which Zeuner called a time-frequency curve, was soon adopted by paleobiologists (Simpson, 1952; Newell, 1952) and has since been widely used (see, for example, papers in Hallam, 1977). The representation of $N(t)$ by a series of straight lines is merely a convention, and its use clearly does not imply the belief that $N(t)$ is really a linear function or a series of linear functions. Commonly used methods of calculating the rate of taxonomic evolution as the slope of the time-frequency curve are, similarly, noninterpretive summary statistics.

By contrast, the logistic and exponential models appear to be based on sound biological theory and to reveal the objective attributes of monophyletic groups of taxa. The analogy between diversification and population growth has been suggested many times (for example: Yule, 1924; Zeuner, 1946; Small, 1950; Cailleux, 1950, 1954; Cisne, 1971; Stanley, 1973, 1974). The usefulness of these diversification models was perceived by Stanley (1975, 1979), who showed that the value of the Malthusian parameter, if it could be estimated, might be a measure of the intrinsic ability of a clade to diversify. Stanley used the simple (undamped) exponential growth equation to specify a method of measuring this parameter (symbolized here by the letter "r") and used calculated r values as a basis for comparing a variety of taxonomic groups. The calculation of r values has since become a growth industry in paleobiology because of the comparative framework erected by Stanley (1979) and also because realistic r values are required as parameters for mathematical and simulation models (see, for example, Walker and Valentine, 1984).

Stanley's method is an important contribution, and his analysis sufficed to bear the weight of his conclusions. However, when we wish to use r values for other than comparative purposes, we must be concerned with the accuracy of individual r values. An important bias is introduced by Stanley's exclusive use of the simple exponential model. As Stanley recognized, "there must be a general tendency for calculated values of [r] to represent underestimates of exponential rates, because some radiations will have followed distinctly sigmoid paths during the interval evaluated. . ." (1979). For those radiations which do follow a sigmoid path, the appropriate diversification function is the damped exponential, or logistic, equation. In this section I will develop an approach to the determination of r which takes account of the possibility that the course of a radiation may be best described by either of these previously proposed diversification functions. The simple exponential equation is a special case of the logistic; I will consider the more general case first.

The logistic model states that the rate of change of N is proportional to N, but that the rate is damped by diversity dependent factors as N approaches some limiting or equilibrium value K. This may be written in the familiar form

$$\frac{dN}{dt} = r(1 - \frac{N}{K})N , \qquad (1)$$

where r, the Malthusian parameter, is the intrinsic rate of diversification. The equation has a solution, which is the familiar S-shaped curve and which is one candidate for N(t). From the solution of Equation 1 we can derive the following formula for r:

$$r = \frac{1}{t} \log \left[\frac{K-N(0)}{N(0)}\right] + \frac{1}{t} \log \left[\frac{N(t)}{K-N(t)}\right] , \tag{2}$$

where "log" is the natural logarithm, and $N(0)$ is the initial condition $N(t)$ at $t=0$, or the number of species at the start of the radiation. The parameter r can be computed from any set of coordinates (t,N) on the logistic curve. However, the equation does not work for $N > K$ because the value of N never reaches K, but only approaches it asymptotically. If Equation 2 is to be used, we must guess what the logistic curve looks like before we can find any set of coordinates on it. It is best to choose the largest value of N which makes Equation 2 computable, and this would generally be the value $N(t=t')=K-1$. Not only will this give the most accurate estimate, it will simplify Equation 2 to the form

$$r = \frac{1}{t'} \log \left[\frac{K-N(0)}{N(0)}\right] + \frac{1}{t'} \log [K-1] , \tag{3}$$

where $(K-1)$ is dimensionless. If the entire radiation proceeds from a single ancestral species, then $N(0) = 1$ and this reduces to

$$r = \frac{2}{t'} \log [K-1] . \tag{4}$$

Although the logistic model is widely thought to be the best model available to describe adaptive radiations, it has never been used to calculate the intrinsic rate of increase. Instead it is becoming customary (following Stanley, 1979) to estimate r by using the simple exponential model under the assumption that the clade is still actively radiating. The idea is that if the term N/K in Equation 1 is small enough, the equation can be simplified to

$$\frac{dN}{dt} = rN , \tag{5}$$

the well-known Malthusian law. The solution $N(t)$ to Equation 5 is an exponential function, $N(t) = N(0) \exp(rt)$, and the formula for the rate of increase is:

$$r = \frac{1}{t} \log \left[\frac{N(t)}{N(0)}\right] . \tag{6}$$

It may be well to emphasize here that the r value of Equation 2 is what we really want to know. The r values found by using Equation 6 are directly comparable to the r values of Equation 2, but only if the "small N/K" assumption is satisfied. If this assumption is not satisfied or (which is more often the case) we don't know whether or not it is satisfied, r values from Equation 6 may be difficult to interpret.

The value of K must be known if we are to make the judgment that N/K is much smaller than 1. For clades which are still actively radiating today, K cannot be known, and criteria which might allow us to assume that the exponential model still applies would be much more difficult to estimate than K generally is. Stanley has used Equation 6 to estimate r values for many clades which he believes are still actively radiating, and has used the latest (Pleistocene or Recent) values for t and N because recent records are usually the most accurate ones. But if my arguments are accepted, these r values might best be calculated using values of t and N from earlier in these radiations, when the "small N/K" assumption is most likely to hold true. If the clade has reached its plateau of diversity, then the value of K is known and there is no reason to use the exponential model to estimate r (it is simply a bad approximation). In this case, r should be calculated using Equation 2, or by some other method based on the logistic model.

Figure 1 is a graphic summary of the above points. For a radiation following a sigmoid path (curve A), the best estimate of r is given by the logistic model. A good estimate of r may be determined by fitting the simple exponential model to the early stage of the radiation, when damping is relatively insignificant (curve B). But if the radiation follows a sigmoid path, and a simple exponential curve is fitted to the point of maximum diversity (curve C), the true relationship of the variables is misrepresented by the model, and the estimate of r will be less than the true value. This error of model misspecification is what Stanley (1979) recognized as an uncorrected bias in his analysis of evolutionary rates. In order to eliminate this bias, we need to specify the right model in each individual case.

If we assume that diversification usually follows a sigmoid path and appears to be undamped only in its early stages, correct model specification may often be possible. This can be illustrated by examining a simple case of an extinct group of organisms whose paleodiversity history seems to indicate a pattern of sigmoid growth. As an example, we will consider the "Llandoverian explosion" of graptolite species at the start of the Silurian. The complex history of Silurian graptolite diversity is discussed by Rickards (1977). It includes many features which cannot be explained by adaptive radiation models, but we need not be concerned with these here. This history also includes two distinct adaptive radiations, the most significant of which occurs in the lower part of the Llandoverian Series.

Figure 2 is a scatter plot of diversity data from the Llandoverian explosion and its immediate sequel. Although the initial period of increasing diversity could be fitted by an exponential curve (or by a straight line, for that matter), the entire data set is better described by a logistic curve. I have generated two possible logistic curves by estimating K directly from the data and calculating r using Equation 2. The lower curve (Figure 2) is based on the mean value K = 52 for the apparent equilibrium phase. Variations around this value do not appear to be random. It is possible that there is a discontinuity at approximately t = 3 Myr and that the upper curve, fitted to the first six data points, best reflects the conditions under which this radiation took place. For

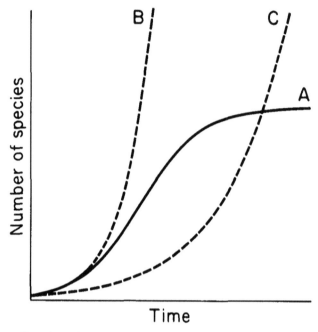

Figure 1. Hypothetical diversification models for an adaptive radiation. A is a sigmoid curve; B and C represent a successful and an unsuccessful attempt, respectively, to use the exponential model to estimate the r value of a radiation which follows curve A.

this curve, K is taken to be 60 species. Values of r for the Llandoverian explosion are 1.28 for the lower curve and 1.82 for the upper curve.

Using this same data set and applying Stanley's method, Carter and others (1980) estimated an r value of 0.57. This is comparable to the highest rates estimated by Stanley (1979). In fact, Stanley's rates seem to level off around this value. The r values calculated using the logistic model may level off at a value two or three times larger than this, as shown by the above example. Comparing Equations 4 and 6, we see that this is about the order of the differences to be expected when these two methods are used for the same data set. This indicates that the range of realistic r values is much greater than has been previously suspected.

While this example is highly suggestive of the degree to which previously published estimates of r are biased, it also demonstrates some of the ambiguities that enter into an analysis of even better-than-average data sets. The Llandoverian radiation of graptolites is used here because the data set features high sampling efficiency and fine-tuned geochronology (see Carter et al., 1980). However, it has other features which make a clear-cut interpretation difficult.

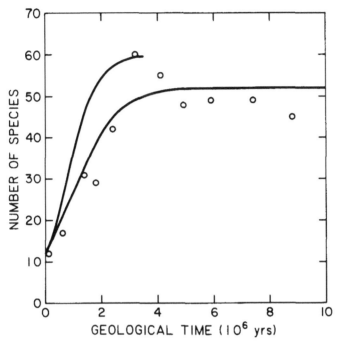

Figure 2. The "Llandoverian explosion" of graptolites. $T = 0$ is the start of the Silurian. Data points are located at the midpoint of each zone. The logistic curves are computer-generated, using r values calculated from Equation 3 (see text). Upper curve: $N(0) = 12$, $N(3) = 59$, $K = 60$, $r = 1.82$. Lower curve: $N(0) = 12$, $N(4) = 51$, $K = 52$, $r = 1.28$. Data are from Carter et al. (1980).

It is unclear whether there is a diversity plateau beginning at approximately $t = 4$ Myr and continuing for 6 Myr, or a diversity peak at $t = 3$ Myr followed by a gradual decline. The difficulty of estimating K accurately is related to the more general problem of distinguishing the influence of intrinsic (diversity dependent) factors from that of extrinsic (diversity independent) factors in observed paleodiversity patterns. If we cannot hope to resolve these, then rate estimates based on both the simple exponential model and the logistic model must be demoted from their theoretically meaningful status to the level of noninterpretive summary statistics. Rather than abandon the modeling effort, we should divide observed paleodiversity patterns among three classes: (1) those which suggest an equilibrium process, (2) those which suggest undamped diversification suddenly cut off by some extrinsic cause, and (3) those which do not definitely suggest either of the above. The data shown in Figure 2 can easily be assigned to the first class. Although much has been said about the inadequacies of the fossil record, especially with regard to estimating diversity at lower taxonomic levels (see, for example, Raup, 1979), many paleodiversity data sets are at least good enough to show whether the exponential or logistic model should be used.

Another problem with the graptolite data set is that it does not define a clade (which is suggested by the fact that both the logistic and the simple exponential model are equally useless in extrapolating back in time from the Silurian). Cladistic unity is generally considered to be a condition for calling an evolutionary event an "adaptive radiation," but this rule is usually honored only in the breach. Species contributing to Llandoverian graptolite diversity belong to several clades, some of which (preeminently the *Monograptus* lineage) were radiating during that time, others of which were in evolutionary decline.

Other data sets do not have these same problems to such a high degree. Some adaptive radiations are known to define clades and do have a more clearly defined equilibrium phase (for example: Cretaceous planktonic foraminifera—see Fisher and Arthur, 1977).

Estimates of r based on the simple exponential model must often have been gross underestimates, and the approach illustrated above may be of substantial help toward the improvement of these estimates. There are other ways of estimating r from the logistic model, and some of these may also be useful. It is not necessary to know K before r can be calculated. A logistic curve is exactly determined by three points, because the equation for the curve has only three constants. If the three points are well chosen and the constants are determined simultaneously, they may define a curve which fits the data very well (see Reiss et al., 1976). The methods of linear regression analysis may also be used when the data have been suitably transformed (see Gilchrist, 1976). These two procedures and the one illustrated above should be the methods of choice in cases where paleodiversity data are of sufficiently high resolution.

DISTANCE STRUCTURED MODELS

THE ADAPTIVE MOSAIC MODEL AND ITS IMPLICATIONS

The logistic and exponential models have exponential increase in common—except that in the logistic equation this increase is modified by a damping function. The idea of exponential increase of species is a scientific theory; it is always open to a challenge from competing scientific theories and (we hope) subject to empirical tests. In this case, a competing theory would have to claim that diversification is not exponential; it would have to say why it is not, and offer an alternative formulation for N(t).

The competing theory presented here is based on the work of Valentine (1980, 1981), whose conceptual model depicts diversification as a process by which an empty adaptive space is filled up with species. The adaptive space is a mosaic of adaptive zones divided into subzones (similar to the idea of Simpson, 1944, 1953); it may be thought of as a template for the multidimensional niche space of Hutchinson (1957). The dimensions of the adaptive space are defined in much the same way as those of niche space—in terms of spectra of environmental conditions and resources. Species can occupy subzones (sometimes called tesserae) in the adaptive space. The mode of speciation is not important for the model, but since new species must be derived from ancestral species, we metaphorically state that a new species can originate only by "jumping" in a random direction from a pre-existing species. There is a modal jump size which usually places a new species on a tessera next to its ancestor. If the new species lands on a tessera already occupied by another species, the speciation attempt is a failure; the new species is competitively excluded by the pre-existing species. In Simpson's (1953) terms, a prospective new species does not have ecological access to tesserae which are already occupied. Speciation attempts and extinctions occur at stochastically constant rates, but the success of speciation depends on the openness of the adaptive space—on how many unoccupied tesserae there are. I will refer to this conceptual model as the adaptive mosaic model.

Adaptive radiations are expected whenever there are large unoccupied regions in the adaptive space; this may occur, for example, following a mass extinction, or an evolutionary breakthrough into a new adaptive zone. A new species may occasionally originate by a large jump and land in an empty region to form the nucleus of a growing cluster of species. Diversification within this cluster will be rapid, but the probability of successfully giving rise to a new species is not the same for all species in the cluster. Those closer to the center will have a harder time finding empty tesserae for their descendants to occupy than will those on the periphery (Walker, 1982). This supposed condition of unequal ability to have descendants contrasts with the exponential and logistic models, which assume complete equality of speciation potential in all members of a clade. This difference distinguishes two general kinds of diversification models, which are referred to here as structured and unstructured models respectively.

The new models are developed by analogy to a class of advanced models in population biology. Populations of species in which the females have long reproductive lives are "age structured." The reproductive potential of the population as a whole is a function of how fertility varies with age and how many individuals there are in each age class. Models of such populations are called age structured models. The adaptive mosaic model suggests that speciation potential in a radiation is a function of distance from the center of the clade, so I will refer to models based on this property as distance structured models.

Distance structured models are justified a priori because of the connections which they establish between the evolutionary process and the pre-existing ecological framework which we know is there. If nothing else, this is at least an advance in theoretical coherence. Unfortunately, structured models are more difficult to test because they require more assumptions than do unstructured models, and also because when we specify any particular structure for the contingencies of speciation in the adaptive space, we describe only a small subset of the possible structures. These considerations suggest that a proposed structure should not only be intuitively reasonable and susceptible to mathematical analysis, but should require the smallest possible number of restrictive assumptions.

ASSUMPTIONS AND CONVENTIONS OF DISTANCE STRUCTURED MODELS

Distance structured models require the following assumptions. All the members of the radiation are descended from a single ancestor (this is not an absolute necessity, but it follows from the adaptive mosaic model). The modal jump size is the only important one; relatively large jumps are invisible to the model, as the descendants land outside the part of the adaptive space available to the clade. We further require that the radiating cluster have spherical symmetry. It is not necessary for the original species to be at the center of the cluster.

Although perfect spherical symmetry is an improbable outcome of the random tessera-filling process described here, it is more probable than any other configuration. In the absence of interference between clades, there will be a strong tendency for all clades in the adaptive space to assume a spherical shape. Consideration must be given to factors which might cause departures from spherical symmetry and the effect of these factors on the predictions of the model. We must assume that anisotropies and discontinuities in the adaptive space (which would induce a nonspherical bias in the shape of the cluster) are invisible to the model or are weak enough in their effects to allow us to neglect them. In the absence of information about the "grain" of the adaptive space, these are minimal assumptions. We must also consider how sensitive the model is to departures from spherical symmetry caused by chance alone, but it will be easier to do this later when we know what the model predictions are.

In order to analyze the geometry of constraints on speciation within the adaptive space, we must specify how many dimensions the space has and how units of distance shall be measured. The number of dimensions will be different for different versions of the model, but the units of measurement will be the same for all versions. Since these units are arbitrary, we will make them as convenient as possible. If we make the modal jump size equal to 1 unit of distance, then the mean distance across a tessera is 1 unit; the mean area of a tessera in 2 dimensions is $1^2 = 1$ unit of area (assuming a rectilinear shape); the mean volume of a tessera in 3 dimensions is $1^3 = 1$ unit of volume, and so on.

THE CIRCUMFERENCE MODEL

To develop a simple distance structured model, we will make two additional assumptions: that there are no extinctions, and that the adaptive space is unbounded and two-dimenstional. Now we can visualize the growing clade as a solid cluster on a plane surface, with only the species on the very edge of the cluster being able to find empty tesserae for their descendant species to occupy. For our purposes here, the "no extinctions" assumption is not as restrictive as it sounds. Random extinctions would have the effect of creating openings within the cluster, but these empty tesserae would then be liable to be filled again, and the proportion of open spaces within the cluster would be maintained at an equilibrium level (for further details see Walker and Valentine, 1984). Species turnover within the cluster would not contribute to the increase in N, the total number of species in the cluster.

If we draw two concentric circles around the edge of the cluster, so as to enclose the outermost ring of species in an annulus 1 unit wide (Figure 3), we will have delimited the only species in the cluster which have any speciation potential. The rate of diversification should be proportional to the number of species in the annulus. If we can find out how many species there are in the annulus, we can write a differential equation similar to Equation 5 and solve it to find the diversification function.

Because of the way the space is scaled, the number of species inside the annulus is equal to the area of the annulus. This area can be approximated by the circumference of a circle with a radius intermediate between the radii of the two circles which define the annulus. Therefore we can write

$$\frac{dN}{dt} = rC = r2\pi x \quad , \tag{7}$$

where C is the circumference of the circle described above, and x is its radius. Equation 7 is referred to here as the circumference model. To make this equation solvable, it is necessary to obtain x in terms of N, and this can be done by using the area formula for a circle, since area and species numbers are measured in the same units. Hence:

$$N = \text{Area} = \pi x^2$$

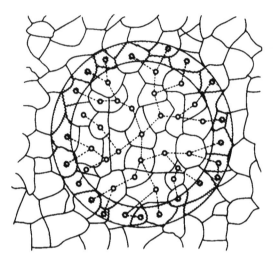

Figure 3. A two-dimensional adaptive space divided into tesserae, with a radiating clade of species. Two circles enclose the outer edge of the clade. Only the species in the annulus (hatched area) have access to empty tesserae for their descendants.

$$x = (\frac{N}{\pi})^{\frac{1}{2}} \ .$$

It should be clear that this x is the radius of the outer circle in Figure 3, and so it is not identical to the x used to calculate the circumference in Equation 7. However it is a good approximation as long as N is not too small, so we do not distinguish between the two x's in our notation. Substituting $x(N)$ into Equation 7 gives

$$\frac{dN}{dt} = r2\pi^{\frac{1}{2}}N^{\frac{1}{2}} \ , \tag{8}$$

which can easily be solved for N(t). Using the initial condition N(0)=1, we get

$$N(t) = [1 + \pi^{\frac{1}{2}}rt]^2 \ . \tag{9}$$

So, the circumference model predicts that N increases as a polynomial of t, and the model meets the criteria for a scientific theory to compete with the theory of exponential increase.

THE CIRCUMFERENCE MODEL IN HIGHER DIMENSIONS

It is possible to extend the circumference model to include radiations in adaptive spaces of any dimensionality. Next we will derive $N(t)$ for $n = 1, 3, 4$, and 5 dimensions; this will suffice to demonstrate trends in $N(t)$ with increasing dimensionality.

In a 3-dimensional adaptive space, N is analogous to the volume of a sphere, and the number of species which can speciate will be approximated by the surface area of the sphere. The analysis proceeds exactly as before, giving the result:

$$N(t) = [1 + (\frac{4\pi}{3})^{1/3} rt]^3 \tag{10}$$

In the case of four or more dimensions the problem is no different, although it is more difficult to visualize. I thank Richard Dodson for showing me that the calculation of volumes of hyperspheres in n-dimensional space is accomplished by integrating the expression

$$dV = J \, d(x^2) \, d\theta_1 \, d\theta_2 \ldots d\theta_{n-1} \quad , \tag{11}$$

where V is the volume; J is the Jacobian of the transformation from Cartesian to spherical coordinates, given by Downes (1966) as

$$J = \frac{1}{2} x^{n-2} \sin^{n-2} \theta_1 \sin^{n-3} \theta_2 \ldots \sin \theta_{n-2} \quad ;$$

and $x, \theta_1, \theta_2, \ldots \theta_{n-1}$ are the variables of the spherical coordinate system. Once the equation for V is derived from Equation 11, it is differentiated with respect to the radius x to give the equation for the surface area of the hypersphere. These equations are then used as before to determine a differential equation in the form of the circumference model. These differential equations for all values of n are collectively referred to here as the generalized circumference model. The solutions (given in Table 1) to these equations for $n = 1$ to 5 dimensions are all polynomials in t. It is evident that the order of the polynomial is the same as the dimensionality of the adaptive space.

Each equation for $N(t)$ in Table 1 defines a whole family of curves. We wish to specify a particular curve from each family so that the various equations can be compared graphically. Since each particular curve is completely determined by two points, we can choose comparable curves from different families by constraining every curve to pass through the same two points. In this way the five circumference equations and the exponential equation are compared in Figures 4 and 5.

In Figure 4, every curve is constrained to pass through the two points $N(0) = 1$ and $N(50) = 50$. $N = 1$ to 50 is chosen as a possible range of the "undamped" phase which will make it possible to distinguish the signal from the noise in paleodiversity data sets. Given the assumptions of exponential

Table 1. Summary of equations compared in Figures 4 and 5.

Exponential equation		$N(t) = \exp(rt)$
Circumference equation,	$n = 1$	$N(t) = 1 + rt$
	$n = 2$	$N(t) = [1 + \pi^{1/2} rt]^2$
	$n = 3$	$N(t) = [1 + (\frac{4}{3})^{1/3} \pi^{1/3} rt]^3$
	$n = 4$	$N(t) = [1 + (\frac{1}{2})^{1/4} \pi^{1/2} rt]^4$
	$n = 5$	$N(t) = [1 + (\frac{8}{15})^{1/5} \pi^{2/5} rt]^5$

growth, it is reasonable to suppose that diversification would still be (relatively) undamped if there is ultimately room for twice this number of species. Numerous examples of adaptive radiations achieving diversities of over 100 species are mentioned by Stanley (1979), so it would seem that $N = 50$ represents a reasonable value to use here. Given the assumptions of the circumference model, it is more difficult to specify a reasonable range of values for N. Diversification is either damped or it is (absolutely) undamped, depending on local conditions in the adaptive space. However, given the large values of N attained by some radiating clades, $N = 50$ is not an unreasonable value for undamped diversification.

One of the most striking of the patterns shown in Figure 4 is that as n increases, the polynomial N(t) approaches closer in appearance to the exponential N(t). This means that if the number of dimensions in the adaptive space is large enough, diversification will always appear to be exponential. The detection of non-exponential kinetics is favored by faster diversification rates and larger values of N. This is illustrated by Figure 5, which is constructed in the same way as Figure 4. These qualities also make it easier to distinguish different versions of the circumference equation from each other.

From each of the diversification functions in Table 1, a different formula for calculating r can be derived. Values of r calculated for the curves in Figures 4 and 5 are given in Table 2. The r parameter has the same significance for all the models, so these r values are all intercomparable in theory. Since r values vary with the number of dimensions assumed in the circumference model, it matters which of these versions is the correct one if we want an accurate estimate of r; but above $n = 3$ the differences between successive r values become as difficult to detect as the differences in the shapes of the curves.

Table 2. Values of r for equations compared in Figures 4 and 5.

		Fig. 4: N(0) = 1 N(50) = 50	Fig. 5: N(0) = 1 N(19) = 844
Exponential equation		0.078	0.35
Circumference equation,	n = 1	0.98	44.37
	n = 2	0.069	0.83
	n = 3	0.033	0.28
	n = 4	0.022	0.16
	n = 5	0.017	0.12

THE GENERAL DISTANCE STRUCTURED MODEL

The generalized circumference model provides us with a glimpse of something different from the overworked exponential equation, but the circumference model involves a number of restrictive assumptions which are not essential to distance structured models in general. I will now show that it is possible to drop the "no extinctions" assumption and the use of a discontinuous distribution of speciation potential, and to place some limits on the ultimate number of species in a clade. A general distance structured model should allow for any pattern of species density to develop, as long as it has spherical symmetry. The general model states that if we take (the number of species at distance x) times (the rate of speciation minus the rate of extinction at distance x), summed up over all values of x, we will have the rate of change of N. In calculus notation:

$$\frac{dN}{dt} = \int_0^{xmax} G(x,t) \{a[1 \cdot D(x,t)] - b\} dx, \qquad (12)$$

where:

$G(x,t)$ = the number of species at distance x at time t
$D(x,t)$ = the relative density of species at distance x at time t. $0 < D < 1$.
xmax = the limiting size of the clade (its maximum radius)
a = the intrinsic rate of speciation, or r+b
b = the intrinsic rate of extinction, or a−r

Although the solution of Equation 12 is not complete, it can be shown that the exponential, logistic and circumference (at least for n=2) models can all be derived from it under the respective sets of assumptions appropriate to them (Walker, 1984). The prospective solution to Equation 12 will be a diversification function with limits of applicability similar to those of the logistic equation.

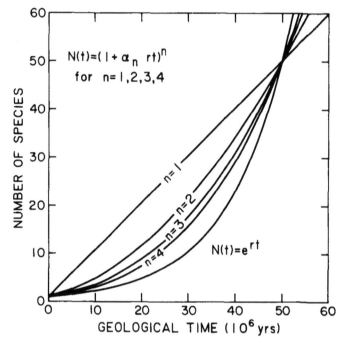

Figure 4. Comparison of solutions to circumference models in 1-, 2-, 3-, and 4-dimensional adaptive space, and to the exponential equation (bottom curve). All curves pass through $N(0) = 1$ and the point $N(50) = 50$. See text for explanation.

TESTING ALTERNATIVE HYPOTHESES

The idea expressed by Stanley (1979) that "speciation is a multiplicative process, and adaptive radiation, whatever its cause, is fundamentally a phenomenon of geometric or exponential increase" has been widely received as beyond question or doubt; but it is as yet unsupported by direct empirical evidence. Indirect evidence which has been adduced in favor of this model (for example: Stanley, 1979, Figure 5.2; Stanley and Newman, 1980, Figure 2) can as easily be explained by the circumference model. It may, however, be possible to test these alternative hypotheses with direct evidence: a series of paleodiversity data points from a single adaptive radiation. Here I consider the problems and requirements of such a test.

It will be recalled that the problem of diversification in an adaptive radiation has been rendered susceptible to mathematical analysis at the cost of considerable simplification. The models discussed here are predictive, but whether they are testable is still open to question. Structured models are

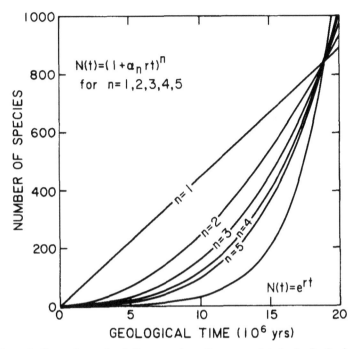

Figure 5. Comparison of solutions to circumference models in 1-, 2-, 3-, 4- and 5-dimensional adaptive space, and to the exponential equation (bottom curve). All curves pass through $N(0)=1$ and the point $N(19)=844$, which corresponds to Stanley's (1979) estimate for the age and modern diversity of the murid rodent clade.

under a special disadvantage because, in general, predictive power and testability will vary inversely with the number of restrictive non-verifiable assumptions. Whether such assumptions are satisfied in any particular case can only be inferred from the conformity of model predictions (diversification functions) to paleodiversity data. Since the assumptions are themselves a part of the model which we are attempting to test by making such a comparison, this involves us in a circular form of reasoning which we would rather avoid if we could. However, any attempt to test diversification models must begin in this way. The fit of diversification functions to paleodiversity data may be statistically tested by regression analysis (techniques are discussed, for example, by Weisberg, 1980). The circumference model has the desirable property that it has only one fitted parameter, r, regardless of the order of the polynomial $N(t)$. Therefore these functions can be compared with the exponential function with no loss of significance due to different degrees of freedom for different models.

A good data set for these purposes would be a clear example of an adaptive radiation; in addition it would have the following properties:

1. Cladistic unity, which is assumed by all models of adaptive radiation.
2. Biogeographic unity, which may be accomplished by considering a radiation within a particular province and attempting to exclude immigrants from adjacent provinces, or by considering organisms with an essentially worldwide distribution such as planktonic foraminifera.
3. Control over sampling problems, which can be attempted by choosing data sets which do not have obviously severe problems. If a constant proportion of the taxa that actually existed are found, the form of the diversification curve will be unchanged. The statistical lack of fit, then, is due to random deviations from this constant proportion.
4. Confidence in geochronology. Random distortions of the absolute time scale can be a very serious problem. However, if uncertainties can be quantified, they may be taken into account by established statistical methods.
5. The clade should rapidly attain a high value of N. This makes the various diversification functions more easily distinguishable, as discussed above.
6. A sufficient number of data points for a statistical test.

It is probably superfluous to note that data sets cannot be readily found in the paleontological literature in the desired form. The literature is full of data collected and tabulated with no particular paleobiological purpose in mind. Some of these data have been used to exemplify adaptive radiations, but seldom if ever has an attempt been made to satisfy more than a few of the criteria listed here. While the attempt to satisfy all of these criteria by the reduction of published data is not necessarily hopeless, it is possible that sensitive tests of alternative diversification functions will be made possible only by data which are collected for that particular purpose.

DISCUSSION

The circumference model fails to predict exponential increase because the number of species capable of speciating successfully is not a constant proportion of the total—it is instead a proportion which grows smaller as the radiating cluster grows larger. This might lead us to suspect that the circumference model is more similar to models of growth by cell proliferation in individuals than it is to models of population growth, and indeed this is the case.

Richards and Kavanagh (1945) counted "well over 150 different mathematical equations" used to describe organic growth; the number is probably much greater today. There are growth equations which have a "family resemblance" to the circumference model, but I have been unable to find a previous formulation of it. The somewhat similar parabolic model (in the notation used here: $N=ct^k$, where c,k are fitted constants) was proposed in the 1920s (see Medawar, 1945) and has been important because of its formal relation to the allometry equation (see Richards and Kavanagh, 1945). A model in some ways similar to the circumference model was described by Eden (1961) and has become known as the Eden growth process. Eden considered growth by simple cell division, with cells regularly located in a two-dimensional lattice, as a stochastic process. His analysis showed, among other things, that clusters of cells tend with high probability to be round and solid. Suggested biological counterparts were: bacterial and tissue cultures in which cells are constrained from moving; the thalli of *Prasiola*; and the sea leattuce *Ulva* (later workers added tumors to this list). The model was modified by Richardson (1973; see also Bramson and Griffeath, 1980) to include cell death, without much change in the conclusions which could be drawn. The model indicates that once the cluster exceeds a certain size it will, with high probability, grow at a radially linear rate. This is also a prediction of the circumference model: $dx/dt=r$ for all n. The probabilistic models make it clear that random departures from spherical symmetry do not substantially alter this conclusion.

The circumference model is basically a deterministic formulation of the Eden growth process. The probabilistic equations may seem more realistic, but in some ways important for this application they are not nearly so informative as the circumference equations. By neglecting random factors we have gained insight into the behavior of the model at small values of N; we have discovered a dependence of model behavior on the assumed dimensionality of the adaptive space, and we have generated quantitative predictions. We hope that random factors neglected in the model formulation may to some extent be taken into account in model testing.

The implications of the adaptive mosaic model depend strongly on the number of dimensions. The case of diversification in a one-dimensional adaptive space is especially interesting. For $n=1$, $N(t)$ becomes the equation for a straight line. Thus a linear diversification curve is, contrary to expectation, an interpretable condition. In an adaptive radiation this would indicate a very high value of the intrinsic rate of increase (see Table 2). However, this diversification pattern might not always be recognized as a "radiation" because the distinctive feature of numerous variations on a single basic type would often be absent. Variations might be few, with descendants looking, on the average, progressively dissimilar to the original species. What this describes is somewhat like Simpson's (1953) "progressive occupation" of adaptive zones, which he contrasts with adaptive radiation. Simpson offers several examples of the progressive occupation mode in vertebrate clades, but it is now a subject of debate whether species selection is not a better explanation for the trends Simpson discusses (see Eldredge and Gould, 1972; Stanley, 1979; Vrba,

1980). We can speculate that the mode of progressive occupation which Simpson thought he saw in the fossil record may, if it occurs, be the result of evolution in an adaptive space of low dimensionality.

As noted above, diversification in a many-dimensional adaptive space is indistinguishable from exponential growth. In fact, an empirically exponential history of diversification is consistent with the adaptive mosaic model if the number of dimensions is great enough. It is not hard to see why this should be so. In a one-dimensional space, the circumference model is equivalent to the exponential model if $N \leqslant 2$ because all the species have the same potential for successfully speciating; but if $N=3$, one of the species is completely surrounded by occupied tesserae, and this equivalence breaks down. In a two-dimensional space, we must reach $N=9$ before a species can be completely surrounded by occupied tesserae. The number of neighboring tesserae increases geometrically with n. In four dimensions every tessera has 80 neighbors, so that any structure which exists in the domain $1 < N < 81$ cannot be expressed by the circumference equation. Considering the numbers of species attained by adaptive radiations, we would not expect the circumference equation to be applicable at all in spaces of more than four dimensions.

Although this may sound like a severe restriction, there are good reasons to think that adaptive space has fewer than four dimensions. Current ideas about the dimensionality of niche space are of obvious relevance to this question. It is generally recognized that, while environmental states may be analyzed into many distinct variables, correlations among these variables reduce the effective number of axes required to fully specify a niche (see review by Pianka, 1981). In practice, niche dimensionality refers to the number of factors which serve to separate species (Levins, 1968) or to demonstrate significant resource partitioning (Schoener, 1974). In the design of ecological field studies, the variables used are few in number (rarely more than 3) and may reflect habitat, food preference, and (least often) time of active feeding. Such studies are highly suggestive of the upper limit on the dimensionality of niche space because they deal with actual biological communities and cover many taxonomic groups "from slime molds to lions" (see review by Schoener, 1974). Niche space is a more or less complete map of adaptive space, and there is no reason to think this conclusion does not carry over to the latter.

There are several unanswered questions.

The logistic model is still the only model of clade growth which incorporates a limit on clade size. The solution of the general distance structured model, however, will provide a competing theory for the logistic model, and will extend the theory presented here to cover many additional cases.

The models discussed here remain to be tested. It is possible to compare the predicted diversification functions to actual paleodiversity patterns, but individual outcomes of such comparisons may be limited in their significance.

Multiple tests and independent verification of assumptions would be required to give firm empirical support to one of these models. Whether it will be possible to do this remains to be seen.

Testability using species level data is uncertain. Paleodiversity data at higher taxonomic levels are much more reliable, and the temptation to use them to test our models is correspondingly great. Therefore, we must ask whether it makes sense to extrapolate distance structured models to explain the proliferation of metazoan families, orders, and other higher taxa. The basic assumptions of distance structured models (a single jump size, spherical symmetry) become increasingly untenable as we consider higher taxa. The adaptive mosaic model implies that, if we look at the level on which the origin of higher taxa becomes visible, the jump size distribution is continuously extended or polymodal; and that chance events can have large effects on the geometrical pattern of diversification in the adaptive space. The complex structure of diversification at this level is determined by jump size distributions, diversification histories, and the modes and timing of mass extinctions. This is a problem which is likely to be amenable to analysis only by computer simulation. Adaptive radiation at the species level has been assumed here to be a relatively orderly process, amenable to a mathematical approach. But the idea that these mathematical models can be extrapolated to the level of higher taxa should be rejected.

ACKNOWLEDGMENTS

I would like to express thanks to James W. Valentine, Dept. of Geological Sciences, University of California, Santa Barbara; and David Jablonski, Dept. of Ecology and Evolutionary Biology, University of Arizona, Tucson, for many helpful and encouraging discussions of the material presented here. I am especially indebted to Richard Dodson, Dept. of Geological Sciences, University of California, Santa Barbara, for stimulating discussion and for invaluable assistance with some aspects of the mathematical analysis. For their helpful comments on the manuscript I thank David Raup, Dept. of Geophysical Sciences, University of Chicago; and Steven Stanley, Dept. of Earth and Planetary Sciences, The Johns Hopkins University. Figures were drafted by Dave Crouch. This research was supported by National Aeronautics and Space Administration grant NAG2-73 to J. W. Valentine. Preston Cloud Research Laboratory Contribution Number 119.

REFERENCES

Bramson, M., and Griffeath, D., 1980. The asymptotic behavior of a probabilistic model for tumor growth. In Jäger, W., Rost, H., and Tautu, P. (eds.), Biological growth and spread, Berlin: Springer-Verlag, 165-172.

Cailleux, A., 1950. Progression géométrique du nombre des espéces et vie en expansion. C. R. S. Géol. Fr. (Paris), 222-224.

Cailleux, A., 1954. How many species? Evolution 8:83-84.

Carter, C., Trexler, J. H., Jr., and Churkin, M., Jr., 1980. Dating of graptolite zones by sedimentation rates: Implications for rates of evolution. Lethaia 13:279-287.

Cisne, J. L., 1971. The evolution of the taxonomic and ecological structure of aquatic free-living arthropods. Geol. Soc. Amer. Ann. Meeting Abstr. 3: 525.

Downes, T. D., 1966. Some relationships among the von Mises distributions of different dimensions. Biometrika 53:269-272.

Eden, M. A., 1961. A two-dimensional growth process. In Neyman, J. (ed.), Contributions to biology and problems of medicine, 4th Berkeley Symp. Math. Stat. Prob. Proc. IV:223-239.

Eldredge, N., and Gould, S. J., 1972. Punctuated equilibria: An alternative to phyletic gradualism. In Schopf, T. J. M. (ed.), Models in paleobiology, San Francisco: Freeman, Cooper & Co., 82-115.

Fisher, A. G., and Arthur, M. A., 1977. Secular variations in the pelagic realm. In Cook, H. E., and Enos, P. (eds.), Deep-water carbonate environments, Soc. Econ. Paleont. Mineral. Spec. Pub. (25):19-50.

Gilchrist, W., 1976. Statistical forcasting. Chichester, England: John Wiley & Sons, 308.

Hallam, A. (ed.), 1977. Patterns of evolution, as illustrated by the fossil record. Amsterdam: Elsevier, 591.

Hutchinson, G. E., 1957. Concluding remarks. Cold Spring Harbor Symp. Quant. Biol. 22:415-427.

Levins, R., 1968. Evolution in changing environments. Princeton, N.J.: Princeton Univ. Press, 120.

May, R. M., 1974. Stability and complexity in model ecosystems. 2nd ed. Princeton, N.J.: Princeton Univ. Press, 265.

May, R. M. (ed.), 1981. Theoretical ecology: Principles and applications. 2nd ed. Sunderland, Mass.: Sinauer Assoc., 489.

Medawar, P. B., 1945. Size, shape, and age. In LeGros Clark, W. E., and Medawar, P. B. (eds.), Essays on growth and form presented to D'Arcy Wentworth Thompson, Oxford: Clarendon Press, 157-187.

Newell, N. D., 1952. Periodicity in invertebrate evolution. J. Paleont. 26:371-385.

Pianka, E. R., 1981. Competition and niche theory. In May, R. M. (ed.), Theoretical ecology: Principles and applications, 2nd ed., Sunderland, Mass.: Sinauer Assoc., 167-196.

Pielou, E. C., 1977. Mathematical ecology. New York: John Wiley Sons, 385.

Raup, D. M., 1979. Biases in the fossil record of species and genera. Carnegie Mus. Nat. Hist. Bull. 13:85-91.

Richards, O. W., and Kavanagh, A. J., 1945. The analysis of growing form. In Le Gros Clark, W. E., and Medawar, P. B. (eds.), Essays on growth and form presented to D'Arcy Wentworth Thompson, Oxford: Clarendon Press, 188-230.

Richardson, D., 1973. Random growth in a tessellation. Cambridge Phil. Soc. Proc. 74:515-528.

Rickards, R. B., 1977. Patterns of evolution in graptolites. In Hallam, A. (ed.), Patterns of evolution, as illustrated by the fossil record, Amsterdam: Elsevier, 333-358.

Reiss, E. L., Callegari, A. J., and Ahluwalia, D. S., 1976. Ordinary differential equations with applications. New York: Rinehart and Winston, 387.

Schoener, T. W., 1974. Resource partitioning in ecological communities. Science 185:27-39.

Simpson, G. G., 1944. Tempo and mode in evolution. New York: Columbia Univ. Press, 237.

Simpson, G. G., 1952. Periodicity in vertebrate evolution. J. Paleont. 26:359-370.

Simpson, G. G., 1953. The major features of evolution. New York: Simon and Schuster, 434.

Small, J., 1950. Quantitative evolution XVI. Increase of species-number in diatoms. Ann. Bot. 14:91-113.

Stanley, S. M., 1973. An explanation for Cope's rule. Evolution 27:1-27.

Stanley, S. M., 1974. Effects of competition on rates of evolution, with special reference to bivalve mollusks and mammals. Syst. Zool. 22:486-506.

Stanley, S. M., 1975. A theory of evolution above the species level. Nat. Acad. Sci. USA Proc. 72:646-650.

Stanley, S. M., 1979. Macroevolution: Pattern and process. San Francisco: W. H. Freeman & Co., 332.

Stanley, S. M., and Newman, W. A., 1980. Competitive exclusion in evolutionary time: The case of the acorn barnacles. Paleobiology 6:173-183.

Valentine, J. W., 1980. Determinants of diversity in higher taxonomic categories. Paleobiology 6:444-450.

Valentine, J. W., 1981. Emergence and radiation of multicellular organisms. In Billingham, J. (ed.), Life in the universe, Cambridge, Mass.: M.I.T. Press, 229-257.

Vrba, E. S., 1980. Evolution, species and fossils: How does life evolve? S. Afr. J. Sci. 76:61-84.

Walker, T. D., 1982. Alternative models of diversification kinetics. 63rd Ann. Meeting Pacific Div. Amer. Assoc. Adv. Sci., Prog. with Abstr.:46.

Walker, T. D., 1984. The evolution of taxonomic diversity in an adaptive mosaic. Ph.D. thesis, Univ. So. Calif., Santa Barbara.

Walker, T. D., and Valentine, J. W., 1984. Equilibrium models of evolutionary species diversity and the number of empty niches. Amer. Nat. 124:887-899.

Weisberg, S., 1980. Applied linear regression. New York: John Wiley & Sons, 283.

Yule, G. U., 1924. A mathematical theory of evolution, based on the conclusions of Dr. J. C. Willis, F.R.S. Phil. Trans. Roy. Soc. Lond. 213(B):21-87.

Zeuner, F. E., 1946. Dating the past: An introduction to geochronology. 1st ed. London: Methuen & Co. Ltd., 444.

Chapter 10

MARINE REGRESSIONS AND MASS EXTINCTIONS: A TEST USING THE MODERN BIOTA

DAVID JABLONSKI

Department of Ecology & Evolutionary Biology, University of Arizona, Tucson*

INTRODUCTION

The coincidence of marine regressions with major episodes of faunal extinction has led many authors to hypothesize a cause-effect relationship (e.g., Chamberlain, 1898a,b; Newell, 1952, 1962, 1967; Hallam, 1981a). While some writers have emphasized physical environmental changes that would accompany regression, such as alterations in climate or oceanic circulation, others have invoked more direct effects, such as increased competitive interactions with decreased shelf area. One set of hypotheses is grounded in the dynamic equilibrium theory of island biogeography, in which decline of diversity with area is regarded as the result of decreased population size and thus increased vulnerability to stochastic extinction processes (Preston, 1962; MacArthur and Wilson, 1967; Simberloff, 1972, 1974). This approach has gained considerable acceptance among paleontologists (see Schopf, 1974; Gould, 1976; Sepkoski, 1976; Hallam, 1981a; Hurst and Watkins, 1981) even as its general explanatory ability has come under question (e.g., Simberloff, 1976, 1981; Connor and McCoy, 1979; Gilbert, 1980). Mass extinction hypotheses based on area effects in continental shelf benthos, however, have not taken into account the great taxonomic richness of the shallow-water biota around oceanic islands. While small drops in relative sea level can eliminate large areas of shallow epicontinental seas, conical islands will

*Current address· Dept. of Geophysical Sciences, University of Chicago, Chicago

actually gain slightly in perimeter and thus in shallow-water area (see Stanley, 1979; Raup and Stanley, 1980). Additionally, destruction of habitat types will be limited in oceanic settings. For example, although some lagoons will be lost due to subaerial exposure, others will replace them as older and more submerged sea mounts come into the photic zone as sea level drops. This paper presents a census of the benthic families represented on Recent oceanic islands, and so provides an estimate of the proportion of the marine biota exempt from extinction by area effects.

MATERIALS AND METHODS

I surveyed the literature on the shallow-water faunas of 22 oceanic islands for members of three major phyla of benthic marine invertebrates: Mollusca (bivalves, prosobranch gastropods), Echinodermata (asteroids, ophiuroids, and echinoids), and Coelenterata (scleractinian corals). These groups were chosen because they have been extensively monographed, constitute a sizeable portion of the marine fossil record, and appear to exhibit different behaviors during mass extinction events (mollusks relatively extinction-resistant, reef-building organisms extinction-prone, and echinoderms intermediate in response; see Newell, 1967, 1971; Valentine, 1973; Sepkoski, 1981, 1982; Hallam, 1981a). Islands were chosen for relative completeness of faunal coverage, and were used only if they were volcanic in origin and surrounded entirely by crust generated at mid-oceanic spreading centers or hot spots. Sites ranged from one of the world's most northerly volcanic islands, Jan Mayen (71°N) to South Georgia and Macquarie Islands (both about 54°S), and include localities from most of the major marine biogeographic regions (see Figure 1).

Faunas were tabulated at the family level in order to increase taxonomic consistency among sources, and to produce data comparable to those recently published on Phanerozoic mass extinctions (e.g., Raup and Sepkoski, 1982). Classification follows the *Treatise on Invertebrate Paleontology* and, for the gastropods, Taylor and Sohl (1962). To compensate for differences in sampling and preparation procedures among islands, families were included only

Figure 1 (facing page). Location of the 22 island faunas surveyed. Abbreviations: ALD, Aldabra; ASC, Ascension; AZO, Azores; BER, Bermuda; CHT, Chatham; CLI, Clipperton; COC, Cocos; COK, Cocos-Keeling; COO, Cook Islands (Raratonga); EAS, Easter; FAE, Faroes; FAN, Fanning; GUA, Guam; HAW, Hawaii; ICE, Iceland; JAN, Jan Mayen; KER, Kerguelen; MAC, Macquarie; MAR, Marquesas; MAS, Mascarenes; MAD, Madeira; SGA, South Georgia.

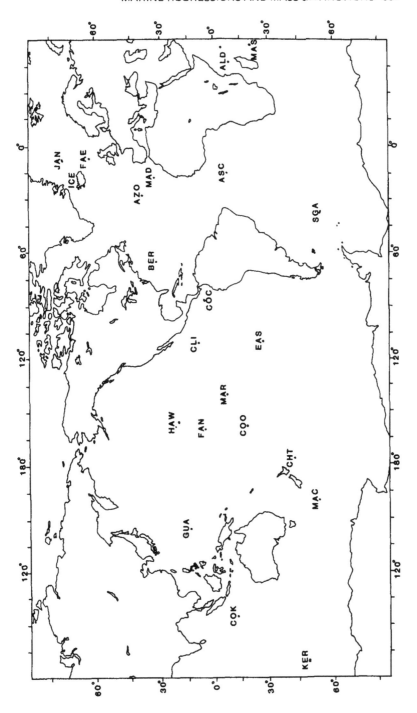

if living representatives occur in depths of 100 m or less, and if maximum adult size exceeds 5 mm. A total of 276 families meet these criteria.

RESULTS

Oceanic islands harbor a remarkably large proportion of today's shallow-water familial diversity (Table 1, Figure 2). Of the 276 families considered here, 239 (87%) have representatives reported from one or more of the 22 oceanic islands in Figure 1; fully 220 (78%) are present on two or more of the islands. The classes are not equally represented on the islands, but their rankings do not conform to their observed vulnerabilities to mass extinctions. All 21 scleractinian coral families are present on at least one island; 19 are recorded from two or more. Also well-represented are the prosobranch gastropods, of which only one out of 82 families is not recorded on the islands, and 78 families are present on two or more islands. Figures are slightly lower for the bivalves and asteroids, each having about 80% of their families on the islands (70% and 64% on two or more islands). Least complete but still extensively represented on the islands are the ophiuroids and echinoids, each with about 76% of their families recorded (70% and 66% respectively on two or more islands).

DISCUSSION

The data suggest that reduction in continental shelf area during marine regression is not sufficient in itself to explain mass extinctions: all other factors being equal (including instantaneous response times), a marine regression eliminating all continental shelf families would cause at most a 13% extinction of Recent families. As will be seen below, even prolonged regression, exceeding faunal relaxation times (cf. Diamond, 1972), would be unlikely to produce a major mass extinction. The biogeographic structure of shallow-water benthos at this taxonomic level is simply not vulnerable to this sort of areal reduction, because too many families are safely established on islands that would not be subject to reductions in habitable area during relative drop in sea level.

The methods of data collection outlined above should have biased against the observed results and in favor of producing a major extinction event, for the following reasons: (1) Only families represented on islands were assumed to survive the regression. Such an extreme pattern is unlikely, since shelf-depth seas have never been entirely eliminated from continental plates. The survivors' totals are thus weighted in favor of decline in familial diversity. (2) Only 22 islands were used out of the thousands actually present in the modern oceans. This constitutes an extremely small sample, yet most families are

Table 1. Percentage of the total, global number of marine invertebrate families actually present on each oceanic island. − = no data available, or data not yet encountered in author's literature search.

	Bivalvia	Gastropoda	Asteroidea	Ophiuroidea	Echinoidea	Scleractinia
Aldabra	32	60	25	59	33	71
Ascension	16	44	11	18	21	−
Azores	20	15	21	29	26	33
Bermuda	14	66	21	53	26	48
Chatham	39	49	32	12	15	0
Clipperton	14	34	7	12	3	14
Cocos	6	28	11	53	21	38
Cocos-Keeling	32	73	18	18	23	67
Cook	26	65	14	53	18	71
Easter	14	46	11	−	15	19
Faroes	38	38	32	41	15	29
Fanning	27	77	18	24	−	57
Guam	18	71	29	12	21	86
Hawaii	48	71	39	47	28	43
Iceland	30	48	36	41	23	14
Jan Mayen	13	26	18	29	5	0
Kerguelen	22	35	32	18	8	0
Macquarie	11	28	14	12	3	0
Madeira	38	63	7	18	5	38
Marquesas	9	32	18	29	−	43
Mascerenes	39	83	46	53	41	71
South Georgia*	9	39	32	41	8	0

* Dates back to the Mesozoic; the nearby South Sandwich Islands might be a more appropriate data source (see Barker et al., 1982, and other papers in that volume).

represented on at least two of the widely scattered islands surveyed. (3) Many of the islands still have not been completely monographed, so that not all families present on the island were recorded. (One general trend that emerged was that the bivalve mollusks have received considerably less attention than gastropods in the course of faunal surveys, particularly in the tropics.) (4) Modern oceans exhibit a high degree of provinciality, so that familial ranges are relatively restricted both longitudinally and latitudinally (as demonstrated for Recent molluscan families by Campbell and Valentine, 1977). This decreases the probability of any given family occurring on one or more islands. In situations of shallower latitudinal gradients or less-dispersed continents, as prevailed during much of the Phanerozoic, families should be more widespread and thus even less vulnerable to extinction due to loss of continental shelf area than in the present day.

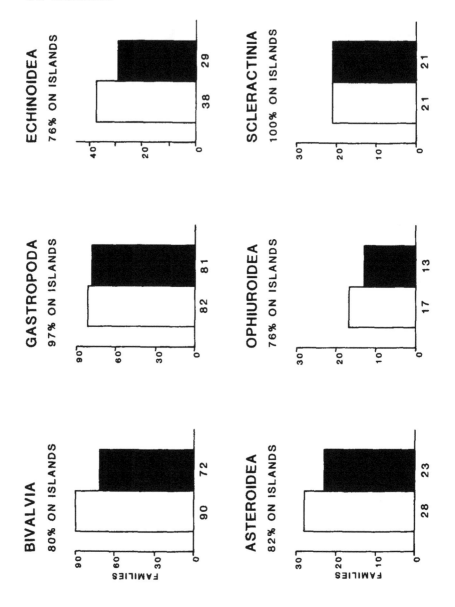

Conclusions similar to these were reached by Raup (1981), who modelled a different kind of biogeographic extinction in terrestrial vertebrates. He tested the modern biota for the magnitude of familial extinction expected for a given radius of total destruction, as might be expected from an extraterrestrial impact, and found that even objects with a lethal radius of 10,000 km would produce an average extinction of only about 12% of the terrestrial vertebrate families. As in the marine biota, even the highly provincial biogeography of the modern terrestrial fauna lacks sufficient endemism at the familial level to be vulnerable to biogeographic extinction.

Several benthic marine families recorded from none or only one of the islands are genuinely limited in distribution and thus would be vulnerable to extinction due to random events during regression. For example, the only shallow-water prosobranch family not recorded on any of the 22 islands is the Diastomatidae, a relict cerithiacean family with only one living species restricted to the southern coast of Australia (Houbrick, 1981). Most of the other prosobranch families are far more widespread. In contrast, the bivalves include a number of low-diversity, geographically-restricted families, including the Parallelodontidae (two living genera, both in Japan); the Pulvinitidae and Trigoniidae (each with one living Australian genus); the Mactromyidae (one genus, in deep water off Australia); Cardinidae (one genus off Mexico); Anatinellidae, Cardiliidae, Cooperellidae, and Glauconomidae (one genus apiece, though apparently widespread); and three anomalodesmatan families (see Morton, 1981), the Pholadomyidae (one living species), the Cleidothaeridae (probably only one living species), and the Laternulidae (one genus, six species, though over a fairly broad region of the Indo-West Pacific). Most of these families are relicts, so that their extirpation during regression would be detected in the fossil record as the final disappearance of a long-dwindling group. In contrast, most families that are diverse on continental shelves have representatives on oceanic islands; abrupt extinction of these large clades would not be expected as a consequence of area effects during regression.

Species-area effects have been most explicitly invoked for the Permo-Triassic event, in which approximately 52% of the skeletonized families became extinct (Newell, 1967; Raup and Sepkoski, 1982; Sepkoski, 1982). Simberloff (1974:267) stated that the close correspondence between shallow continental shelf area and number of marine invertebrate families during the Permo-Triassic "can indeed be construed as a result of a causal connection between

Figure 2 (facing page). Number of shallow-water families in the world biota (white bars) represented on 22 oceanic islands (black bars). Representation of modern families on islands among the six classes range from 76-100%.

area and biotic diversity, and that Schopf's [1974] hypothesis that the . . . extinctions were the consequence of changing sea level is biologically plausible." Schopf (1974:140) viewed the Permo-Triassic extinctions "as merely another manifestation of this relationship"–i.e., the "equilibrium model in which faunal diversity is largely predictable from area for colonization" (but see Schopf, 1979). Schopf's measurements of habitable area necessarily did not include Permo-Triassic oceanic islands, and judging from the distribution of the Recent biota those islands would have supported the great majority of shallow-water families extant at that time. The Permo-Triassic extinction, then, is far too large to be accounted for by simple area effects due to contraction of shelf seas (Figure 3).

Regression near the Permo-Triassic boundary extended over several geologic stages, comprising several million years at least (e.g., Newell, 1973; Schopf, 1974). Consequently, it could be argued that area effects might still have played an important role in the extinction if the families sequestered on islands had sufficiently small population sizes that random, attritional extinction removed a high proportion of them during the prolonged interval of lowered sea level. This duration argument is one of the standard explanations for the association of extinctions with regressions in the geologic past despite the lack of mass extinctions during the rapid glacio-eustatic oscillations of the Pleistocene (e.g., Hallam, 1981a:227). The modern distributional data do not support this argument, however, because they suggest that most families will be resistant even to prolonged low stands of sea level. At least 78% of the families are represented on two or more of the islands surveyed, indicating that most shallow-water families are spread over many habitat patches, possess large population sizes, and maintain that population over many localities, well-distributed both longitudinally and latitudinally. Thus while individual islands would be subject to habitat alteration or random extinction, a rich and stable pool of taxa could persist in spite of local extinctions even during prolonged regression, if all other factors remained constant; an analogy to Pleistocene paleobiogeography would not be inappropriate under such circumstances (see Stoddart, 1976; Taylor, 1978; and Rosen, 1981, on persistence of reef-dwelling species during the Pleistocene).

Taking a different approach but reaching similar conclusions to the present paper, Wise and Schopf (1981) further maintain that reduction in shelf sea area during the Pleistocene would not have eliminated enough species to produce significant extinctions at higher taxonomic levels. Their conclusions should be applied to the rest of the Phanerozoic with caution, however, in light of the important distinction between vertical sea level fluctuations (the Pleistocene oscillations are apparently among the largest in geologic history) and expansion and contraction of shelf seas (even maximum interglacial

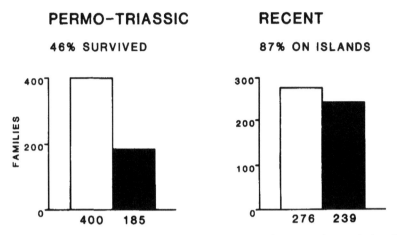

Figure 3. Comparison of the number of skeletonized families before and after the Permo-Triassic extinction (Raup and Sepkoski, 1982) with the number of families in the Recent biota that would persist on oceanic islands even if the continental shelf biota were completely eliminated. So many families are present on oceanic islands that area effects during regression of shelf seas would not be sufficient to produce such a massive extinction at high taxonomic levels.

transgressions did not create shelf seas comparable to those lost at the close of the Paleozoic and Mesozoic).

The molluscan fossil record during the Permo-Triassic interval also suggests that the correlation between continental shelf area and familial diversity was not a directly causal one. In an important but little-cited paper, Batten (1973) found that there are three times as many Paleozoic gastropod families in the upper Mid-Triassic (Ladinian) as in the lower Triassic (Scythian). Furthermore, *none* of the 16 Paleozoic families recorded from the Ladinian are known from the Upper Permian! Some families, then, completely disappear from the Late Permian-Early Triassic fossil record only to reappear unscathed in the Mid-Triassic—a pattern of apparent extinction and return also observed in Permo-Triassic bivalves (Nakazawa and Runnegar, 1973; see also Newell, 1973). Clearly, refugia were indeed available for some benthic organisms, and were effective at the familial level for millions of years. But many families did suffer terminal extinction, despite the fact that present-day biogeography suggests that they would have been represented in island refugia as well.

One reviewer argued that reduced rates of seafloor spreading, a commonly postulated mechanism for the end-Permian regression (Schopf, 1974; Bambach et al., 1980), would have removed most oceanic islands from the photic zone due to crustal cooling. However, lithospheric heat loss proceeds at a constant exponential rate, so that depth is a simple function of crustal age, not spreading rate (Parsons and Sclater, 1977; Pitman, 1978; Anderson and Kilbeck, 1981; and references therein). Islands created at hot spots (or at midoceanic ridges) would be transported laterally at diminished rates, but their initial elevation and subsidence rates would be unaffected by a seafloor spreading slowdown. Only a total cessation of seafloor spreading (as suggested by Schopf, 1974) could produce the complete failure of heat flow and consequent contraction of the oceanic lithosphere needed to drown the world's oceanic islands. This is highly unlikely, however, in light of the major role of seafloor spreading in planetary heat flux (e.g., Sclater et al., 1980); at any rate, there is considerable evidence for active subduction through this interval (e.g., Ridd, 1980; McElhinny et al., 1981; Morel and Irving, 1981; Wright, 1982; Cox and Gordon, 1982), indicating that while plate tectonic processes may have slowed, they did not grind to a halt. (In fact the global Permo-Triassic regression need not have been a consequence of any alteration of seafloor spreading rate or change in mid-ocean ridge volume. As the continental plates accreted into Pangaea, the enormous world ocean would have contained "more percentage of old ocean crust lying around ready to be subducted than at any other time before or since [in] the Phanerozoic" [Coney et al., 1980: 332]. The end-Paleozoic regressive trend thus could have resulted at least in part from the increased average depth and enlarged capacity of the world ocean basin.)

Direct extrapolation of the modern biodistributional data to the fossil record would, of course, be inappropriate if the biogeographic structure of the marine benthos in ancient seas were radically different from that of the present day. For example, if a very large proportion of the shallow-water families were endemic to epicontinental seas, area effects during regression might become important. However, this would require greater than 40% endemism at the familial level (not to individual continents or biogeographic regions, but to inland seas) to produce a Permo-Triassic scale event, and this figure seems excessive; most epicontinental endemic centers are distinguished on the basis of genera, not families (e.g., papers in Hallam, 1973). A more significant variable may be long-term changes in species/family ratios (see Valentine, 1970, 1973), and an assessment of this factor, along with a test for differences in species/family ratios between shelf and island faunas, will be the subject of another paper (Jablonski and Flessa, 1985; Flessa and Jablonski, 1985).

Conclusions drawn from the Recent biota might also be questioned if the dispersal capabilities of marine benthos have changed significantly. The broad distribution of modern families could be in part a reflection of larval dispersal capability, with high-dispersal larval stages enabling propagules to maintain communication between populations over broad geographic ranges, and to recolonize islands (or the mainland) in the event of local extinctions. While many members of the post-Paleozoic fauna (see Sepkoski, 1981) have planktotrophic larval stages, living representatives of the mainstays of the Paleozoic benthos–the articulate brachiopods, crinoids, and bryozoans–are almost exclusively nonplanktotrophic and thus generally have restricted larval dispersal capability (see Jablonski and Lutz, 1983). It could thus be argued that high taxa most vulnerable to extinction at the close of the Permian were those with low dispersal capabilities (e.g., Carr and Kitchell, 1980:440), which would be unlikely to maintain broad geographic ranges to persist on islands because of low probabilities of recolonization after local extinction.

Unfortunately, little is known of the reproductive modes of dispersal capabilities of Paleozoic taxa, and modern relatives may be unrepresentative of Paleozoic adaptations. Strathmann (1978) and Valentine and Jablonski (1983) have suggested that at least some Paleozoic crinoids and articulate brachiopods were planktotrophic, and that only after the Permian extinction were these groups exclusively nonplanktotrophic in development (see also Jablonski and Lutz, 1983, and references therein on Paleozoic bryozoans). The shift could be attributed either to preferential extinction of planktotrophic clades, or chance loss due to the magnitude of the extinction. Furthermore, although species' ranges correlate well with larval dispersal capability (Hansen, 1980; Jablonski, 1980a, 1982), many benthic families whose members are exclusively nonplanktotrophic have attained broad geographic distributions encompassing oceanic islands. These families include: the bivalves Nuculidae, Nuculanidae, Carditidae, Montacutidae, Erycinidae; gastropods Acmaeidae, Patellidae, Trochidae, Turbinidae, Volutidae (see Jablonski and Lutz, 1980, 1983 for references); scleractinian corals, so far as is known (e.g., Heck and McCoy, 1978); and many bryozoan families (see Lagaaij and Cook, 1973). Therefore, although it is possible that some Paleozoic groups suffered losses at the Permo-Triassic boundary for reasons related to their mode of reproduction, such selectivity probably could not have been directly mediated by differences in familial geographic range.

An alternative to the equilibrium model for the species-area relationship is the "habitat-diversity" model: large areas include more kinds of habitats than small areas (see Connor and McCoy, 1979). This alternative approach might be more appropriate than the equilibrium model in considering marine invertebrate diversity changes, particularly since epicontinental seas apparently did include unique benthic environments (see Hallam, 1981a for review). Although more quantitative data are needed before definitive conclusions can be reached, reduction in the number of habitat patches during regression was probably insufficient to produce an extinction of Permo-Triassic magnitude because: (1) Many of the unique epicontinental sea habitats were extreme environments (e.g.,high-salinity or low-oxygen settings) that supported relatively few taxa (see Bambach, 1977). (2) As mentioned previously, endemism in epicontinental seas, including unique habitat types, was manifest primarily at generic rather than familial levels. (3) Family-area curves have very low slopes (see, for example, Simberloff, 1974 for several terrestrial data sets; Flessa, 1975 for terrestrial mammals; and Hallam's 1977 plot for Jurassic marine bivalves). This trend can also be inferred from the flattening of taxon-area frequency distributions ("hollow curves") at increasingly higher taxonomic levels (see Anderson, 1974; Flessa and Thomas, this volume). If these curves are simply a reflection of the number of habitats available with changing area, such low slopes would require loss of more habitable area than was originally available if a diversity decline of the magnitude of the Permo-Triassic extinction is to result (a conclusion consistent with the results of Wise and Schopf, 1981).

By anyone's estimate, the Permo-Triassic extinction was a major event in the history of marine life, removing over 50% of the families (Newell, 1967; Schopf, 1974; Raup and Sepkoski, 1982) and from 75% to a staggering 96% of the species (Valentine et al., 1978; Raup, 1979); the extinction is equivalent at the generic level to some 85 m.y. of background extinction (Raup, 1978). In addition, the extinction does at least roughly coincide with a marine regression, as do the rest of the major extinction events recognized by Raup and Sepkoski (1982; see also Newell, 1967; Sepkoski, 1982): the terminal Ordovician (e.g., Berry and Boucot, 1973); the Frasnian (e.g., Johnson, 1974; but cf. Schlager, 1981 and House, 1975 for a more complex picture); the terminal Triassic (Hallam, 1981b); and the terminal Cretaceous (e.g., Matsumoto, 1980). The question remains then, that if area effects are insufficient to produce such major declines at high taxonomic levels in marine organisms, what is the connection, if any, between regression and mass extinction?

The lag of Triassic rediversification relative to re-expanded continental shelf area (Newell, 1967; Schopf, 1974; Waterhouse and Bonham-Carter, 1976) suggests a diversity-dependent mechanism for the Permo-Triassic event, or at least for the recovery (Valentine, 1972, 1973; Carr and Kitchell, 1980). As Carr and Kitchell (1980) are careful to point out, many kinds of perturbations other than reduction in shelf area could lower global carrying capacity and produce the observed diversity pattern. The most attractive hypotheses are still those related to changes in global geography: suturing of continents would have altered the configurations of ocean basins and reduced the global number of discrete shallow-water provinces, and exposure and coalescence of landmasses would have shifted global climate away from a maritime to a continental regime (Valentine, 1972, 1973; Valentine and Moores, 1972, 1973; Schopf, 1979; see also Newell, 1971; Flessa and Sepkoski, 1978; Bambach et al., 1980; Barron et al., 1980; Hambrey and Harland, 1981).

These are not new arguments, but they are consistent with the biogeography of modern seas and with the paleontological data. The physical environmental changes would have been felt most strongly at low latitudes, and there is indeed a latitudinal pattern of extinction in the Permo-Triassic faunas, with the tropical biota including reef dwellers being much more severely affected than high-latitude and cosmopolitan taxa (Newell, 1971; Valentine, 1972, 1973; Valentine and Moores, 1973; Bretsky, 1973; Waterhouse, 1973; Waterhouse and Bonham-Carter, 1976). (The Tethyan biota was not totally eradicated, however, and it is interesting that a number of Mesozoic families first appear in Permian rocks of this region [Dagis and Ustritsky, 1973; Batten, 1979].) Similarly pronounced low-latitude extinctions have been reported in the Ordovician (Sheehan, 1979; Berry, 1979), Frasnian (Copper, 1977), Cretaceous (Kauffman, 1979; Thierstein, 1981; Hallock, 1982), and, to a less marked degree, in the Triassic (Hallam, 1981b) extinctions.

An extinction approximately equal in magnitude to the Permo-Triassic can be generated in the Recent island data by eliminating families restricted to the tropics (here taken as occurring exclusively on islands within the belt of coral reef construction). By this criterion, 18% of molluscan families, 47% of echinoderm families, and 67% of scleractinian families are lost. Unlike the area-effect test, these magnitudes and rankings conform to those observed in mass extinction events in the fossil record.

CONCLUSION

The biogeography of the modern world suggests that species-area effects during regression are not sufficient to produce an extinction at the family level of the magnitude of the Permo-Triassic event: the great majority of marine families would persist in the undiminished shallow-water regions around oceanic islands. These results, then, help to explain why not every regression is accompanied by a mass extinction. More generally, the data presented here demonstrate again that evolutionary and biogeographic patterns or processes at the species level are not always readily extrapolated to familial or higher taxonomic levels; nor is backtracking from higher levels always appropriate. This is in part a function of the degree of inclusiveness of the taxonomic units; for example, two provinces or geologic periods may share few species but many families. Patterns will not simply be damped going up the taxonomic hierarchy, however. Different taxonomic levels appear to have some degree of evolutionary independence, with rates and directions of change not totally predictable from those of lower-level components. Every family of course originates with a species, and every family becomes extinct when the last of its species is lost. But while speciation frequencies may have remained roughly steady through much of the Phanerozoic (e.g., Paleozoic brachiopod durations and rates are comparable to Cenozoic bivalves; Stanley, 1979), the frequency of origination of new orders clearly has not (e.g., Valentine, 1973;

Jablonski, 1980b). This raises the question of whether the biogeographic deployment of higher taxa has changed through geologic time along with constraints on their rates of origination.

The sheer magnitude of the Permo-Triassic event indicates that additional factors must have come into play, most likely involving reduction in the number of shallow-water provinces and a disruption of tropical marine ecosystems. This biogeographic mechanism does not involve probabilistic extinction among a global biota whose inhabitants can be treated as identical particles, or extinction centered around a point source of disturbance, but extinction depending on membership in particular biogeographic regions. In this sense the Permian extinction was selective rather than random, but a search for the constellation of traits that doomed a given clade to extinction must include biogeographic affinity, with all the coadapted biological features that this entails. Taxa characterized by a particular reproductive mode, for example, or body size, or membership in a particular community, may have survived not because these traits were specifically selected for, but because these traits were characteristic of a province (or latitude) that suffered relatively little disturbance. Biogeography and macroevolution are probably more closely linked than is generally assumed.

ACKNOWLEDGMENTS

I thank Peter J. Coney, Karl W. Flessa, Susan M. Kidwell, Earl D. McCoy, David M. Raup, J. John Sepkoski, Jr., and James W. Valentine for valuable discussions and reviews. I am also grateful to John D. Taylor and Geerat J. Vermeij for providing unpublished data on the mollusks of Aldabra and the Marianas, respectively, and to David J. Bottjer (and the redoubtable Hancock Library, U.S.C.), J. Wyatt Durham, and Philip W. Signor for advice and assistance regarding published sources. Supported in part by NSF grant EAR 81-21212.

REFERENCES

Anderson, R. N., and Kilbeck, J. N. S., 1981. Oceanic heat flow. In Emiliani, C. (ed.), The sea. Vol. 7, The oceanic lithosphere, New York: Wiley, 489-523.

Anderson, S., 1974. Patterns of faunal evolution. Q. Rev. Biol. 49:311-332.

Bambach, R. K., 1977. Species richness in marine benthic habitats through the Phanerozoic. Paleobiology 3:152-167.

Bambach, R. K., Scotese, C. R., and Ziegler, A. M., 1980. Before Pangaea:

The geographies of the Paleozoic world. Am. Sci. 68:26-38.

Barker, P. F., Hill, I. A., Weaver, S. D., and Pankhurst, R. J., 1982. The origin of the eastern South Scotia Ridge as an intraoceanic island arc. In Craddock, C. (ed.), Antarctic geoscience, Madison, Wis.: Univ. Wisconsin Press, Internat. Union Geol. Sci. Ser. B. (4):203-211.

Barron, E. J., Sloan, J. L., II, and Harrison, C. G. A., 1980. Potential significance of land-sea distribution and surface albedo variations as a climatic forcing factor; 180 m.y. to the present. Palaeogeog. Palaeoclimat., Palaeoecol. 30:17-40.

Batten, R. L., 1973. The vicissitudes of the gastropods during the interval of Guadalupian-Ladinian time. In Logan, A., and Hills, L. V. (eds.), The Permian and Triassic systems and their mutual boundary, Can. Soc. Petrol. Geol. Mem. 2:596-607.

Batten, R. L., 1979. Permian gastropods from Perak, Malaysia, Pt. 2, The trochids, patellids, and neritids. Am. Mus. Novit. 2685, 26.

Berry, W. B. N., 1979. Graptolite biogeography: A biogeography of some Lower Paleozoic plankton. In Gray, J., and Boucot, A. J. (eds.), Historical biogeography, plate tectonics, and the changing environment, Corvallis, Or.: Oregon State Univ. Press, 105-115.

Berry, W. B. N., and Boucot, A. J., 1973. Glacio-eustatic control of Upper Ordovician-Early Silurian platform sedimentation and faunal changes. Geol. Soc. Amer. Bull. 84:275-284.

Boucot, A. J., 1975. Evolution and extinction rate controls. Amsterdam: Elsevier, 427.

Bretsky, P. W., 1973. Evolutionary patterns in the Paleozoic Bivalvia: Documentation and some theoretical considerations. Geol. Soc. Amer. Bull. 84:2079-2096.

Campbell, C. A., and Valentine, J. W., 1977. Comparability of modern and ancient faunal provinces. Paleobiology 3:49-57.

Carr, T. R., and Kitchell, J. A., 1980. Dynamics of taxonomic diversity. Paleobiology 6:427-433.

Chamberlin, T. C., 1898a. The ulterior basis of time divisions and the classification of geologic history. J. Geol. 6:449-462.

Chamberlin, T. C., 1898b. A systematic source of evolution of provincial faunas. J. Geol. 6:597-608.

Coney, P. J., Jones, D. L., and Monger, J. W. H., 1980. Cordilleran suspect terranes. Nature 288:329-333.

Connor, E. F., and McCoy, E. D., 1979. The statistics and biology of the species-area relationship. Am. Nat. 113:791-833.

Copper, P., 1977. Paleolatitudes in the Devonian of Brazil and the Frasnian-Fammenian mass extinction. Palaeogeogr., Palaeoclimat., Palaeoecol. 21: 165-207.

Cox, A., and Gordon, R. G., 1982. Paleomagnetic Euler poles for the absolute motion of Laurasia during the Late Paleozoic and Mesozoic (Abstr.). Geol. Soc. Amer. Abstr. 14:468.

Dagis, A. S., and Ustritsky, V. I., 1973. The main relationships between the changes in marine fauna at the close of the Permian and the beginning of the Triassic. In Logan, A., and Hills, L. V. (eds.), The Permian and Triassic systems and their mutual boundary, Can. Soc. Petrol. Geol. Mem. 2:647-654.

Diamond, J. M., 1972. Biogeographic kinetics: Estimation of relaxation times for avifaunas of southwest Pacific islands. Natl. Acad. Sci. USA Proc. 69: 3199-3203.

Flessa, K. W., 1975. Area, continental drift and mammalian diversity. Paleobiology 1:189-194.

Flessa, K. W., and Jablonski, D., 1985. Declining Phanerozoic extinction rates: Effects of taxonomic structure? Nature 313:216-218.

Flessa, K. W., and Sepkoski, J. J., Jr., 1978. On the relationship between Phanerozoic diversity and changes in habitable area. Paleobiology 4:359-366.

Gilbert, F. S., 1980. The equilibrium theory of island biogeography: Fact or fiction? Biogeogr. J. 7:209-235.

Gould, S. J., 1976. Palaeontology plus ecology as palaeobiology. In May, R. M. (ed.), Theoretical ecology: Principles and applications, Philadelphia: Saunders, 218-236.

Hallam, A. (ed.), 1973. Atlas of palaeobiogeography. Amsterdam: Elsevier; 531.

Hallam, A., 1977. Jurassic bivalve biogeography. Paleobiology 3:58-73.

Hallam, A., 1981a. Facies interpretation and the stratigraphic record. Amsterdam: Elsevier, 291.

Hallam, A., 1981b. The end-Triassic bivalve extinction event. Palaeogeogr., Palaeoclimat., Palaeoecol. 35:1-44.

Hallock, P., 1982. Evolution and extinction in larger Foraminifera. 3rd N. Amer. Paleont. Conv. Proc. I:221-225.

Hambrey, M. J., and Harland, W. B., 1981. The evolution of climates. In Cocks, L. R. M. (ed.), The evolving earth, Cambridge: Brit. Mus. (Nat. Hist.) and Cambridge Univ. Press, 137-152.

Hansen, T. A., 1980. Influence of larval dispersal and geographic distribution on species longevity in neogastropods. Paleobiology 6:193-207.

Heck, K. L., Jr., and McCoy, E. D., 1978. Long-distance dispersal and the reef-building corals of the eastern Pacific. Mar. Biol. 48:349-356.

Houbrick, R. S., 1981. Anatomy of *Diastoma melanoides* (Reeve, 1849) with remarks on the systematic position of the family Diastomatidae (Prosobranchia: Gastropoda). Biol. Soc. Washington Proc. 94:598-621.

House, M. R., 1975. Faunas and time in the marine Devonian. Yorkshire Geol. Soc. Proc. 40:459-490.

Hurst, J. M., and Watkins, R., 1981. Lower Paleozoic clastic, level-bottom community organization and evolution based on Caradoc and Ludlow comparisons. In Gray, J., Boucot, A. J., and Berry, W. B. N. (eds.), Communities of the past, Stroudsburg, Pa.: Hutchinson Ross Co., 69-100.

Jablonski, D., 1980a. Apparent versus real biotic effects of transgressions and regressions. Paleobiology 6:397-407.

Jablonski, D., 1980b. Adaptive radiations: Fossil evidence for two modes (Abstr.). 2nd Internat. Congr. Syst. Evol. Biol. Abstr.:243.

Jablonski, D., 1982. Evolutionary rates and modes in Late Cretaceous gastropods: Role of larval ecology. 3rd N. Amer. Paleont. Conv. Proc. I:257-262.

Jablonski, D., and Flessa, K. W., 1985. The taxonomic structure of shallow-water marine faunas: Implications for Phanerozoic extinctions. Malacologia, in press.

Jablonski, D., and Lutz, R. A., 1980. Molluscan larval shell morphology; Ecological and paleontological applications. In Rhoads, D. C., and Lutz, R. A. (eds.), Skeletal growth of aquatic organisms, New York: Plenum, 323-377.

Jablonski, D., and Lutz, R. A., 1983. Larval ecology of benthic marine invertebrates: Paleobiologic implications. Biol. Rev. 58:21-89.

Johnson, J. G., 1974. Extinction of perched faunas. Geology 2:479-482.

Kauffman, E. G., 1979. The ecology and biogeography of the Cretaceous-Tertiary extinction event. In Christensen, W. K., and Birkelund, T. (eds.), Cretaceous-Tertiary boundary events symposium, Univ. Copenhagen, II: 29-37.

Lagaaij, R., and Cook, P. L., 1973. Some Tertiary to Recent Bryozoa. In Hallam, A. (ed.), Atlas of palaeobiogeography, Amsterdam: Elsevier, 489-498.

MacArthur, R. H., and Wilson, E. O., 1967. The theory of island biogeography. Princeton, N.J.: Princeton Univ. Press, 203.

Matsumoto, T., 1980. Inter-regional correlation of transgressions and regressions in the Cretaceous period. Cret. Res. 1:359-373.

McElhinny, M. W., Embleton, B. J. J., Ma, X. H., and Zhang, Z. K., 1981. Fragmentation of Asia in the Permian. Nature 293:212-216.

Morel, P., and Irving, E., 1981. Paleomagnetism and the evolution of Pangea. J. Geophys. Res. 86:1858-1872.

Morton, B., 1981. The Anomalodesmata. Malacologia 21:35-60.

Nakazawa, K., and Runnegar, B., 1973. The Permian-Triassic boundary: A crisis for bivalves? In Logan, A., and Hills, L. V. (eds.), The Permian and Triassic systems and their mutual boundary, Can. Soc. Petrol. Geol. Mem. 2:608-621.

Newell, N. D., 1952. Periodicity in invertebrate evolution. J. Paleontol. 26: 371-385.

Newell, N. D., 1962. Paleontological gaps and geochronology. J. Paleontol. 36:592-610.

Newell, N. D., 1967. Revolutions in the history of life. Geol. Soc. Amer. Spec. Paper 89:63-91.

Newell, N. D., 1971. An outline history of tropical organic reefs. Amer. Mus. Novit. 2465; 37.

Newell, N. D., 1973. The very last moment of the Paleozoic Era. In Logan, A., and Hills, L. V. (eds.), The Permian and Triassic Systems and their mutual boundary. Can. Soc. Petrol. Geol. Mem. 2:1-10.

Parsons, B., and Sclater, J. G., 1977. An analysis of the variation of ocean floor heat flow and bathymetry with age. J. Geophys. Res. 82:803-827.

Pitman, W. C., III, 1978. Relationship between eustacy and stratigraphic sequences of passive margins. Geol. Soc. Amer. Bull. 89:1389-1403.

Preston, F. W., 1962. The canonical distribution of commonness and rarity. Ecology 43:185-205, 410-432.

Raup, D. M., 1978. Cohort analysis of generic survivorship. Paleobiology 4: 1-15.

Raup, D. M., 1979. Size of the Permo-Triassic bottleneck and its evolutionary implications. Science 206:217-218.

Raup, D. M., 1981. Extinction: Bad genes or bad luck? Acta Geol. Hispanica 16:25-33.

Raup, D. M., and Sepkoski, J. J., Jr., 1982. Mass extinctions in the marine fossil record. Science 215:1501-1503.

Raup, D. M., and Stanley, S. M., 1980. Principles of paleontology. 2nd ed. San Francisco: Freeman, 481.

Ridd, M. F., 1980. Possible Paleozoic drift of SE Asia and Triassic collision with China. J. Geol. Soc. London 137:635-640.

Rosen, B. R., 1981. The tropical high diversity enigma—the corals'-eye view. In Forey, P. L. (ed.), The evolving biosphere, Brit. Mus. (Nat. Hist.) and Cambridge Univ. Press, 103-129.

Schlager, W., 1981. The paradox of drowned reefs and carbonate platforms. Geol. Soc. Amer. Bull. 92:197-211.

Schopf, T. J. M., 1974. Permo-Triassic extinction: Relation to sea-floor spreading. J. Geol. 82:129-143.

Schopf, T. J. M., 1979. The role of biogeographic provinces in regulating marine faunal diversity through geologic time. In Gray, J., and Boucot, A. J. (eds.), Historical biogeography, plate tectonics, and the changing environment, Corvallis, Or.: Oregon State Univ. Press, 449-457.

Sclater, J. G., Jaupart, C., and Galson, D., 1980. The heat flow through oceanic and continental crust and the heat loss of the earth. Rev. Geophys. Space Phys. 18:269-311.

Sepkoski, J. J., Jr., 1976. Species diversity in the Phanerozoic: Species-area effects. Paleobiology 2:298-303.

Sepkoski, J. J., Jr., 1981. A factor analytic description of the Phanerozoic marine fossil record. Paleobiology 7:36-53.

Sepkoski, J. J., Jr., 1982. Mass extinctions in the Phanerozoic oceans: A review. Geol. Soc. Amer. Spec. Paper 190:283-289.

Sheehan, P. M., 1979. Swedish Late Ordovician marine benthic assemblages and their bearing on brachiopod zoogeography. In Gray, J., and Boucot, A. J. (eds.), Historical biogeography, plate tectonics, and the changing environment, Corvallis, Or.: Oregon State Univ. Press, 61-73.

Simberloff, D. S., 1972. Models in biogeography. In Schopf, T. J. M. (ed.), Models in paleobiology, San Francisco: Freeman, Cooper, 160-191.

Simberloff, D. S., 1974. Permo-Triassic extinctions: Effects of area on biotic equilibrium. J. Geol. 82:267-274.

Simberloff, D. S., 1976. Species turnover and equilibrium island biogeography.

Science 194:572-578.

Simberloff, D. S., 1981. Community effects of introduced species. In Nitecki, M. H. (ed.), Biotic crises in ecologic and evolutionary time, New York: Academic Press, 53-81.

Stanley, S. M., 1979. Macroevolution, pattern and process. San Francisco: Freeman, 332.

Stoddart, D. R., 1976. Continuity and crisis in the reef community. Micronesica 12:1-9.

Strathmann, R. R., 1978. The evolution and loss of feeding larval stages of marine invertebrates. Evolution 32:894-906.

Taylor, D. W., and Sohl, N. F., 1962. An outline of gastropod classification. Malacologia 1:7-32.

Taylor, J. D., 1978. Faunal response to the instability of reef habitats: Pleistocene molluscan assemblages of Aldabra Atoll. Palaeontology 21:1-30.

Thierstein, H. R., 1981. Late Cretaceous nannoplankton and the change at the Cretaceous-Tertiary boundary. Soc. Econ. Paleont. Mineral. Spec. Pub. 32:355-394.

Valentine, J. W., 1970. How many marine invertebrate fossil species? A new approximation. J. Paleontol. 44:410-415.

Valentine, J. W., 1972. Conceptual models of ecosystem evolution. In Schopf, T. J. M. (ed.), Models in paleobiology, San Francisco: Freeman, Cooper, 192-215.

Valentine, J. W., 1973. Evolutionary paleoecology of the marine biosphere. Englewood Cliffs, N.J.: Prentice-Hall, 511.

Valentine, J. W., Foin, T. C., and Peart, D., 1978. A provincial model of Phanerozoic marine diversity. Paleobiology 4:55-66.

Valentine, J. W., and Jablonski, D., 1983. Larval adaptations and patterns of brachiopod diversity in space and time. Evolution 37:1052-1061.

Valentine, J. W., and Moores, E. M., 1972. Global tectonics and the fossil record. J. Geol. 80:167-184.

Valentine, J. W., and Moores, E. M., 1973. Provinciality and diversity across the Permian-Triassic boundary. In Logan, A., and Hills, L. V. (eds.), The Permian and Triassic systems and their mutual boundary, Can. Soc. Petrol. Geol. Mem. 2:759-766.

Waterhouse, J. B., 1973. The Permian-Triassic boundary in New Zealand and New Caledonia and its relationship to world climatic changes and extinction of Permian life. In Logan, A., and Hills, L. V. (eds.), The Permian and Triassic systems and their mutual boundary, Can. Soc. Petrol. Geol. Mem. 2:445-464.

Waterhouse, J. B., and Bonham-Carter, G., 1976. Range, proportionate representation, and demise of brachiopod families through Permian Period. Geol. Mag. 113:401-428.

Wise, K. P., and Schopf, T. J. M., 1981. Was marine faunal diversity in the Pleistocene affected by changes in sea level? Paleobiology 7:394-399.

Wright, J. E., 1982. Permo-Triassic accretionary subduction complex, southwestern Klamath Mountains, Western California. J. Geophys. Res 87:3805-3818.

Chapter 11

MODELING THE BIOGEOGRAPHIC
REGULATION OF EVOLUTIONARY RATES

KARL W FLESSA

Department of Geosciences, University of Arizona, Tucson

RICHARD H. THOMAS

Department of Ecology & Evolutionary Biology, University of Arizona, Tucson

> Next to determining the question whether species have real existence, the consideration of the laws which regulate their geographical distribution is a subject of primary importance to the geologist. It is only by studying these laws with attention, by observing the positions which groups of species occupy at present, and inquiring how these may be varied in the course of time by migrations, by changes in physical geography, and other causes, that we can hope to learn whether the duration of species be limited, or in what manner the state of the animate world is affected by the endless vicissitudes of the inanimate.
>
> Charles Lyell, *Principles of Geology*, 1832, p. 66.

INTRODUCTION

The geographical distribution of organisms formed one of the major foundations of Darwin's demonstration of the fact of evolution. Since that time, however, biogeography has played a largely tangential role in the development of the theory of evolution. With some notable exceptions, much of the biogeographic literature became occupied with the delineation of faunal and floral zones and the explication of the historical factors that led to their establishment. Even Mayr's development of the allopatric model (Mayr, 1942), despite its obvious geographic focus, did not emerge from the traditional biogeographic research of the times.

Recently, however, there has been a rekindling of interest in the role of geography in the origin and disappearance of species. The complementary approaches of vicariance biogeography and phylogenetic systematics are generating a re-examination of biogeographic and systematic patterns (see, for example, Nelson and Platnick, 1980). The role of a dynamic geography in regulating many of the major features of Phanerozoic diversity has become increasingly apparent (Valentine and Moores, 1972; Flessa and Imbrie, 1973). And the MacArthurian approach to geographic ecology (see, for example, MacArthur, 1972; Cody and Diamond, 1975) has added immeasurably to our understanding of the fine geographic structure of adaptation, speciation and extinction.

In the spirit of this biogeographic revival of interest in the evolutionary process, we seek to analyze the evolutionary implications of the frequency distribution of taxa among regions. We term the frequency distribution of taxa in increasing numbers of regions (or some other quantitative measure of geographic range), a *biogeographic frequency distribution*. The shape of such frequency distributions is the result of the interaction of the probabilities of speciation, extinction, and range expansion. In our analyses, we hope to gain some insight into the biogeographic regulation of evolutionary rates. To what extent are the probabilities of range expansion, extinction, and speciation governed by the geographic range of the species?

THE BIOGEOGRAPHIC FREQUENCY DISTRIBUTION
OF GENERA OF MARINE BIVALVE MOLLUSCS

We shall illustrate the derivation (and some of the problems inherent in its construction) of a biogeographic frequency distribution by reference to the genera of marine bivalve molluscs. The resulting biogeographic frequency distribution is, we feel, typical of a variety of different animals and plants, and forms the basis for subsequent modeling.

MATERIALS AND METHODS

Lists of the genera of marine bivalve molluscs present in each of the 71 regions shown in Figure 1 were assembled from the published literature. Literature sources are given in the Appendix. Although many of the faunas correspond to conventional faunal provinces, some faunas are samples from geopolitical rather than biogeographic units. While no precise criteria were employed in the assembly of the data, efforts were made to include only the results of comprehensive sampling and reasonably modern systematic study. These general guidelines were occasionally relaxed in order to include faunal lists from more remote areas. Most of the studies included are less than 25 years old and most concentrate on the bivalve faunas of the shelf regions; few include deep-sea forms. The generic level was chosen to lessen the effects of poor or eccentric taxonomic work. The generic names were set to the standards of Vokes (1967). This procedure removed objective synonyms from further analysis. One thousand and thirty-two genera remained after this step. Problems associated with varying taxonomic practices and different systematic philosophies undoubtedly remain. These difficulties are considered below.

The data used here were assembled as part of a project on patterns of molluscan faunal similarity and will be published in a future paper. The lists of genera are also available on request from the senior author.

DISCUSSION

The frequency with which genera occur in increasing numbers of regions is shown in Figure 2. This distribution shows that most genera are confined to one or two regions and that the numbers of genera that occur in increasing numbers of regions fall off rapidly and then persist at very low values. In other terms, the proportion of bivalve genera that are cosmopolitan is very small, while the endemic genera make up the vast majority of the bivalves. This biogeographic frequency distribution is another expression of the fact that bivalve faunal provinces tend to be quite distinct even at the generic level (see Campbell and Valentine, 1977). Two genera share the title of "champion cosmopolitan"–*Nucula* and *Hiatella*.

Despite our efforts at nomenclatorial standardization, many features of this biogeographic frequency distribution are influenced by the mixture of taxonomic practices and systematic philosophies represented by our compilation of data. Consider, for example, the effect of genera that are broadly defined (i.e., "wastebasket" genera)–*Ostrea* may be a good example. Such genera will tend to elongate the tail of the distribution. An opposite effect is caused by genera that are erected as a consequence of parochial systematic practices. This is the problem caused by subjective synonyms. Such genera will tend to elevate the peak of the distribution. An additional effect results from our use of the generic level. Genera with many species will tend, on the average, to be widespread, even if their component species are endemic. These collectively cosmopolitan genera will tend to elongate the tail of the distribution.

The basic pattern of the strongly concave distribution of genera among regions may be exaggerated by the biasing factors mentioned above, but we doubt that the pattern is generated solely by the biases. This conclusion stems not only from our faith in the quality of the data but from the similarity of the biogeographic frequency distribution of other groups. Anderson (1977) presents a frequency distribution of species of mammals among areas of increasing size within North America. The distribution is strongly concave. Anderson used a single compilation of species and we judge that the problems of splitting and lumping are less severe among students of North American mammals than they are among students of the world ocean's bivalve molluscs. Hansen (1980) presents biogeographic frequency distributions of species of volutid gastropods from the Tertiary of the Gulf Coast (shown here as Figure 5) and Jablonski (1982) makes a similar plot for Late Cretaceous gastropods of the Gulf and Atlantic Coastal Plain. In both instances, the frequency distributions of the species with nonplanktotrophic larval stages are strongly concave.

A disquieting exception to the pattern we describe here is provided by Jablonski and Valentine (1981). They present biogeographic frequency distributions of species from 13 families of marine bivalves and gastropods from the eastern Pacific shelf. The distributions are quite varied in shape and few even approach the strongly concave pattern exhibited in Figure 2. Their measure of geographic range (kilometers of shoreline rather than number of regions) plus the strong provinciality of the eastern Pacific shelf may explain the discrepancy. Little intraprovincial geographic restriction will tend to produce species ranges whose lower limits will be no less than the geographic extent of a single province. Under such circumstances, the characteristic peak of the strongly concave curve is difficult to produce. Indeed, the tendency will be toward a polymodal biogeographic frequency distribution with each mode corresponding to the geographic breadth of a particular province.

Substantial assurance that the distinctively concave biogeographic frequency distribution is not only real, but characteristic of a variety of animals and plants comes from the work of Willis (1922). In fact, our strongly concave distribution is simply the "hollow curve" of distribution discovered by Willis.

THE HOLLOW CURVE

In 1922 the botanist J. C. Willis proposed his hypothesis of "age and area" in a book of that title (see also Willis and Yule, 1922). Willis' hypothesis was that, as a general rule, endemic species were geologically young while widespread species were relatively old. Willis came to this conclusion as a result of

---

Figure 1 (facing page). Location of marine bivalve faunas. Numbers refer to literature sources listed in Appendix.

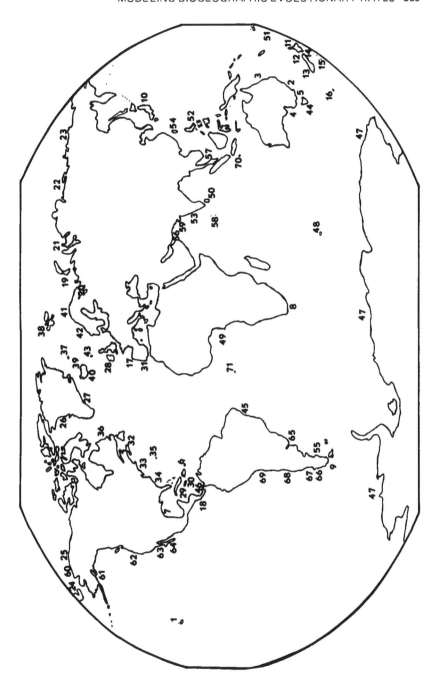

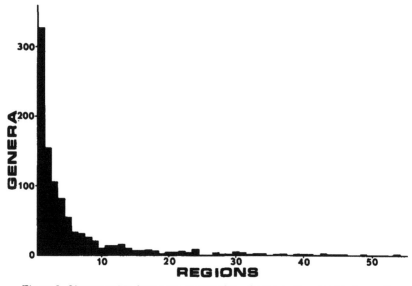

Figure 2. Biogeographic frequency distribution of genera of marine bivalve molluscs.

his studies of the distribution of species among genera and of species and genera among areas. In both cases, Willis found a characteristic distribution: most genera were monotypic and the numbers of those that were increasingly polytypic rapidly decreased and then slowly declined; most species were endemic and the numbers of those that were increasingly widespread rapidly decreased and then slowly declined. To Willis, the prevalence of the hollow curve among many groups was evidence for some mechanical process of evolution that generated new species at a more-or-less regular rate. In Willis' view, the longer a species persisted, the more widespread it would become (thus the correlation between age and area). The regularity of the distribution of species among genera argued against the idea then current that speciose genera were somehow better adapted and thus more successful than monotypic genera. Indeed, Yule (1924) demonstrated mathematically that "luck" was all that was necessary to become a polytypic genus. To Willis, all this was evidence against the importance of natural selection and adaptation in the formation of new species and new genera. Willis argued for an evolutionary mode driven by the pressures of mutation—"mutations that were at times of rank sufficient to give rise to Linnean species, genera, or even families" (Willis and Yule, 1922: 179.)

Willis' conclusions were not well received. (For a more comprehensive review of the impact of Willis' ideas, see Anderson, 1974.) Willis' argument that endemic species tended to be newcomers rather than ancient relicts was severely criticized (Sinnott, 1917; Bateson, 1923). The hollow curve of species among genera was shown to be susceptible to the vagaries of systematic practice (Chamberlain, 1924). Wright (1941) demonstrated that this hollow curve did not require an evolutionary mode that was driven by mutation. Bateson (1923) deemed the hollow curve of geographic distribution a simple consequence of the distribution of species among genera. This last criticism is worthy of special attention. As mentioned above, speciose genera will tend to be more widespread than paucispecific ones. Indeed, "What else could we expect?" (Bateson, 1923:43). Although many of Willis' hollow curves are based on the distribution of genera among areas, a significant number (ten) are based on the distribution of *species* among areas occupied. Bateson's (1923) criticism was not entirely fair.

Efforts to simulate the hollow curve with simple models of speciation and extinction have focused entirely on the frequency distribution of species within genera. Yule (1924) and Anderson and Anderson (1975) developed models based on random rates of speciation and extinction. Their models easily generated the characteristic hollow curve. The hollow curve of the distribution of species among genera needs no special explanation: genera with many species do not necessarily differ in evolutionary properties from genera with one or only a few species. The distribution of the sizes of genera in nature is what is to be expected from stochastic speciation and extinction. This conclusion lends credence to the assumptions used in the simulations of Raup et al. (1973). If, as seems likely, the distribution of species among areas is independent of the distribution of species among genera, what additional insights can be provided by simulating the hollow curve of geographic distribution? Do biogeographic patterns regulate the probability of certain evolutionary events?

THE MODEL

Consider the first species of a clade. Given that its mode of origin involves a small population, it will be present, initially, only within some small area. Within some subsequent interval of time, there will be a certain probability that the species will: a) extend its geographic range to include more areas, b) contract its geographic range (local or terminal extinction), and c) produce a new species. Although the sum of the probabilities of range expansion and extinction must be less than or equal to one, the probability of speciation is independent of either expansion or extinction. The probability that the species does nothing at all is simply the difference between 1.0 and the sum of the probabilities of expansion and extinction times the difference between 1.0 and the probability of speciation.

The essential components of our simulation are shown in Figure 3. Each of the transitions indicated by the arrows is assigned a probability. Note that our model considers terminal extinction (the transition from one to zero regions) as simply another expression of range contraction (local, "ecological" extinction). The model permits only one step expansion or contraction per time interval. Skipping regions (from 2 to zero, for example) is not allowed. Speciation (EVO in Figure 3) is modeled to proceed by adding new species to the category of one region only. This was done not simply for computational convenience. It reflects our view that most speciation involves branching and is most often accomplished in small populations. A phyletic style of speciation is permitted by our model to the extent that branching is involved and that the gradual transition from the ancestral species (one that persists in other regions) is restricted to a single region. While phyletic speciation might occur over several regions in nature, it is not recognized by our model. Phyletic speciation without branching is irrelevant here because that process adds a "new" species at the expense of an "old" one. No change in the total number of species or in their distribution among areas need result.

AN EXAMPLE

Figure 4 illustrates the workings of the model through four time intervals. At Time 1, probabilities are assigned to each transition and the simulation is "seeded" with 100 species. As shown in Figure 4, this particular simulation specifies decreasing probabilities of extinction (EXT) and speciation (EVO) with increasing geographic range. The probability of range expansion (EXP) is set to increase with increasing geographic range.

At Time 2, the 0.1 probability of range expansion has caused 10 species to extend their range to include two regions. Fifty species have gone extinct (0.5 x 100), while 60 new species (0.6 x 100) make their first appearance in a single region. Time 2, in this example, fortuitously results in no net change for the number of species restricted to one region $(100 - 10 - 50 + 60 = 100)$. There has been substantial turnover, however.

At Time 3, two species (0.2 x 10) have extended their range from two to three regions, ten have spread from one to two (0.1 x 100), while 4 (0.4 x 10) have contracted from two to one. This results in a total of 14 species that occur in two regions $(10 - 2 - 4 + 10 = 14)$. Meanwhile, 50 species that were restricted to one region have gone extinct (0.5 x 100), ten have expanded to two regions (0.1 x 100), four species have retreated from two regions to one (0.4 x 10) while 60 (0.6 x 100) plus five (0.5 x 10) species have been added by speciation from one and two regions respectively. All this shuffling about results in a sum of 109 species now restricted to one region $(100 - 10 - 50 + 4 + 60 + 5 = 109)$.

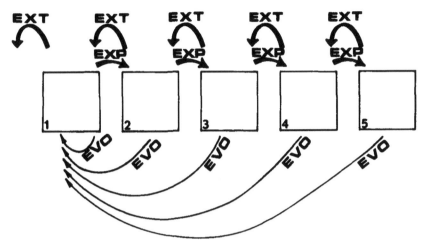

Figure 3. Essential components of model for the biogeographic regulation of evolutionary probabilities. Boxes indicate the number of regions occupied. EXP is the probability of range expansion, EXT is the probability of extinction (range contraction), and EVO is the probability of speciation. See text for further explanation.

By Time 4, the shape of the biogeographic frequency distribution begins to emerge. One species (0.3 x 2, rounded off) has spread from three to four regions. One species (0.3 x 2) has contracted from three to two regions, while three species (0.2 x 14) have expanded from two to three regions. This leaves a total of three species that occur in three regions $(2 - 1 - 1 + 3 = 3)$. Three species have expanded from two to three regions (we will spare the reader the details of the arithmetic at this point), one has retreated from three regions back to two, six species have withdrawn from two regions back to one, while eleven have spread from one region to two. This leaves a total of 18 species that occur in two regions $(14 - 2 - 6 + 1 + 11 = 18)$. Finally, the number of species that occur in only one region now totals 120. This results from the loss of 11 species that expanded their range to include two regions, the loss of 54 species through terminal extinction, the contribution of 3 species formerly found in two regions, and the addition of 65, 7, and 1 species from the evolution among species in one, two, and three regions, respectively $(109 - 11 - 54 + 3 + 65 + 7 + 1 = 120)$.

Computers are well-suited for this sort of tedious and repetitive arithmetic. One was used in all our simulations.

THE SIMULATION PROGRAM

The model was programmed to generate a distribution of species among 25 regions. Simulations were terminated upon the attainment of either a stable distribution (one in which the ratios of species in adjacent categories were constant from time interval to time interval) or the passage of 200 time intervals. The effects of varying the probabilities of range expansion, extinction, and speciation on the attainment and rate of approach to stability are currently under study and will not be considered here. A different combination of rates of expansion, extinction, and speciation was specified for each run. A listing of the program is available from the senior author.

The simulations were done to determine the effects of different combinations of evolutionary probabilities on biogeographic frequency distributions. We assigned each transition (expansion, extinction, and speciation) a probability that either increased, decreased, or remained constant with an increase in geographic range. Thus, each probability function can have a positive, negative, or zero slope. For simplicity's sake, we consider only linear functions. In the simulations, each positive slope specifies an increasing probability with increasing geographic range. Probabilities increase linearly from 0.30 to 0.49. Each negative slope specifies a decreasing probability with increasing geographic range. Probabilities decrease linearly from 0.49 to 0.30. Each zero slope specifies a constant probability with increasing geographic range. Probabilities are set at 0.40. We consider only the 27 basic combinations of slope shown in Table 1. Given these slopes and ranges of probabilities, only models two through nine and model 18 attained stability within 200 time intervals. Obviously, the actual number of possible combinations of slopes, intercepts, and non-linear functions approaches infinity. Examination of the effects of all the possibilities not considered here must await an analytical solution to the generation of the biogeographic frequency distribution. In the meantime, we hope these 27 simulations are instructive in their own right. We simply pose the question: what combination of slopes of probabilities are necessary for the production of the biogeographic frequency distributions seen in nature? We should note that our model requires us to pose the question in this fashion. A more realistic question would avoid the assumption of cause and effect between geographic range and evolutionary probability. Geographic range may, after all, simply be influenced by the same factors that govern evolutionary events. We suspect that this is at least partially the case, as will be seen in the subsequent discussions.

Figure 4 (facing page). Example of a model through four time intervals following a "seeding" with 100 species in Time 1. See text for explanation.

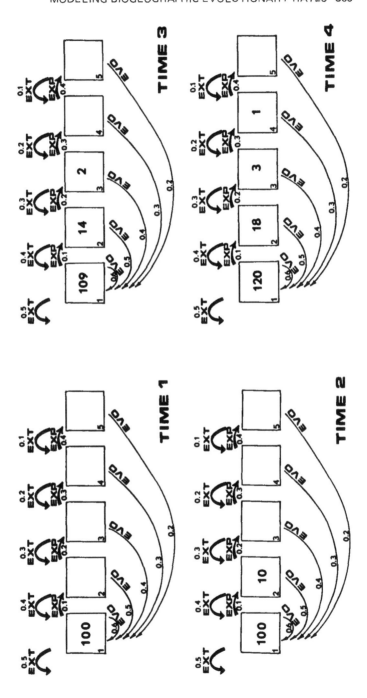

Table 1. Combinations of slopes used in simulations and results of tests of concavity and fossil record. + indicates probability increases with increasing geographic range; − indicates decrease; 0 indicates constant probability.

| | SLOPE OF PROBABILITY OF | | | PASSES TEST OF | |
| | Range | | | | |
Model	Expansion	Extinction	Speciation	Concavity	Fossil Record
1	+	+	+	yes	no
2	+	+	0	yes	no
3	+	+	−	yes	no
4	+	0	+	yes	no
5	+	0	0	yes	no
6	+	0	−	yes	no
7	+	−	+	yes	yes
8	+	−	0	yes	yes
9	+	−	−	yes	yes
10	0	+	+	no	no
11	0	+	0	no	no
12	0	+	−	no	no
13	0	0	+	no	no
14	0	0	0	no	no
15	0	0	−	no	no
16	0	−	+	no	yes
17	0	−	0	no	yes
18	0	−	−	no	yes
19	−	+	+	no	no
20	−	+	0	no	no
21	−	+	−	no	no
22	−	0	+	no	no
23	−	0	0	no	no
24	−	0	−	no	no
25	−	−	+	no	yes
26	−	−	0	no	yes
27	−	−	−	no	yes

GENERATING THE HOLLOW CURVE:
THE TEST OF CONCAVITY

The feature of the biogeographic frequency distribution that gives the hollow curve its name is the pronounced upward concavity of the curve. Thus, the principal criterion by which we evaluate the fit of our simulated distributions to the natural hollow curve is concavity. Our application of the test of concavity is deliberately visual and thus, somewhat subjective. We feel, however, that the variation within natural hollow curves and the constraints of our simulations (linear functions, etc.) make application of elaborate curve-fitting procedures inappropriate. As will be seen, the results of the test are rather clear.

Table 1 shows that the only combinations that pass the test of concavity are those in which the probability of range expansion increases with increasing geographic range. From the point of view of the frequency distribution, this is an eminently reasonable result: how else could a long, low tail of cosmopolitan species be produced?

The biological justification for the increasing probability of range expansion with increasing geographic range is also easily arrived at. It seems reasonable that those species that are widespread will have a larger proportion of good dispersers among them than those species that are endemic. Like the rich getting richer, the cosmopolitans tend to get more cosmopolitan. The association of a wide geographic range with high dispersal ability is certainly part of the conventional wisdom of biogeography. Direct documentation comes from a variety of sources, only some of which will be mentioned here. Shuto (1974), Scheltema (1977), and Hansen (1980) document the tendency for gastropods with relatively long-lived planktic larval stages to have extensive geographic ranges. Species of gastropods with non-planktic larval stages, on the other hand, tend to be more narrowly distributed. Jablonski (1979) has detailed a similar pattern among Cretaceous bivalve molluscs. Flessa (1981) shows that the degree of intercontinental mammalian faunal similarity increases when bats are included in the analysis. Clearly, dispersal mode is important in determining a species' range.

The effect of the mode of dispersal on the shape of the biogeographic frequency distribution is shown dramatically by Hansen (1980) and Jablonski (1982). Hansen's figure, reproduced here as Figure 5, demonstrates that the long, low tail of the hollow curve of early Tertiary volutid gastropods is generated by species with a planktic mode of larval dispersal. The tall peak of endemics is characterized by species with a non-planktic mode of larval dispersal.

Thus, the hollow curve of geographic distribution simply results from the increasing probability of range expansion with increasing geographic range. Any combination of speciation and extinction probabilities may be associated with a positive slope of range expansion. Models one through nine pass the test of concavity.

FURTHER CONSTRAINTS ON EVOLUTIONARY ODDS:
THE TEST OF THE FOSSIL RECORD

Although a concave biogeographic frequency distribution results from any combination of speciation and extinction probability functions associated with an increasing probability of range expansion, evidence from the fossil

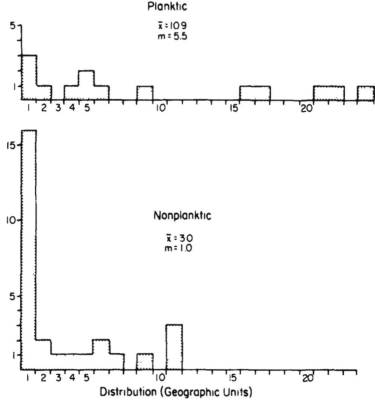

Figure 5. Biogeographic frequency distribution of Lower Tertiary volutid gastropods, Gulf Coast, U.S.A. Species with planktic larval stage shown in upper histogram, species with nonplanktic larval stage below. $\bar{x}$ indicates mean and m is median geographic range, expressed as the number of geographic units occupied. Reproduced from Hansen (1980) and used with the kind permission of *Paleobiology* and Thor Hansen.

record appears to further constrain the number of models remaining. More than fifty years after its formulation, Willis' age and area hypothesis finds increasing support. Widespread species tend to be geologically long-lived, endemic species tend to be geologically short-lived.

Willis (1922) pictured the age and area relationship as a consequence of the regular, "clockwork" production and dispersal of species. The longer a species persisted, the more area it was likely to occupy. Thus, new species were endemics; old species became cosmopolitan through mere persistence. Some evidence for Willis' postulated relationship between age and area was provided by Small (1922), but support from the fossil record was scanty and weak.

Recent work by many authors suggests that a species may be long-lived because it is widespread—a fascinating reversal of the interpretation of cause and effect. Ordovician through Jurassic bivalves (Bretsky, 1973), Siluro-Devonian brachiopods (Boucot, 1975), Recent, tropical Venerid bivalves (Jackson, 1974), Tertiary neogastropods (Hansen, 1980), Cretaceous molluscs (Jablonski, 1980; Koch, 1980), and Tertiary bivalves (Hoffman and Szubzda-Studencka, 1982) display a positive correlation between geographic range and geologic duration. Geologically long-lived species do tend to be more widely distributed than short-lived species. The cause of this relationship is taken to be the evolutionary "security" provided by a broad geographic distribution and the biological properties associated with that distribution. Jackson (1974), for example, points out that widespread species, by virtue of their apparent powers of dispersal, are capable of recolonizing areas affected by regional environmental catastrophes, while endemic species run the risk of terminal extinction. Species may be also widespread because they are eurytopic, regardless of their means or effectiveness of dispersal. Both Jackson (1974) and Jablonski and Valentine (1981) demonstrate a tendency for bivalve species inhabiting waters less than one meter's depth to have a greater geographic range than deeper water species. The shallow water species, it is argued, have broader environmental tolerances. This association of eurytopy with cosmopolitanism provides double protection from many of the causes of extinction. Boucot (1975) suggests that cosmopolitan species have large populations and that it is the larger size of the population that provides protection from extinction and inhibits speciation. Koch (1980) provides the only test of this hypothesis and finds it unsupported by the data.

That the relationship between geographic distribution and evolutionary longevity is one of cause and effect, respectively, is supported by the predominance of planktic larval stages among widely distributed species of gastropods (Figure 5). If a broad distribution was simply a consequence of longevity, larval types would be distributed randomly along a scale of increasing geographic range. Furthermore, Hoffman and Szubzda-Studencka (1982) point out that broad geographic distributions were attained quite rapidly in 20 out of 31 species of bivalves that originated in the Middle Miocene of eastern Europe.

The fossil record suggests that the probability of extinction decreases with increasing geographic range. Models seven, eight, and nine (Table 1) pass both the test of concavity and the test of the fossil record.

A POSITIVE CORRELATION OF PROBABILITIES
OF EXTINCTION AND SPECIATION?

Is there any reason to choose one of the remaining models rather than another? What is the nature of the relationship between the probability of speciation and geographic range? Little evidence is available, and what there is is often characterized by reasoning many steps removed from the data.

Again, Hansen's (1980) analysis of the duration and distribution of volutid gastropods in the Gulf Coast Tertiary provides some insight. Species with a nonplanktic larval stage show a slight tendency toward greater speciation rates than do the species with a planktic larval stage (Hansen, 1980:Fig. 10). Insofar as those species with a planktic larval stage are widespread, the probability of speciation seems to decrease with increasing geographic range. More evidence for this correlation, though even more indirect, is provided by Stanley (1980:260-265).

The lack of much direct evidence notwithstanding, the logic of the decreasing probability of speciation with increasing geographic range has not escaped many authors. Jackson (1974), Shuto (1974), Scheltema (1977), Vermeij (1978), and Hansen (1980) have all pointed out that those biological properties that result in wide geographic ranges (effective dispersal phases, eurytopy) will act to retard speciation. Populations of species with formidable powers of dispersal will be less likely to become isolated than populations of species with limited dispersal abilities. There is an obvious counter-argument. As the size of a species' range increases, the chances that that range will become dissected by some vicariant event will also increase. Thus, the probability of speciation would *increase* with increasing geographic range. This is the basis of Rosenzweig's (1975) argument for continental steady-state diversity. But this range-dissection argument has force only if the assumption is made that geographic events affect all species equally. A species that is widespread because of an effective dispersal phase may not be sensitive to the event that caused the geographic isolation of a species without an effective dispersal phase. If geographic range was the only variable and the biological properties of species were unimportant, then a case could be made for a negative correlation of extinction and speciation probabilities. We doubt that this will be possible.

However tentatively, we suspect that as geographic range increases, the probability of speciation will likely decrease. Model nine (Table 1) seems to be the most likely combination of evolutionary probabilities.

CONCLUSIONS

1. The distribution of genera of marine bivalve molluscs among 71 regions follows a "hollow curve." Most genera are restricted to one or a few regions while the numbers of those that occur in more and more areas fall off rapidly and then persist at a low level. Endemics are most frequent while cosmopolitans are few in number.

2. The similarity of this hollow curve to that displayed by species of mammals, gastropods, and many other animals and plants suggests that it is a common biogeographic pattern.
3. Simulations in which the probability of range expansion increases with increasing geographic range produce the hollow curve. The frequency distribution is not sensitive to the probabilities of extinction or speciation.
4. The fossil record of molluscs and brachiopods indicates that cosmopolitan species tend to have longer geologic ranges. The probability of extinction appears to decline with increasing geographic range.
5. Because those properties that enhance geographic spread tend to retard speciation, the probability of speciation may also decline with increasing geographic range.

ACKNOWLEDGMENTS

We thank the University of Arizona for its support of this research and Tony Murer for his help in assembling the faunal lists. We are grateful to Sydney Anderson, Thor Hansen, Antoni Hoffman, and David Jablonski for their helpful reviews of an earlier draft of this paper. R. Thomas was supported in part by an NSF Graduate Fellowship.

REFERENCES

Anderson, S., 1974. Patterns of faunal evolution. Quart. Rev. Biol. 49:311-332.

Anderson, S., 1977. Geographic ranges of North American terrestrial mammals. Amer. Mus. Novit. (2629).

Anderson, S., and Anderson, C. S., 1975. Three Monte Carlo models of faunal evolution. Amer. Mus. Novit. (2563).

Bateson, W., 1923. Area of distribution as a measure of evolutionary age. Nature 111:39-43.

Boucot, A. J., 1975. Evolution and extinction rate controls. Amsterdam: Elsevier.

Bretsky, P. W., 1973. Evolutionary patterns in the Paleozoic Bivalvia: Documentation and some theoretical considerations. Geol. Soc. Amer. Bull. 84:2079-2096.

Campbell, C. A., and Valentine, J. W., 1977. Comparability of modern and ancient marine faunal provinces. Paleobiology 3:49-57.

Chamberlain, J. C., 1924. The hollow curve of distribution. Amer. Nat. 58: 350-374.

Cody, M. L., and Diamond, J. M. (eds.), 1975. Ecology and evolution of Communities. Cambridge, Mass.: Belknap/Harvard Univ. Press.

Flessa, K. W., 1981. The regulation of mammalian faunal similarity among the continents. J. Biogeogr. 8:427-437.

Flessa, K. W., and Imbrie, J., 1973. Evolutionary pulsations: Evidence from Phanerozoic diversity patterns. In Tarling, D. H., and Runcorn, S. K. (eds.), Implications of continental drift to the earth sciences, New York: Academic Press, 1:247-285.

Hansen, T. A., 1980. Influence of larval dispersal and geographic distribution on species longevity in neogastropods. Paleobiology 6:193-207.

Hoffman, A., and Szubzda-Studencka, B., 1982. Bivalve species duration and ecologic characteristics in the Badenian (Miocene) marine sandy facies of Poland. N. Jb. Geol. Palaont. Abh. 163:122-135.

Jablonski, D., 1979. Paleoecology, paleobiogeography, and evolutionary patterns of Late Cretaceous Gulf and Atlantic Coastal Plain mollusks. Unpubl. Ph.D. dissert., Yale Univ.

Jablonski, D., 1980. Apparent versus real biotic effects of transgression and regressions. Paleobiology 6:397-407.

Jablonski, D., 1982. Evolutionary rates and modes in Late Cretaceous gastropods: Role of larval ecology, 3rd North Amer. Paleontol. Conv. Proc. I: 257-262.

Jablonski, D., and Valentine, J. W., 1981. Onshore-offshore gradients in Recent eastern Pacific shelf faunas and their paleobiogeographic significance. In Scudder, G. G. E., and Reveal, J. L. (eds.), Evolution today, 2nd Internat. Congr. System. Evolu. Biol. Proc., Pittsburgh, Pa.: Hunt Institute, 441-453.

Jackson, J. B. C., 1974. Biogeographic consequences of eurytopy and stenotopy among marine bivalves and their evolutionary significance. Amer. Nat. 108:541-560.

Koch, C. F., 1980. Bivalve species duration, areal extent and population size in a Cretaceous sea. Paleobiology 6:184-192.

Lyell, C., 1832. Principles of geology, vol. 2. London: John Murray.

MacArthur, R. H., 1972. Geographical ecology. New York: Harper & Row.

Mayr, E., 1942. Systematics and the origin of species. New York: Columbia Univ. Press.

Nelson, G. J., and Platnick, N. I., 1980. Systematics and biogeography. New York: Columbia Univ. Press.

Raup, D. M., Gould, S. J., Schopf, T. J. M., and Simberloff, D. S., 1973. Stochastic models of phylogeny and the evolution of diversity. J. Geol. 81: 525-542.

Rosenzweig, M. L., 1975. On continental steady states of species diversity. In Cody, M. L., and Diamond, J. M. (eds.), Ecology and evolution of communities, Cambridge, Mass.: Belknap/Harvard Univ. Press, 121-140.

Scheltema, R. S., 1977. Dispersal of marine invertebrate organisms: Paleobiogeographic and biostratigraphic implications. In Kauffman, E. G., and Hazel, J. E. (eds.), Concepts and methods of biostratigraphy, Stroudsburg, Pa.: Dowden, Hutchinson and Ross, 73-108.

Shuto, T., 1974. Larval ecology of prosobranch gastropods and its bearing on biogeography and paleontology. Lethaia 7:239-256.

Sinnott, E. W., 1917. The "age and area" hypothesis of Willis. Science 46: 457-459.

Small, J., 1922. Age and area, and size and space, in the Compositae. In Willis, J. C., Age and Area, London: Cambridge Univ. Press, 119-136.

Stanley, S. M., 1979. Macroevolution. San Francisco: W. H. Freeman.

Valentine, J. W., and Moores, E. M., 1972. Global tectonics and the fossil record. J. Geol. 80:167-184.

Vermeij, G. J., 1978. Biogeography and adaptation. Cambridge, Mass.: Harvard Univ. Press.

Vokes, H. E., 1967. Genera of the Bivalvia: A systematic and bibliographic catalogue. Amer. Paleont. Bull. 51(232):103-394.

Willis, J. C., 1922. Age and area. London: Cambridge Univ. Press.

Willis, J. C., and Yule, G. U., 1922. Some statistics of evolution and geographical distribution in plants and animals, and their significance. Nature 109: 177-179.

Wright, S., 1941. The "age and area" concept extended. Ecology 22:345-347.

Yule, G. U., 1924. A mathematical theory of evolution, based on the conclusions of Dr. J. C. Willis, F.R.S. Phil. Trans. Roy. Soc. London, Ser. B 213:21-87.

APPENDIX

Literature sources used. Numbers refer to areas shown in Figure 1.

(29)   Abbott, R. T., 1958. The marine molluscs of Grand Cayman Island, British West Indies. Mon. Acad. Nat. Sci. Philadelphia, 138.

(32-36)   Abbott, R. T., 1974. American seashells, the marine Mollusca of the Atlantic and Pacific coasts of North America. New York: Van Nostrand Rheinhold, 663.

(8)   Barnard, K. H., 1974. Contributions to the knowledge of South African marine Mollusca, Pt. VII. Revised faunal list. Ann. South African Mus., 47(Pt. 5):663-781.

(55)   Carcelles, A. R., 1950. Catalogo de los Moluscos marinos de Patagonia. Anales Mus. Nahuel Huapi Perito Francisco P. Moreno, II: 41-99.

(9)   Carcelles, A. R., and Williamson, S. I., 1951. Catalogo de los Moluscos marinos de la Provincia Magallanica. Rev. Inst. Nac. Investigacion Ciencias Naturales, Ciencias Zoologicas, II (5), Buenos Aires, 383.

(4)   Cotton, B. C., and Godfrey, F. K., 1938. The molluscs of South Australia, Pt. 1, The Pelecypoda. So. Australian Branch Brit. Sci. Guild, Adelaide, 314.

(1)   Dall, W. H., Bartsch, P., and Rehder, H. A., 1938. A manual of

the Recent and fossil marine pelecypod mollusks of the Hawaiian Islands. Bernice P. Bishop Mus. Bull. 153; 233.

(65)   Figueras, A., and Sicardi, D. E., 1968-1970. Catalogo de los molluscos marinos del Uruguay, Pt. II, III, IV. Communicaciones de la Sociedad Malacologica del Uruguay, II (15, 16, 17, 18): 255-275, 355-378, 407-424.

(19-24)   Filatova, Z. A., 1957. General review of the bivalve mollusks of the northern seas of the U.S.S.R. In Nikitin, B. N. (ed.), Marine biology, Moscow: Inst. Oceanology Trans. 20 (Am. Inst. Biol. Sci. trans., 1959), 302.

(3)   Hedley, C., 1910. The marine fauna of Queensland. Australian Assoc. Adv. Sci. Mtg. 12th, Brisbane, 1909, Rept., sec. D.:329-371.

(52)   Hidalgo, J. G., 1904-1905. Catalogo de los Moluscos Testaceos de las Ilas Filipinas, jo lo y Marianas, I: Moluscos marinos. Madrid: Gaceta de Madrid, 408.

(17)   Hidalgo y Rodrigues, J. G., 1917. Fauna malacologica de España, Portugal y las Baleares. Serie Zoologica Trabajos del Mus. Nacional Ciencias Naturales, Madrid (30), 752.

(30)   Humfrey, M. 1975. Sea shells of the West Indies, a guide to the marine molluscs of the Caribbean. London: William Collins Sons & Co. Ltd., 351.

(2)   Iredale, T., and McMichael, D. F., 1962. A reference list of the marine Mollusca of New South Wales. Sydney: Australian Mus. Mem. II, 109.

(54)   Kaneko, S., 1943. Shells from Formosa (lists). Nat. Hist. Soc. Taiwan Trans. 33:660-677.

(18)   Keen, A. M., 1958. Sea shells of tropical west America; Marine mollusks from lower California to Columbia. Stanford, Calif.: Stanford Univ. Press, 624.

(56)   Khan, M. D., and Dastagir, S. G., 1971. On the Mollusca, pelecypod fauna of Pakistan. Karachi, Pakistan: Agr. Res. Council Pakistan Zool. Surv. Dept., 17-54.

(10)   Kira, T., 1955. Coloured illustrations of the shells of Japan. Osaka, Japan: Hoikusha, 204.

(50)   Kirtisinghe, P., 1978. Sea shells of Sri Lanka. Rutland, Vt.: Charles E. Tuttle Co., 202.

(59)   Kunda, H. L., 1965. On the marine fauna of the Gulf of Kutch, Pt. III Pelecypods. Bombay Nat. Hist. Soc. J. 62 (2):209-236.

(57)   Lynge, H., 1909. The Danish expedition to Siam 1899-1900, IV Marine Lamellibranciata. Copenhagen: D. KGL. Danske Vidensk. Selsk. Skrifter, 7, Rackke, Naturvidensk, O. G. Mathem. Afd., 3, 299.

(25)   MacGinitie, N., 1959. Marine mollusca of Point Barrow, Alaska. U.S. Nat. Mus. Proc. 109 (3412):59-208.

(5)   MacPherson, J. H., and Gabriel, C. J., 1962. Marine molluscs of Victoria. Melbourne: Victoria Natl. Mus. Handb. 2, 475.

(70)   Maes, V. O., 1968. The littoral marine mollusks of Cocos-Keeling Islands (Indian Ocean). Acad. Nat. Sci. Phila. Proc. 119 (4):93-217.

(44)   May, W. L., 1923. An illustrated index of Tasmanian shells. Tasmania: John Vail, Govt. Printer, 100.

(6)   McLean, R. A., 1951. The Pelecypoda or bivalve mollusks of Porto Rico and the Virgin Islands. N.Y. Acad. Sci., Sci. Survey of Porto Rico and Virgin Islands 17 (Pt. 1), 183.

(6)   Menon, P. K. B., Sareen, M. L., and Tandon, K. K., 1967. On the marine fauna of Goa: A preliminary survey, Pt. 2 Mollusca, Pelecypoda. Res. Bull. Punjab Univ. (N.S.) Sci., 18 (Pt. III-IV):315-320.

(49)   Marche-Marchad, I., 1958. Nouveau catalogue de la collection de mollusques Testaces marin de L'Ifan. Ifan, Dakar: Inst. Fr. D'Afrique Noire, XVI, 64.

(31)   Nobre, A., 1938-1940. Fauna malacologica de Portugal: I Molluscos marinhos e das aquas salabras. Barcelos, Companhia Editora do Minho, 806.

(26-27, 37-43)   Ockelman, W. K., 1958. Marine Lamellibranchiata. In The zoology of East Greenland, Medd. om Gronland, 122; 256.

(46)   Olsson, A. A., and McGinity, T. L., 1958. Recent marine mollusks from the Caribbean coast of Panama with the description of some new genera and species. Amer. Paleont. Bull. 39 (177):1-58.

(51)   Ostergaard, J. M., 1935. Recent and fossil mollusca of Tongatabu. Bernice P. Bishop Mus. Bull. 131; 59.

(11-16)   Powell, A. W. B., 1957. Shells of New Zealand: An illustrated handbook. Christchurch, N.Z.: Whitcombe and Tombs, Ltd., 202.

(47,48)   Powell, A. W. B., 1960. Antarctic and subantarctic mollusca. Auckland Inst. & Mus. Rec. 5 (3, 4):117-193.

(7)   Pulley, T. E., 1952. An illustrated checklist of the marine mollusks of Texas. Texas J. Sci. 4:167-199.

(45)   Rios, E. C., 1970. Coastal Brazilian seashells. Rio Grande, Brazil: Mus. Oceanogr. de Rio Grande, Fundacao Cidade do Rio Grande, 255.

(71)   Rosewater, J., 1975. An annotated list of the marine mollusks of Ascension Island, South Atlantic Ocean. Smithsonian Contr. Zool. (189):41.

(58)   Smith, E. A., 1906. The fauna and geography of the Maldive and Lackadive Archipelago, marine mollusca. Cambridge Univ. Press, II:1046-1057.

(66-69)   Soot-Ryen, T., 1959. Reports of the Lund University Chile Expedition 1948-1949; 35, Pelecypoda. Lunds Univ. Arsskriften N.F. avd. 2, bd. 55, n. 4, 6 Kungl. Fysiografiska Sallskapets Handlinger N.F.

(57)   Suvatti, C., 1950. Fauna of Thailand, mollusca.Bangkok, Thailand: Dept. Fish. Sec. 10:32-126.

(28)   Tebble, N., 1976. British bivalve seashells, a handbook for identification. Edinburgh, Scotland: Her Majesty's Stationery Office, 212.

(60-64) Valentine, J. W., 1978. Pers. commun.: compiled from sources listed in Campbell, C. A., and Valentine, J. W., 1977, Comparability of modern and ancient marine faunal provinces, Paleobiology 3:49-57.

Part IV

DIVERSITY PROFILES OF INDIVIDUAL CLADES

During the late nineteenth century, numbers of leading paleontologists attempted to infer evolutionary processes from the fossil record. They met with limited success, partly because they were looking for regularities in the history of clades which could be regarded as "laws" and used as definitive components of an evolutionary theory. Youthful, mature, elderly and senescent stages in the history of clades were sometimes claimed as a regularity; even second childhood was sometimes detected. Although this analogy between clade history and life history has been abandoned, there are still occasional suggestions as to how the spindle diagram of an average clade should look, given certain models of evolution and ecology.

In this section, two of the likeliest data sets available are analyzed for signs of agreement with the expectations of some predictive model. Hardy examines the trilobites in Upper Cambrian biomeres. These faunas are well-defined, having been studied in great detail systematically and stratigraphically. Ward and Signor, by contrast, review the record of ammonites through the entire Mesozoic Era. Although this is necessarily a more general study, vast labors have been expended on the stratigraphy and systematics of this group also. Strikingly, neither of these groups has yielded clade diversification patterns that are consistent with any particular predictions. Was this predictable?

A final paper (by Valentine) attempts to account for the volatility or jumpiness of clade diversities in terms of an ecological model which involves a biosphere kept open to evolutionary opportunity through extinction and the inhibiting effects on speciation of adaptive minima which act as barriers to diversification.

Chapter 12

TESTING FOR ADAPTIVE RADIATION:
THE PTYCHASPID (TRILOBITA)
BIOMERE OF THE LATE CAMBRIAN

MARGARET C HARDY

Dept. of Ecology & Evolutionary Biology, University of Arizona, Tucson*

INTRODUCTION

"It is at the lower taxonomic levels where evolutionary and ecologic theory is most robust and it is at this level where interpretive analysis may be the most fruitful—given the proper data base" (Raup, 1979).

Recently several authors (Gould et al., 1977; Sepkoski, 1978, 1979; Stanley, 1979; Valentine, 1980, 1981; Walker, 1985) have presented models of taxonomic diversity patterns during adaptive radiations. In order to avoid problems arising from the incomplete nature of the fossil record, the primary emphasis of these studies is on patterns of radiation in higher taxa, over long periods of time. An important assumption of these studies is that species diversity patterns follow those of higher taxa. While Sepkoski's (1978) simulations appear to support this assumption, documentation using species level data is lacking. This paper examines a trilobite radiation at the species level for those characteristics considered important in these previous studies.

The Ptychaspid Biomere is generally regarded as a well-documented adaptive radiation of Late Cambrian trilobites (Stitt, 1971a,b, 1975, 1977; Palmer, 1979). The data from this biomere were chosen because many of the notorious biases in the fossil record (poor or inconsistent preservation, variability due to geographic location, and the pull of the Recent; Raup, 1979) are not thought to be important factors. Due to the detailed nature of these data (see

*Current address: 220 Clarence St., No. 2, Richmond, CA 94801

below), the dynamics of adaptive radiation at the species level can be examined with regard to the models cited above.

An adaptive radiation may be defined as a divergent cluster of taxa of monophyletic origin whose proliferation is due primarily to the opening of a new ecospace (Valentine, 1973; Eldredge and Cracraft, 1980). This ecospace can either be a previously unoccupied habitat or a new "adaptive zone" (Simpson, 1944, 1953; Valentine, 1973; Eldredge and Cracraft, 1980). Eldredge and Cracraft (1980) point out that this proliferation of species should be rapid relative to a taxon's diversification rate before and after the radiation.

The literature on adaptive radiation falls into two basic categories: 1) empirical descriptions of overall patterns of various adaptive radiations; and 2) predictive models based upon the hypothesized mechanisms underlying adaptive radiation. The empirical patterns can be summarized as follows: Rapid speciation occurs at the beginning of a radiation with increasing extinction as the radiation proceeds (Simpson, 1944, 1953; Gould et al., 1977; Stanley, 1979; Eldredge and Cracraft, 1980). Thus, species durations are shortest during the early part of the radiation when evolutionary rates are high (Simpson, 1944, 1953; Stanley, 1979; Eldredge and Cracraft, 1980) and hollow (concave) survivorship curves are produced (Simpson, 1944, 1953; Stanley, 1979). Most of the higher taxa also originate at the beginning of the radiation (Valentine, 1977, 1980). Due to this early origination of taxa, a "heavy-bottomed" clade should characterize adaptive radiations—one with a center of gravity of less than 0.5 (Gould et al., 1977). Finally, some diversity-dependent factor(s) act(s) to limit the radiation so that a maximum diversity is attained and speciation then decreases relative to extinction (Simpson, 1944, 1953; Gould et al., 1977; Sepkoski, 1978, 1979; Stanley, 1979; Valentine, 1972, 1973).

With these basic patterns in mind, a number of models have been used to further predict overall patterns of adaptive radiation. Most of the models proposed so far postulate a sigmoidal pattern of diversity increase—a pattern which is directly analogous to the (Lokta-Volterra) logistic growth curve of populations (see Pearl, 1927, for original application). In the case of taxa, originations are assumed to increase geometrically early in the radiation. An environmental carrying capacity for taxa is also assumed (Sepkoski, 1978, 1979; Stanley, 1979). However, a sigmoidal diversity pattern for adaptive radiation can also be attained using a non-geometric increase of taxa. Such models do not require a diversity equilibrium (Valentine, 1980, 1981; Walker, 1985). Sigmoidal diversity increases were also used by Carr and Kitchell, (1980) to model the recovery of taxonomic diversity from saturated and unsaturated conditions. Diversity patterns of marine metazoan orders and families do exhibit a sigmoidal pattern (Sepkoski, 1978, 1979) and Stanley (1979) presents various clades which are consistent with his model.

Palmer (1965a:149) defined a biomere as "a regional biostratigraphic unit bounded by abrupt non-evolutionary changes in the dominant elements of a single phylum." He originally used this term to describe a Late Cambrian trilobite assemblage in the Great Basin. Within this unit, the Pterocephalid Biomere, Palmer found a pattern of increasing trilobite diversity towards the top of the biomere with an abrupt extinction event marking the upper boundary. The lower boundary was postulated to be diachronous and was defined by the appearance of key species within the trilobite families Pterocephalidae and Elvinidae. Each biomere was considered to be the result of repeated invasions of the shelf environment by ancestral stocks of oceanic species which then evolved *in situ*. These newly evolved trilobite species were later exterminated by a global change in oceanic conditions (Palmer, 1965a,b), perhaps an influx of cold water (Lochman and Duncan, 1944; Taylor, 1977). The biomere boundaries do not appear to be coincident with major lithologic changes, nor with other faunal changes (Palmer, 1965a,b).

Trilobite biomeres have also been recognized outside of Palmer's originial study area. Two biomeres have been described in Texas (Longacre, 1970) and Oklahoma (Longacre, 1970; Stitt, 1971b, 1977): the original Pterocephalid Biomere and overlying it, the Ptychaspid Biomere. It is this younger biomere, the Ptychaspid Biomere, which is the subject of this paper.

On the basis of his work with the Ptychaspid Biomere, Stitt (1971b, 1975, 1977) further divided biomeres into four stages, each representing a phase of adaptive radiation. The first stage contains relatively few species, typically with short stratigraphic ranges and high morphological variability. The next two stages have successively increasing species diversity, longer species ranges, and less morphological variability. The final stage has the highest species diversity and also has high familial diversity, but here species ranges are short. This final stage has been characterized as a time of "evolutionary desperation" for the trilobite species present (Stitt, 1975:385). Palmer (1979) agreed with this interpretation of biomeres, although he did feel it would be more appropriate to redefine the end of a biomere to coincide, not with the end of the fourth stage, but with the end of the third stage where the initial invasion of the subsequent biomere occurs.

Because the Ptychaspid Biomere presents a well-studied adaptive radiation, it provides an excellent opportunity to examine at the species level those characteristics and patterns considered important for recognizing adaptive radiations at higher levels.

MATERIALS AND METHODS

The data used in this study were collected by Stitt (1971a, 1977) from the

Western Arbuckle Mountains and the Wichita Mountains in southern Oklahoma. Each site comprises approximately 1800 feet of limestone and dolomite that make up the Timber Hills and Arbuckle groups. At both localities a basal sandstone unit rests disconformably on the underlying group. Also present in each section are algal-stromatolites, located at about mid-section (Figure 1). These deposits record shallow cratonic sedimentation during the Late Cambrian and Early Ordovician epochs. A transgression, possibly of Franconian age, is thought to have deposited the basal sandstone unit. A gradual deepening, shallowing, and then deepening again appears to have occurred, with the shallowest interval being recorded by the algal-stromatolite deposits (Stitt, 1971a, 1977).

Stitt (1971a, 1977) sampled a total of four sections: three in the Western Arbuckle Mountains and one in the Wichita Mountains. Collection intervals equaled one foot, producing extremely well-sampled sections from these areas. Although trilobites are not present in every sample, they are found consistently throughout each section and the exception of the dolomite lithologies.

A composite of all four sections was made, for the purposes of this study. I used Shaw's (1964) technique of graphic correlation (this removed all sampling gaps caused by the dolomite). The composite section was created using all common species' originations and extinctions (Figure 2). It was not necessary to remove any species with short or aberrant ranges because the initial correlation using the entire trilobite assemblage was extremely high, with a correlation coefficient equal to 0.98 ($N = 146$, $p < 0.001$). While only the composite section is used here, all analyses were done on the original sections as well, and all conclusions drawn apply to both forms of the data.

The species diversity curve (Figure 3) was calculated using total species' ranges as determined by the first and last appearances of a given species within the composite section. This method assumes that the species was always present somewhere within the depositional basin, even if it was not found in all of the samples lying within its range (a simple range-through assumption). Familial diversity (Figures 4 and 5) was also determined using this method. In this case, the assumption was made that the family was present, even if genera and species of that family were not present in all samples throughout that family's range.

Species survivorship was calculated for the Biomere as a whole (Figure 6). This was done by dividing stratigraphic ranges into ten foot intervals. Percent survivorship was then determined by measuring the proportion of species surviving from one interval to the next. Due to the trimodality of the species diversity curve (see Results), species survivorship was also calculated for each 'mode' of the curve (Figure 6). Net change in speciation (Figure 7) was determined for 100 foot intervals, starting at the base of each section.

The two spindle diagrams (Figures 8 and 9) were created by placing the

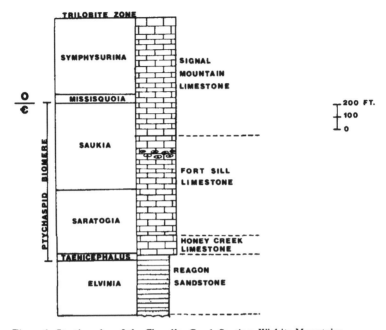

Figure 1. Stratigraphy of the Chandler Creek Section, Wichita Mountains.

diversity curves such that time (here measured in feet) was on the y-axis. A mirror image of the curve was then drawn. A number of "clade statistics," based on shape, were proposed by Gould et al. (1977) as important indicators of evolutionary patterns within clades. Two of these statistics were utilized here: center of gravity and uniformity. Of the original statistics, these two appear to provide the best basis for interclade comparisons. The other statistics are highly clade-dependent for their values (Stanley, et al., 1977; Ward and Signor, 1985).

The center of gravity locates mean diversity in time. It indicates where the average diversity occurred for a given clade. The center of gravity was calculated by setting the time axis on a scale of zero to one. This interval was then divided into twenty equal parts and the species diversity present at each of those divisions was weighted by the proportion represented by that division.

Uniformity is a measure of how constant a particular clade's diversity has been throughout its duration: the more variable the diversity, the lower the uniformity. Uniformity was determined by enclosing the clade in the smallest rectangle possible and computing the percent area of that rectangle that the clade covered (for a more detailed explanation of these statistics and their computation, see Gould et al., 1977).

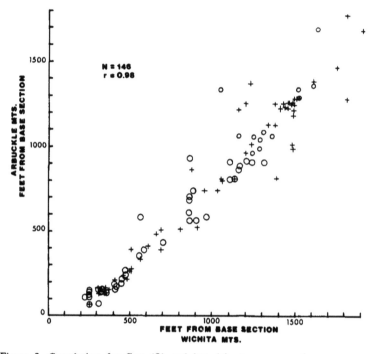

Figure 2. Correlation for first (O) and last (+) appearances of common species.

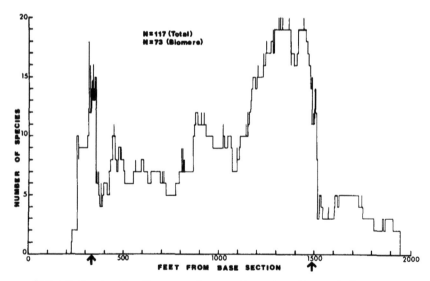

Figure 3. Species diversity curve of composite section. (Arrows indicate beginning and end of Biomere.)

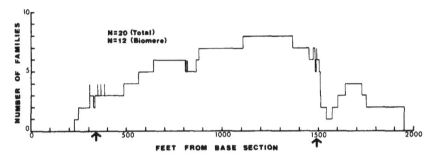

Figure 4. Familial diversity curve of composite section.

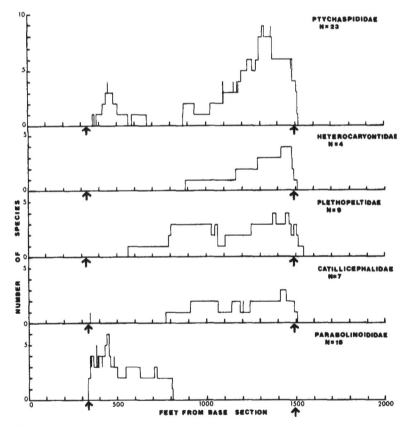

Figure 5. Within-family species diversity curves.

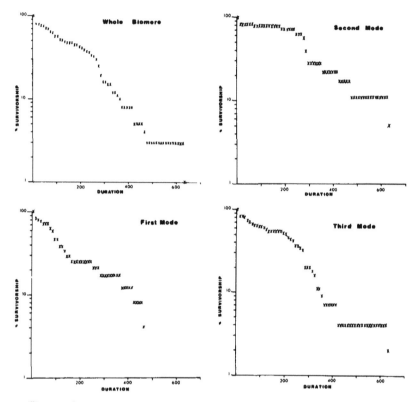

Figure 6. Species survivorship curves.

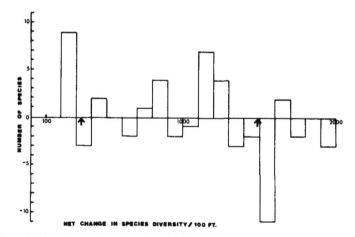

Figure 7. Net changes in species diversity/100 feet.

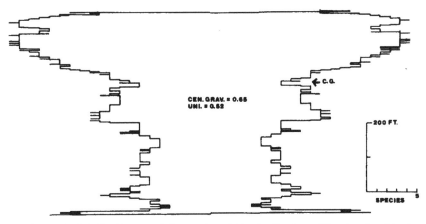

Figure 8. Spindle diagram of the species within the Ptychopariid order found in the Ptychaspid Biomere. (C.G. = center of gravity.)

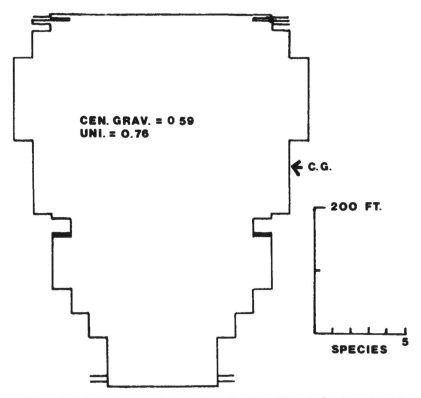

Figure 9. Spindle diagram of the familial diversity within the Ptychaspariid order.

RESULTS

Figure 3 indicates that the species diversity curve for the Ptychaspid Biomere does not have a sigmoidal shape nor does the increase from beginning to end appear to be geometric, as would have been predicted by the models discussed above. In fact, the pattern is trimodal: there are two minor, rapid increases, followed by gradual declines, and then a final, relatively gradual increase which peaks at maximum diversity for the Biomere. The extinction event occurs quickly after this peak and diversity drops very rapidly.

In order to control for artifacts which might be due to taxonomic classification, two other analyses were done: familial diversity (Figure 4) and within-family species diversity (Figure 5). The familial diversity curve does not show a sigmoidal pattern of increase. It is interesting to note, however, the differences between the species diversity curve and the familial one. At the familial level there is a gradual increase of diversity until the end of the Biomere but the species curve lacks its smoothness. The high peak of the species curve is also lost at the familial level. This suggests the kind and amount of information that is lost when higher taxonomic levels are used to infer species level patterns.

The within-family species diversity curves (Figure 5) also fail to exhibit sigmoidal patterns. Five of the twelve families present account for the majority of the area under the species diversity curve. Two families, the Ptychaspididae and the Parabolinodidae, are responsible for the first diversity peak (or mode). The second mode is composed primarily of the Catillicephalidae and the Plethopeltidae, although the Heterocaryontidae and the Ptychaspididae do contribute a small amount. The final mode is composed of four families: the Catillicephalidae, the Plethopeltidae, the Heterocaryontidae, and the Ptychaspididae. Once again, none of these individual family diversity curves indicates either a sigmoidal or geometric pattern.

Hollow (concave) survivorship curves, resulting from a large number of short-lived species early in the radiation, are predicted. The survivorship curve for the Biomere as a whole (Figure 6) does not provide much insight because it is neither extremely concave nor is it convex. However, survivorship curves from each of the three modes give some indication of the variation in species survivorship that occurs throughout the Biomere. Those species with the shortest durations are indeed found at the beginning of the Biomere (the first mode), while those with the longest durations are in the second mode. The third mode's survivorship is lower than that for the second mode but higher than that of the first mode. This may indicate that survivorship at the end of the Biomere would have been longer but that species ranges were truncated by the extinction event. This pattern of short-lived species at the beginning of the Biomere followed by longer survivorship as the Biomere continues and potentially as it ends, is what has been observed in other adaptive radiations (Simpson, 1944, 1953; Stanley, 1979).

Several authors (Simpson, 1944, 1953; Sepkoski, 1978, 1979; Stanley, 1979) have observed that speciation rates are relatively high compared to extinction rates at the beginning of an adaptive radiation. As the radiation progresses, the extinction rates supposedly either equilibrate with speciation rates or overtake and exceed them. A plot of net changes in species number (Figure 7) does not follow such a pattern; rather, episodes of net speciation and extinction tend to alternate throughout the Biomere. These alternations coincide with the increases and decreases in species diversity.

Gould et al. (1977) predicted, based upon the pattern of species durations described above, that adaptive radiations should generate "heavy-bottomed" clades. They also observed that Cambrian and Ordovician clades have lower centers of gravity than clades within the same taxonomic group found in later systems. This suggests that the Ptychaspid Biomere would have a low center of gravity, probably below 0.5. However, for families within the order Ptychopariida, the center of gravity is higher than those for Cambrian and Ordovician trilobites found by Gould et al. (1977) (Table 1). It is also higher than for trilobites in general. In fact the Ptychopariida clade is not heavy-bottomed, but is top-heavy instead.

Uniformity for families within the Ptychoparid order (Table 1) was much higher than that found by Gould et al. (1977) for all families within trilobite orders. The uniformity for species within the order was much lower than for families within the order. This is consistent with a trend noted by Gould et al. (1977) for uniformity to decrease as more disparate taxonomic levels are grouped (species being more disparate from orders than families in the taxonomic hierarchy). Gould et al. (1977) also suggested that uniformity may be lower in certain cases in which poor preservation creates variability. Because preservation is considered consistent for the data presented here (there are few large gaps and they are small relative to the entire section), this may be a possible explanation for the high uniformity exhibited by these data.

Table 1. Centers of gravity and uniformities for trilobite clades.

| | *This Paper* | | *Gould et al., 1977* | |
	Species Within Ptychopariida	Families Within Ptychopariida	All Trilobites	Cambrian-Ordo. Trilobites
Center of Gravity	0.65	0.59	0.410	0.488
Uniformity	0.52	0.76	0.588	--

DISCUSSION

Many problems are encountered when analyzing adaptive radiations because the term "adaptive radiation" is used to describe two very different phenomena. The first and perhaps most common use of adaptive radiation occurs as a general description of any divergent group of taxa. Typically these taxa show many variations on a theme and thus monophyly is inferred. The second use appears to be more theoretical in nature and contains many implicit and explicit assumptions about the processes that produce an adaptive radiation. It is basically this second type of adaptive radiation that is summarized at the beginning of this paper.

Huxley (1953) has pointed out that all of life is an adaptive radiation and Gould et al. (1977:25) "argue that steadily increasing diversity marks the history of clades in 'normal' times. . ." Should all clades then be classified as adaptive radiations or is the term to be applied to phenomena that are not "normal?" If adaptive radiation is to be used in its most general sense, then any clade which has diversified becomes an adaptive radiation. Because the models presented above assume a more special case, the general definition is useless for studies of the type presented here. To analyze an adaptive radiation of the first type using models which are based on adaptive radiations of the second type gives results that are inconsistent with predictions and makes clear explanations difficult indeed. Much of the disagreement between the Biomere data and the models examined in this analysis is probably due to this lack of a consistent use (or definition) of the term "adaptive radiation."

Throughout the remainder of this paper, the term " adaptive radiation" will refer to the second more restricted definition. An adaptive radiation will be defined as a monophyletic group that undergoes rapid speciation at the beginning of the radiation and then shows a gradual decline in its speciation rate. This is the definition which most closely matches those used in the models examined in this paper and, in my opinion, is the more interesting application of the term under these circumstances. If we are to continue to use this term for both phenomena, then some distinction must be made. A preliminary discussion on types of adaptive radiations has been given by Jablonski (1980).

At least two explanations can be offered to explain the lack of agreement between the predictions made by these models of adaptive radiation (Gould et al., 1977; Sepkoski, 1978, 1979; Stanley, 1977; Valentine, 1980, 1981; Walker, 1985) and the Biomere data. Either the Ptychaspid Biomere represents an adaptive radiation of trilobites that differs from the models or the Biomere is not an adaptive radiation and is the result of other natural processes.

THE BIOMERE AS AN ADAPTIVE RADIATION

Because some of the assumptions of the models are violated by the Biomere data, their lack of agreement may (at least in part) be caused by this. An important assumption of all of the models examined here is one of monophyly for the clade undergoing the adaptive radiation. Eldredge (1977) does not feel this is the case for the trilobite order, Ptychopariida, used here. The importance of this assumption can not be determined but, at least for the Metazoa, it does not appear to be a necessary condition. Sepkoski's (1978, 1979) data on marine metazoan orders and families do show a logistic pattern as predicted, even though this group as a whole is paraphyletic. This indicates that a sigmoidal curve can be generated when the assumption of strict monophyly is relaxed. The situation is the same for the Ptychoparids used in this study— they are also paraphyletic and possibly even polyphyletic (Eldredge, 1977).

Sepkoski's (1978, 1979) and Stanley's (1979) models of adaptive radiation both assume that the ecospace being invaded is empty and free of competitors. In the Oklahoma sections, brachiopods, conodonts, and crinoids are all present (Stitt, 1971a, 1977) and are potential competitors. Once again, Sepkoski's (1978, 1979) data on marine metazoan orders and families also violates this assumption. Many phyla of soft-bodied organisms are present during the Cambrian explosion and also throughout the rest of the Phanerozoic. They too are potential competitors who are not included in the diversity curves. However, their presence does not appear to affect Sepkoski's results.

Since these assumptions do not appear to be necessary conditions for the production of a sigmoidal pattern of diversity increase during an adaptive radiation, further explanation is needed to explain why the Biomere data does not follow the predicted pattern if, indeed, it is an adaptive radiation. One possibility is that these disparities are the result of taxonomy. This study is at the species level while other studies (Sepkoski, 1978, 1979; Stanley, 1979) utilize data comprised primarily of higher taxa. Sepkoski's (1978) simulations do indicate that a direct relationship exists between higher taxa and species, but there is no direct evidence to support this result. In fact, the dynamics of an adaptive radiation at the species level need not be as simple as those at higher taxonomic levels. Two pieces of evidence support this conclusion.

One is Sepkoski's (1978, 1979) data sets for marine metazoan orders and families. While the overall diversity pattern is basically the same for his two sets, multiple equilibria are necessary to generate the familial pattern rather than the single equilibrium needed to describe the ordinal pattern. This variability hints at an increasingly complex pattern as lower and lower taxonomic units are analyzed within an adaptive radiation. The second piece of evidence comes from the Ptychaspid Biomere. While the overall pattern of increasing diversity is apparent in both diversity curves and diversity peaks are coincident for both, the species level curve is far more complex than the familial one.

To a certain extent, these differences between patterns at various taxonomic levels (Sepkoski's and this paper's) are a consequence of taxonomic classification. Unless all the species present in the adaptive radiation belong to a monospecific higher taxon, it is unlikely that the higher taxon will be as variable through time as the species. This is because species ranges within a higher taxon are rarely coincident and lend greater stability (i.e., longevity) to the higher taxon. These differences may also be caused by the fact that species are far more sensitive to environmental fluctuations than are higher taxa (unless the higher taxa are monospecific). Each species within a higher taxon has its own set of unique resources and tolerances, some, but not all of which, may be shared with other species. Therefore, the more species a higher taxon contains the more resources it relies on and the lower the probability that all of its resources will become unavailable at any given time. This sensitivity could cause species diversity patterns during adaptive radiations to be obscured by environmental fluctuations or changes, even though they might be present at higher taxonomic levels. However, for the Ptychaspid Biomere, at least, this does not appear to be the case since families do not show the predicted pattern either.

These data might still represent an adaptive radiation despite the absence of predicted patterns, if they result from three pulses of adaptive radiation. This is suggested by the trimodality of the species diversity curve (Figure 3). However, none of the three modes shows geometric increase or a sigmoidal pattern nor do the species survivorship curves (Figure 6) indicate three separate events. Nonetheless, if each sampling interval represents a sufficiently large period of time, the initial stages of these adaptive radiations would have been obscured, either through time averaging of the deposits or due to lack of preservation of the presumably small initial populations of the newly arisen and rapidly evolving trilobites (Rosenzweig, personal communication).

A final explanation may be that the models used here are not adequate representations of adaptive radiations, especially at the species level. Analyses of the dynamics of adaptive radiations are a relatively recent development (with the notable exception of Simpson's work, 1944 and 1953). Because of this, very little data exist which can be used to confirm or deny many of the assumptions that have been made about adaptive radiation.

Final evidence to support the conclusion that the Ptychaspid Biomere is not an adaptive radiation can be found if the global family durations of the families present in the Ptychaspid Biomere (Sepkoski, in press) are examined. These data show that only four of the twelve families originate during the Biomere—the other eight originate earlier. Of the four originating during the Ptychaspid Biomere, only one, the family Ptychaspididae, is a prominent family within the Biomere (Figure 5); the other three are very minor components of the overall species diversity (data not shown). Because the majority of the families originate earlier, the Ptychaspid Biomere can not be a record of oceanic migration and *in situ* evolution of these families from a single ancestral stock. Each of these families could be undergoing their own adaptive radiation but the within-family species diversity curves (Figure 5) give no indication that this is occurring.

None of the explanations given above appear to be sufficient to reconcile the differences present between the patterns exhibited by the Biomere data and those predicted by the models presented here. Therefore, the Biomere data do not appear to be an adaptive radiation. This conclusion is based upon analogy with Sepkoski's data (1978, 1979) which have many of the same difficulties and yet, conform to the predictions of the models.

AN ALTERNATIVE EXPLANATION OF THE BIOMERE DATA

The most obvious alternative explanation for the nature of the Ptychaspid Biomere is that it records a biofacies migration: Palmer (1982) has recently presented data that indicates that this may be the situation for the older Pterocephalid Biomere. A habitat change from shallow to deep water conditions, with minor fluctuations within a general transgressive sequence, could be the cause of the patterns observed. A long period of time relative to trilobite evolution is represented in this section (73 species in 5-10 million years). Because of this, recognition of similar habitats, on the basis of faunal composition, may be difficult because faunal composition does not remain constant long enough to be preserved more than once. Rapid trilobite evolution obscures patterns of faunal composition within a habitat because replacement of most species within a given habitat has occurred before that habitat can be sampled again. Therefore, this section may represent changes in both time (trilobite evolution) and space (habitat migration). With both of these situations changing, the resulting pattern would be expected to be very complex and the deciphering of the important generating factors necessarily difficult.

The hypothesis of changing habitat could be tested by a detailed paleoenvironmental study of the Oklahoma sections. If the fluctuations in diversity are the result of changes in environment, then a correlation between depositional environment and diversity is predicted. If shallow cratonic seas were similar to modern shelf regions, then the highest within-habitat species diversity should be found in deeper waters. Also a morphological analysis might prove helpful. If morphology is being influenced by local habitats, then a correlation between habitat and morphology should be possible.

SUMMARY AND CONCLUSIONS

A well-documented assemblage of Late Cambrian trilobites, the Ptychaspid Biomere, is examined at the species level for characteristics considered diagnostic of adaptive radiations at higher taxonomic levels. Diversity curves of species, families, and species within a family show neither a sigmoid pattern nor geometric increase, two characteristics considered important in the dynamics of adaptive radiations. Net changes in species diversity are calculated and do not show the predicted inverse relationship. The clade statistics, center of gravity and uniformity, also differ from those predicted for adaptive radiations. Division of the data into three modes and species survivorship curves of each of these divisions give the only set of results consistent with those predicted for adaptive radiations: species with the shortest durations are found early in the Biomere while those with longer durations occur later.

This data set violates some of the assumptions of the models of adaptive radiation examined in this paper (most importantly assumptions of monophyly and empty ecospace). One possible explanation is that the data do not show the predicted patterns because of these violations. However, Sepkoski's (1978, 1979) marine metazoan data sets, which also violate these assumptions, do exhibit the patterns predicted. Thus, adaptive radiations can produce the predicted patterns when these assumptions are violated. This suggests that an alternative hypothesis is necessary to explain the trilobite data. Further data to support this conclusion are global family durations of the trilobite families present in the Biomere. Their durations do not coincide with an *in situ* origin and subsequent adaptive radiation.

The most obvious alternative hypothesis is that the trilobite data represent diversity patterns produced by biofacies migration. Habitat recognition, within a general transgressive sequence, on the basis of faunal composition is made difficult by the high faunal turnover. Two analyses would test this conclusion: 1) Detailed paleoenvironmental work could provide data which would allow species diversity to be correlated with habitat. 2) Morphological analyses of the trilobites could indicate habitat types and again allow for correlation of diversity and habitat.

ACKNOWLEDGMENTS

I am very indebted to Jim Stitt whose data I used and who kindly and patiently explained many aspects of it to me. He has offered many valuable comments and criticisms even though some of my conclusions differed from his. I am also extremely grateful to Dave Jablonski who has read innumerable versions of this paper while still managing to maintain a fresh and positive attitude towards the ideas presented in it. For their input and guidance throughout the progress of this study, I would also like to thank Karl Flessa, Susan Kidwell, and Mike Rosenzweig. Special thanks go to J. J. Sepkoski, Jr. who provided me with galley proofs of his compendium before its publication. Mike Sanderson and Michael Edson both provided invaluable help in the preparation of this manuscript. This project was supported by numerous and timely grants from my parents, Billie and David Hardy. This study was submitted in partial completion of a Master of Science degree at the University of Arizona.

REFERENCES

Carr, T. R., and Kitchell, J. A., 1980. Dynamics of taxonomic diversity. Paleobiology 6:427-443.

Eldredge, N., 1977. Trilobites and evolutionary patterns. In Hallam, A. (ed.), Patterns of evolution, Amsterdam: Elsevier, 27-57.

Eldredge, N., and Cracraft, J., 1980. Phylogenetic patterns and the evolutionary process. New York: Columbia Univ. Press, 349.

Gould, S. J., Raup, D. M., Sepkoski, J. J., Jr., Schopf, T. J. M., and Simberloff, D. S., 1977. The shape of evolution: A comparison of real and random clades. Paleobiology 3:23-40.

Huxley, J., 1953. Evolution in Action. New York: The New American Library of World Literature, 141.

Jablonski, D., 1980. Adaptive radiation: Fossil evidence for two modes. 2nd Internat. Cong. Syst. Evol. Biol. Abstr.:243.

Lochman, C., and Duncan, D., 1944. Early Upper Cambrian faunas of central Montana. Geol. Soc. Amer. Spec. Paper (54):1-181.

Longacre, S. A., 1970. Trilobites of the Upper Cambrian Ptychaspid Biomere, Wilberns Formation, central Texas. Paleont. Soc. Mem. 4:1-70.

Palmer, A. R., 1965a. Biomere—A new kind of stratigraphic unit. J. Paleont. 39:149-153.

Palmer, A. R., 1965b. Trilobites of the Late Cambrian Pterochephalid Biomere in the Great Basin, United States. U.S. Geol. Surv. Prof. Paper (493):1-105.

Palmer, A. R., 1979. Biomere boundaries re-examined. Alcheringa 3:33-41.

Palmer, A. R., 1982. Biomere boundaries: A possible test for extraterrestrial perturbation of the biosphere. In Silver, L., and Schultz, P. (eds.), Geological implications of impacts of large asteroids and comets on the earth, Geol. Soc. Amer. Spec. Paper (190):469-475.

Pearl, R., 1927. The growth of populations. Quart. Rev. Biol. 2:532-548.

Raup, D. M., 1979. Biases in the fossil record of species and genera. Carnegie Mus. Nat. Hist. Bull. (13):85-91.

Sepkoski, J. J., Jr., 1978. A kinetic model of Phanerozoic taxonomic diversity,

I Analysis of marine orders. Paleobiology 4:223-251.

Sepkoski, J. J., Jr., 1979. A kinetic model of Phanerozoic taxonomic diversity, II Early Phanerozoic families and multiple equilibria. Paleobiology 5: 222-251.

Sepkoski, J. J., Jr., 1982. A compendium of fossil marine families. Milwaukee Public Museum Contr. Biol. Geol. (51), 125.

Shaw, A. B., 1964. Time in stratigraphy. New York: McGraw-Hill, 365.

Simpson, G. G., 1944. Tempo and mode in evolution. New York: Columbia Univ. Press, 237.

Simpson, G. G., 1953. The major features of evolution. New York: Columbia Univ. Press, 434.

Stanley, S. M., 1979. Macroevolution. San Francisco:W. H. Freeman, 332.

Stanley, S. M., Signor, P. W., III, Lidgard, S., and Karr, A. T., 1981. Natural clades differ from "random" clades: Simulations and analyses. Paleobiology 7:115-127.

Stitt, J. H., 1971a. Late Cambrian and earliest Ordovician trilobites, Timbered Hills and Lower Arbuckle groups, Western Arbuckle Mountains, Murray County, Oklahoma. Okla. Geol. Surv. Bull. 110:1-70.

Stitt, J. H., 1971b. Repeating evolutionary pattern in Late Cambrian trilobite biomeres. J. Paleont. 45:178-181.

Stitt, J. H., 1975. Adaptive radiation, trilobite paleoecology and extinction, Ptychaspid Biomere, Late Cambrian of Oklahoma. Fossils and Strata 4: 381-390.

Stitt, J. H., 1977. Late Cambrian and earliest Ordovician trilobites, Wichita Mountains area, Oklahoma. Okla. Geol. Surv. Bull. 124:1-62.

Taylor, M. E., 1977. Late Cambrian of western North America: Trilobite biofacies, environmental significance and biostratigraphic implications. In Kauffman, E. G., and Hazel, J. E. (eds.), Concepts and methods of biostratigraphy, Stroudsburg, Pa.: Dowden, Hutchinson and Ross, 397-425.

Valentine, J. W., 1972. Conceptual models of ecosystem evolution. In Schopf, T. J. M. (ed.), Models in paleobiology, San Francisco: Freeman, Cooper & Co., 192-215.

Valentine, J. W., 1973. Evolutionary paleoecology of the marine biosphere. Englewood Cliffs, N.J.: Prentice-Hall, 511.

Valentine, J. W., 1977. General patterns of Metazoan evolution. In Hallam, A. (ed.), Patterns of evolution, Amsterdam: Elsevier, 27-57.

Valentine, J. W., 1980. Determinants of diversity in higher taxonomic categories. Paleobiology 6:444-450.

Valentine, J. W., 1981. Emergence and radiation of multicellular organisms. In Billingham, J. (ed.), Life in the universe, Cambridge, Mass.: The MIT Press, 229-257.

Walker, T. D., 1985. Diversification functions and the rate of taxonomic evolution. In Valentine, J. W. (ed.), Phanerozoic diversity patterns: Profiles in macroevolution, Princeton, N.J.: Princeton Univ. Press and Amer. Assoc. Adv. Sci. (this volume).

Ward, P. D., and Signor, P. W., III, 1985. Evolutionary patterns of Jurassic and Cretaceous ammonites: An analysis of clade shape. In Valentine, J. W. (ed.), Phanerozoic diversity patterns: Profiles in macroevolution, Princeton, N.J.: Princeton Univ. Press and Amer. Assoc. Adv. Sci. (this volume).

Chapter 13

EVOLUTIONARY PATTERNS OF
JURASSIC AND CRETACEOUS AMMONITES:
AN ANALYSIS OF CLADE SHAPE

PETER D. WARD* and PHILIP W. SIGNOR, III

Department of Geology, University of California, Davis

INTRODUCTION

Clade diversity (spindle) diagrams graphically portray changes in the diversity of monophyletically derived taxa through time. They are often used to depict examples of evolutionary events, such as radiation or extinction, or to infer results of environmental events in evolutionary time, such as competition or predation. Recently, clade diversity diagrams and their uses have been critically reexamined, in efforts to reduce the hazards inherent in the uncritical interpretation of such diagrams and to propose appropriate null hypotheses against which deterministic explanations may be tested (Raup et al., 1973; Raup and Gould, 1974; Gould et al., 1977; Schopf, 1979; Stanley et al., 1981). Through production of computer generated clade diversity diagrams, Gould et al. (1977) were able to show that cladograms derived from random models often produce shapes which appear to have deterministic meaning (but see Stanley et al., 1981 and Raup, 1981). These workers introduced simple clade statistics that allow quantitative evaluation of different clade diagrams, so that evolutionary difference intrinsic to the component lineages could be examined.

 In this paper, we continue in this general theme of clade shape analysis. Rather than using simulated cases, however, we will analyze the clade diagrams of a well known, well documented group: Jurassic and Cretaceous

*Current address· Dept. of Geological Sciences, University of Washington, Seattle

ammonites. We will examine these clades of ammonite genera within families using the clade statistics proposed by Gould et al. (1977) in an attempt to differentiate evolutionary tempo within the ammonite families. We will then evaluate the efficacy of Gould et al.'s statistical analysis, and propose an alternative index which we feel may be equally useful.

Many ammonite family clades have a characteristic shape, previously noted by various workers (Gould et al., 1977, Stanley, 1979): short and wide, indicative of rapid radiation followed by precipitous decline and final extinction. These shapes result from high rates of extinction and origination. It is, of course, this rapid evolution that make most ammonites such useful biostratigraphic tools. Not all ammonite clades have these shapes, however.

Some workers have viewed the characteristic ammonite clade shapes as perhaps due to taxonomic practice, rather than reflecting real ammonite evolutionary tempo, and have ignored the presence of other clade shapes within the ammonites. For example, "Clades of genera within families of Mesozoic ammonites, for example, are extremely short and fat. We do not deny that evolution may have been very rapid within this group, but prodigious oversplitting inspired by stratigraphic utility may be the primary cause of these unusual shapes" (Gould et al., 1977:29). From the same source comes the following: "...we may be quite sure that mammals and ammonites are oversplit at the family level—ammonites for their stratigraphic utility as Mesozoic guide fossils, mammals because we tend to make finer distinctions between animals that look more and more like us" (Gould et al., 1977:31). Such taxonomic practice would tend to reduce the size of clades (Gould et al., found mammal and ammonite family clades to be the smallest in average generic richness per family of the twelve higher taxa they examined). Such reasoning, however, must be used consistently. The presence of long, thin ammonite clades along with the short, fat clades must be reconciled with these assertions.

Are the characteristic clade shapes of ammonites simply the result of taxonomic practice or is there information about evolutionary tempo in these groups that is mirrored in the taxonomy? Certainly, not all ammonite clades are short and fat. Elsewhere (Ward and Signor, 1983), we found a wide variety of significantly different mean longevities for genera within families of ten or more genera. While most families were composed of genera with ranges between 1 and 10 million years, giving a mean generic longevity of about 7 m.y., some families had mean generic ranges far in excess of this figure. That some families are composed of much longer ranging genera suggests that the longer ranging taxa are not randomly distributed among the families. For example, while only one in ten ammonite genera last more than 12 m.y., 11 of 13 tetragonitid genera exceed that duration. The prospect of this occurring by chance is exceedingly remote. These families with long-ranging genera also tend to be at lower average diversity for any given time than the shorter-ranging families, thereby producing long, thin clade shapes.

Schopf et al. (1975) suggested that evolutionary range may be a function of morphological complexity of taxa; more complex taxa might tend to have shorter ranges since evolutionary change would be more readily detected than in morphologically simple taxa. Although some of the long-ranging ammonite families, most notably the Phylloceratidae and Lytoceratidae, are composed of many morphologically simple genera, other long-ranging ammonite families, such as most of the Upper Cretaceous heteromorphic ammonite families, are composed of long ranging, morphologically complex genera. Furthermore, virtually all long ranging Jurassic and Cretaceous ammonite genera, irrespective of taxonomic affinity, share a common morphological trait: siphuncles of low diameter but high wall thickness (Ward and Signor, 1983). Those genera with small, thick siphuncles (interpreted to be deeper water varieties by Westermann, 1971) had lower extinction rates than those taxa with weaker connecting rings.

These findings suggest that a continuum of evolutionary tempos existed within Mesozoic ammonites, where at one end clades show high origination and extinction rates, producing short, fat clades and the other where clades show the opposite, producing long thin clade shapes. We will apply the clade statistics to Jurassic and Cretaceous ammonites to ask two questions: first, are there predictable relationships between clade shape and evolutionary tempo and, second, what are the limitations of clade statistics in evolutionary analysis?

MATERIALS AND METHODS

Gould et al. (1977) addressed the "general and obvious concerns" of paleontologists studying the morphology of clade diversity diagrams: How big is a clade? How wide is it? How much does its diversity fluctuate through time? Where is it widest? To answer these questions, they proposed the following clade statistics:

1. Size, a measure of persistence and diversity (the total number of lineage-time units).
2. Duration (DUR)—the number of time intervals that a clade persists.
3. Maximum Diversity (MAX)—the highest diversity during any one time period of the clade's duration.
4. Center of Gravity (CG)—a measure of relative position in time of the median time-diversity unit.
5. Uniformity (UNI)—a measure of the uniformity of diversity through time.

With the exception of CG, all of these statistics are readily calculated from raw data, and so can be derived from a clade diversity diagram. We modified the size statistic to a measure of total lineage components; in our treatment SIZE refers to only the total number of genera in a family. We did not use the descriptor CG because it was invented as a measure of diversity trends (Gould et al., 1977), which is not a concern here.

We compiled stage and substage level data for 981 genera of Jurassic and Cretaceous ammonites, distributed among 83 families. All data for Jurassic genera comes from the *Treatise on Invertebrate Paleontology* (Part 4) and data for Cretaceous genera is primarily from the *Treatise* but has been updated according to more recent literature. Ammonite genera found in any part of a stage (or substage for the Bajocian, Albian and Campanian stages) were considered to range throughout that stage or substage. (This procedure was used instead of assigning some fraction of the stage duration to an ammonite occurring in a stage, as suggested by Sepkoski [1975], because the stages are defined on the basis of the ammonites.) Absolute durations for the stages and substages were taken from Obradovitch and Cobban (1975), Van Hinte (1975a, b) and Van Eysinga (1975). Clade statistics were then calculated treating the genera as the "lineages" within clades (families), for families with five or more genera.

The clade diversity diagrams of Jurassic and Cretaceous ammonite genera within families as derived from our sources are shown in Figure 1. From this

Figure 1 (facing page). Clade shapes of Jurassic and Cretaceous ammonites. Diagrams are presented as half-spindles to conserve space. The key to the family names is as follows: 1, Phylloceratidae; 2, Tissotiidae; 3, Hamitidae; 4, Schloenbachiidae; 5, Lytoceratidae; 6, Anisoceratidae; 7, Tetragonitidae; 8, Diplomoceratidae; 9, Nostoceratidae; 10, Baculitidae; 11, Scaphitidae; 12, Pachydiscidae; 13, Collignoniceratidae; 14, Sphenodiscidae; 15, Placenticeratidae; 16, Desmoceratidae; 17, Kossmaticeratidae; 18, Coilopoceratidae; 19, Acanthoceratidae; 20, Labeceratidae; 21, Heteroceratidae; 22, Deshayesitidae; 23, Trochleiceratidae; 24, Brancoceratidae; 25, Vascoceratidae; 26, Binneyitidae; 27, Muniericeratidae; 28, Engonoceratidae; 29, Forbesiceratidae; 30, Flickiidae; 31, Douvilleiceratidae; 32, Silesitidae; 33, Lyeliceratidae; 34, Turrilitidae; 35, Leymerielidae; 36, Hoplitidae; 37, Oppeliidae; 38, Olcostephanidae; 39, Protetragonitidae; 40, Ptychoceratidae; 41, Bochianitidae; 42, Holcodiscidae; 43, Ancyloceratidae; 44, Pulchelliidae; 45, Hemihoplitidae; 46, Oosterellidae; 47, Berriasellidae; 48, Kosmoceratidae; 49, Cardioceratidae; 50, Clydoniceratidae; 51, Macrocephalitidae; 52, Reineckeiidae; 53, Tulitidae; 54, Perisphinctidae; 55, Aspidoceratidae; 56, Mayaitidae; 57, Haploceratidae; 58, Pachyceratidae; 59, Stephanoceratidae; 60, Strigoceratidae; 61, Sphaeroceratidae; 62, Psiloceratidae; 63, Amaltheidae; 64, Juraphillitidae; 65, Derolytoceratidae; 66, Spiroceratidae; 67, Parkisoniidae; 68, Oxynoticeratidae; 69, Otoitidae; 70, Arietitidae; 71, Morphoceratidae; 72, Hildoceratidae; 73, Pleurocanthitidae; 74, Eoderoceratidae; 75, Sonniniidae; 76, Polymorphitidae; 77, Dactylioceratidae; 78, Schlotheimiidae; 79, Graphoceratidae; 80, Liparoceratidae; 81, Hammatoceratidae; 82, Echioceratidae.

diagram, it is evident that a variety of shapes are present. (To save space, we have turned the normal "spindles" of diversity and time into half-spindles.)

RESULTS

The Jurassic and Cretaceous ammonite families used in this study are listed in Table 1. Included for each family are overall diversity of the family (SIZE), duration of the entire family clade (DUR), mean generic duration for the family (GDUR), maximum diversity at any given time (MAX), and a new measure which we will describe below, PROFILE (PRO).

CLADE DURATION AND DIVERSITY

The variety of clade morphologies shown in Figure 1 leads to a number of questions concerning clade structure and evolutionary tempo of the basic clade units, in this case the genera. One such question concerns the possibility of a relationship between clade duration and total clade diversity. Restated, we may ask: is there a significant correlation between clade duration and total number of component genera in families of Jurassic and Cretaceous ammonites? If ammonite evolution resembled a random branching process, with equal probabilities of extinction and origination, one would expect such a correlation to exist (Gould et al., 1977).

In Figure 2 we have plotted family duration (DUR) against the total number of genera (SIZE) for each of the families used in the study. A wide scatter of points is evident; the correlation coefficient for the linear regression is not significant ($r = -.283$). This finding indicates that clade duration is not a function of clade diversity in Jurassic and Cretaceous ammonites. Short-ranging clades may be of high or low diversity. However, there does appear to be a relationship between total clade duration and clade diversity in those clades with many long-ranging genera. In Figure 2, those families with mean generic durations of greater than 10 million years are circled. If their duration is plotted separately against diversity, a significant correlation is found ($r = .636$); suggesting that when generic longevity is long the predicted relationship between SIZE and DUR does occur. The remaining clades, excluding those with long-ranging genera, still show a non-significant correlation ($r = .560$).

DUR and GENERIC LONGEVITY

If clade duration is not related to clade diversity, can we isolate the factors which do control its fate? As pointed out in a variety of papers, origination and extinction rates are important factors (Raup et al., 1973; Raup and

Gould, 1974; Gould et al., 1977; Stanley, 1979; Stanley et al., 1981; Raup, 1981). A measure of extinction rates for ammonite genera is mean generic duration per family, since Stanley (1977) has shown that mean duration is inversely proportional to the extinction rate, assuming survivorship rates are approximately constant. We have plotted mean generic duration for the clades against family duration (DUR) in Figure 3. There is a statistically significant correlation between these two parameters (r = .802). It appears that duration of the families is in some way controlled by mean generic duration; the longer lived the genera the longer lived the family itself. To some extent, this result is not surprising; long-lived genera cannot occur in short-ranging families. But it is interesting to note that long-lived ammonite families are normally not composed of many short-ranging genera.

MEAN GENERIC DURATION and UNI

Gould et al. (1977) proposed UNI as a measure of the 'evenness' of diversity through time within a clade; clades with nearly constant diversity through time had high UNI values. High UNI clades were found to be those with low probabilities of branching and extinction. In Figure 4 we have plotted UNI values of the family clades against mean generic duration per family, which we have shown to be a good predictor of clade longevity. As can be seen in Figure 4, no correlation appears to exist (r = -.205). The long, thin ammonite clades, such as those of the phylloceratids, lytoceratids, tetragonitids, and diplomoceratids appear to have enough fluctuation in diversity to yield a variety of UNI values (Table 1). For the ammonite clades, UNI appears to be of little value in characterizing clades of different evolutionary tempos.

MAX

Gould et al. (1977) found high MAX values to occur in clades with high values of origination and extinction. Likewise, we expected that high values of MAX should be found in clades with short-lived genera. We were surprised to find that this was not the case; there was no significant relationship between mean generic longevity and MAX (Figure 5; r = -.223). Families with long-lived genera tended to remain small, as predicted, but families with short-lived genera varied substantially in their MAX values. Here again, MAX seems to be of questionable value in characterizing ammonite clades of different evolutionary tempos.

PROFILE STATISTIC

The ammonite family clade shapes appear distinct from stochastic clade

Table 1. Clade statistics of Jurassic and Cretaceous ammonite families.
All statistics were calculated with the time scales of Van Hinte
(1976a, b). GDUR is the mean duration of genera assigned to
the family; other statistics are as defined in text.

FAMILY	SIZE	DUR	GDUR	MAX	UNI	PRO
Acanthoceratidae	21	7	4.1	14	.88	.5
Anisoceratidae	8	22	6.4	3	.77	7.3
Ancyloceratidae	25	25	8.3	15	.55	1.7
Arietitidae	25	10	5.2	25	.52	.4
Aspidoceratidae	33	24	7.1	16	.61	1.5
Baculitidae	6	35	10.3	2	.86	17.5
Berriasellidae	44	28	8.2	18	.71	1.6
Brancoceratidae	14	18	6.9	9	.59	2.0
Cardioceratidae	32	19	6.7	17	.66	.9
Collignoniceratidae	20	31	5.4	13	.27	2.3
Craspeditidae	11	70	5.6	5	.17	14.0
Dactylioceratidae	9	7	2.7	7	.50	1.0
Desmoceratidae	25	69	14.2	8	.64	8.6
Diplomoceratidae	7	25	14.0	5	.78	5.0
Douvilleiceratidae	13	20	6.9	8	.56	2.5
Echinoceratidae	7	5	5.0	7	1.0	.7
Engonoceratidae	7	21	8.0	4	.67	5.3
Eoderoceratidae	17	10	5.3	9	1.0	1.1
Graphoceratidae	17	6	3.2	11	.82	.6
Hamitidae	5	28	9.2	4	.41	7.0
Hammatoceratidae	13	8	2.4	8	.23	1.0
Haploceratidae	10	52	10.4	4	.5	13.0
Hildoceratidae	39	13	2.9	25	.34	.5
Hoicodiscidae	5	11	6.0	4	.68	2.8
Hoplitidae	23	30	7.2	11	.50	2.7
Juraphyllitidae	8	15	6.3	5	.67	3.0
Kosmoceratidae	13	5	5.0	13	1.0	.6
Kossmaticeratidae	10	39	7.4	5	.38	7.8
Liparoceratidae	9	5	5.0	9	1.0	.6
Lyelliceratidae	7	23	7.1	3	.72	7.7
Lytoceratidae	10	114	21.8	4	.48	28.5
Macrocephalidae	14	9	4.9	12	.64	.8
Mayatidae	6	8	8.0	8	1.0	1.3
Nostoceratidae	12	30	9.9	7	.57	4.3
Olcostephanidae	16	32	9.1	7	.65	4.6
Oppelliidae	60	77	7.5	19	.31	4.1
Otoitidae	8	3	3.0	8	1.0	.4
Oxynoticeratidae	11	10	5.0	8	.69	1.3
Pachyceratidae	6	13	5.5	5	.51	2.6
Pachydiscidae	11	44	14.2	8	.44	5.5
Parkinsoniidae	14	13	4.3	11	.42	1.2
Perisphinctidae	105	37	6.2	32	.55	1.2
Phylloceratidae	14	132	36.0	8	.48	16.5

Table 1.   Continued.

FAMILY	SIZE	DUR	GDUR	MAX	UNI	PRO
Placenticeratidae	6	30	9.3	3	.62	10.0
Pleurocanthidae	7	15	5.0	4	.58	3.8
Polymorphitidae	9	5	5.0	9	1.0	.6
Psiloceratidae	17	10	5.0	11	.77	.9
Pytchoceratidae	5	31	10.6	4	.42	.8
Pulchellidae	6	18	7.5	6	.42	3.0
Reineckiidae	6	5	5.0	6	1.0	.8
Scaphitidae	8	39	10.3	4	.53	9.8
Schloenbachidae	5	12	4.8	4	.50	3.0
Schlotheimiidae	8	10	5.0	4	1.0	2.5
Sonninidae	16	13	3.6	14	.32	.8
Sphenodiscidae	5	9	5.8	5	.64	1.8
Stephanoceratidae	10	10	3.8	5	.76	2.0
Tetragonitidae	13	55	20.5	7	.69	7.9
Tissotiidae	6	8	3.0	4	.56	2.0
Turrilitidae	10	13	6.1	6	.78	2.2
Tulitidae	7	9	4.3	5	.67	1.8
Vascoceratidae	8	3	3.0	8	1.0	.4

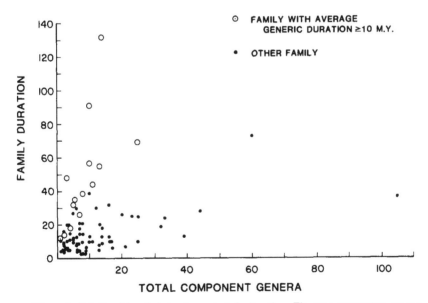

Figure 2. Relationship of clade size and clade duration. The two parameters are not significantly related ($r = .263$). Among long-lived families (hollow circles) there does tend to be a relationship between clade size and duration but this does not extend to the whole data set.

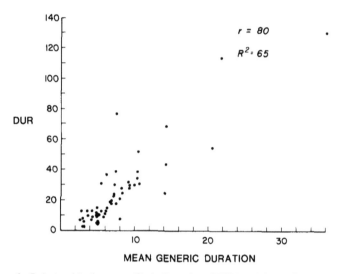

Figure 3. Relationship between Clade Duration (DUR) and Mean Generic Duration (GDUR). The two parameters are highly correlated, indicating that long-lived families are indeed composed of long-lived genera.

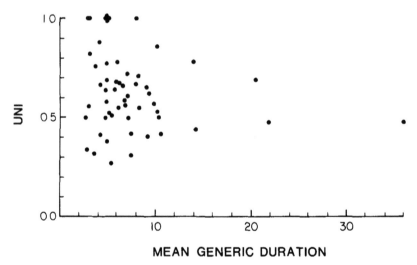

Figure 4. Relationship between Mean Generic Duration (GDUR) and Uniformity (UNI). The two parameters are poorly correlated ($r = -.205$), in contrast to the predictions based on random branching models (Gould et al., 1977).

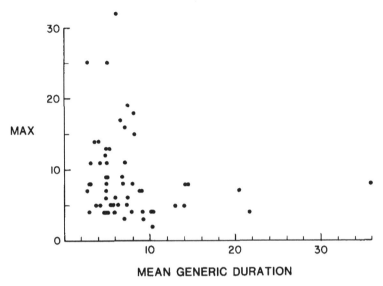

Figure 5. Relationship between Maximum Diversity (MAX) and Mean Generic Duration (GDUR). MAX and GDUR are not correlated (r = -.223), although the clades with the highest MAX values do tend to be those with short-lived genera, as predicted by Gould et al. (1977).

shapes since SIZE, MAX, and duration of the ammonite clades (DUR) do not appear to be correlated, and since long-lived genera are not randomly distributed among the clades, but are concentrated in a few, long-lived clades. These findings lead to an important question: do the morphologies of the ammonite family clades form a continuum between short, fat clades and the long and thin clades, or is there a bimodality in clade shape between these two extremes? Essentially, we are asking if evolution within the ammonites occurred at two distinct tempos, or as a continuum between groups with high and low rates of origination and extinction.

To approach this question, we would like to employ a single descriptor of clade shape. To this end, we propose a new clade statistic to describe clade shape: DUR (clade duration) divided by MAX (the highest diversity attained during any unit of time during the existence of the clade). We will call this new clade descriptor "profile" (PRO, for profile of the clade). Long thin clades will tend to have high PRO values; increasingly short, wide clades will have lower PRO values.

The PRO values for the ammonite clades are listed in Table 1. There is a strong relationship between mean generic longevity and clade shape (PRO) (r = .73) (Figure 6). Clades with long-lived genera tend to be long and thin.

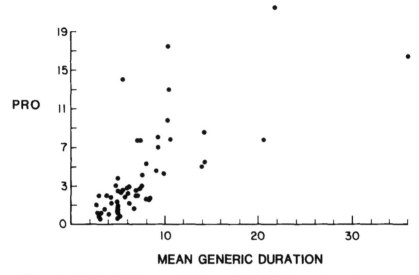

Figure 6. Relationship between clade length/width (PRO) and Mean Generic Duration (GDUR). PRO and GDUR are highly correlated (r = .725), indicating that clades with long-lived genera tend to remain at relatively smaller diversities. The ecological and evolutionary reasons for this present an interesting question: why should longevity trade off with diversity?

Even given the high correlation, however, the PRO statistic does not necessarily reflect only underlying evolutionary tempo; only about half the variance in PRO is accounted for by the correlation with mean generic duration (R ≈ .53).

To examine the question of bimodality we made a histogram of PRO values for the family clades (Figure 7). The resulting pattern shows that no obvious bimodality exists; the histogram is continuous and skewed to the right. This pattern is similar to the histograms for average generic duration per family and for the total generic duration of the entire sample of ammonites used in this study.

JURASSIC VERSUS CRETACEOUS AMMONITE CLADE SHAPES

We have shown that a variety of clade shapes exist for clades of Jurassic and Cretaceous ammonites. A further question of interest relates to the possibility of differences between the Jurassic and Cretaceous clades. Are there consistent patterns which distinguish one group of clades from the other?

Combined UNI, PRO, and mean generic longevity values have been

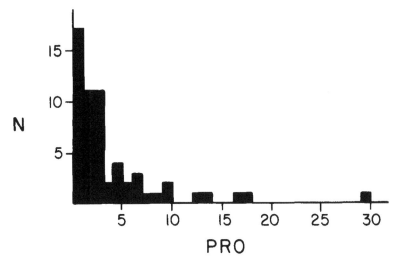

Figure 7. Histogram of PRO values. The PRO values tend to be skewed to the right, where relatively few long, skinny clades exist.

tabulated for families of the Lower, Middle and Upper Jurassic and the Lower and Upper Cretaceous (Table 2). If a family occurred in any part of a series, it was considered to range through all the series. In this crude fashion, we have arrived at estimates of average clade shapes in these five series.

UNI varies very little through the time interval, ranging from .67 in the Lower Jurassic to .57 in the Lower Cretaceous. All of these figures are below the UNI value found by Gould et al. (1977) for the entire sample of ammonite families used in their study. The high UNI values for ammonites in general are derived from their characteristically short clades, which reach maximum diversity quickly and then go extinct. Because this entire life history often takes place within one stage it leads to a large number of family clades which are rectangular, giving UNI values of 1.0.

In contrast to UNI, the PRO values for the Jurassic and Cretaceous differ markedly. PRO values show a progressive increase in the Jurassic, and then markedly increase in the Cretaceous. The equality of PRO values for the Lower and Upper Cretaceous is probably an artifact of the sampling method, since many short-ranging ammonite families of the Albian extend slightly upward into the Cenomanian and were therefore included among the Upper Cretaceous clades. It is our observation that clade shapes during later parts of the Upper Cretaceous show even higher PRO values than the Lower Cretaceous.

The differences in the distribution of clade shapes between the Jurassic and

Table 2. Mean values of UNI, PRO and GDUR for families present in
five intervals of the Jurassic and Cretaceous. As the end of
the Mesozoic approached, ammonite clades tended to become
long and thin. UNI values remained nearly constant, despite
the increase in mean generic longevity per family.

TIME INTERVAL	UNI	PRO	GDUR
Upper Cretaceous	.61	7.2	10.7
Lower Cretaceous	.57	7.2	11.8
Upper Jurassic	.63	6.0	9.0
Middle Jurassic	.65	4.7	8.0
Lower Jurassic	.67	3.7	8.0

Cretaceous can best be illustrated by comparing percent frequencies of various PRO clade values (Figure 8). This figure illustrates the progressive shift to higher PRO values through time of the Jurassic and Cretaceous ammonite clades, indicating that average clade shapes became progressively longer and thinner. Part of this could well be due to differences in taxonomic treatment. Part, however, must reflect real differences in evolutionary tempo.

The increase of the PRO values during the Jurassic and Cretaceous may in part be due to difference in taxonomic practice and effort. Far more paleontological effort has gone into ammonite biostratigraphy of the Jurassic, leading to a finer taxonomy and ever smaller families. The differences in PRO values, however, cannot be due entirely to this factor. There simply are more long-ranging taxa in the Cretaceous than in the Jurassic. This is demonstrated by a comparison of the mean generic duration of Jurassic and Cretaceous genera. These values range between 8 and 9 million years per genus in the Jurassic and between 11 and 12 million years per genus in the Cretaceous. The Cretaceous evolution of long-ranging taxa belonging to the Desmoceratidae, Tetragonitidae, Baculitidae, and various Upper Cretaceous heteromorphic ammonite families accounts for this.

DISCUSSION

Our compilations and analyses of ammonite clade shapes for the Jurassic and Cretaceous suggest that, at least for the clades examined, decisive (and as one reviewer put it, "delightful") differences occur between real and random clades. Ammonites appear to have evolved in decidedly non-random ways. In particular, two of the clade statistics proposed by Gould et al. (1977), UNI and MAX, do not behave as predicted when applied to ammonite clades. As other workers have already noted that the clade simulations do not mimic the

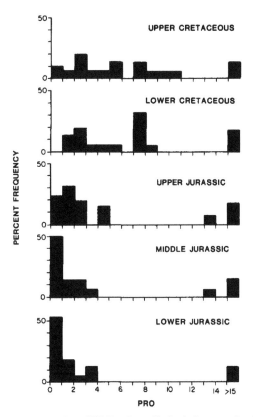

Figure 8. Percent frequencies of PRO values. Each clade occurring in a time interval was added to that period's clades, hence, many clades were added to the clades of several different time intervals.

evolution of all real clades, this is not unduly surprising (Stanley et al., 1981; Raup, 1981).

Despite the failure of these two statistics to behave as expected (when applied to ammonites), it is clear that some aspects of clade shape reflect variations in evolutionary tempo of the component lineages (genera). In particular, the correlation between ammonite family longevity and mean generic duration suggests that some aspects of clade shape can be used as evidence in estimating the evolutionary rates of component taxa.

As noted above, clade duration is not correlated with total clade size. This stands in contrast to some predictions based upon a random model, where size should be highly correlated with the total number of component lineages. The role played by random factors in evolution remains a subject of contention but the results presented here suggest that ammonite evolution did not produce patterns which can reasonably be expected from a random process. In contrast, Arnold (1982) found variation in numbers of component species to be an important factor in determining generic longevity of globigerinid forams. Whether this reflects a difference in evolutionary pattern or a taxonomic difference remains to be determined.

Elsewhere (Ward and Signor, 1983), we have shown that ammonite generic duration appears to be correlated with siphuncular thickness, which is believed by some workers to indicate depth tolerance (Westermann, 1971), but which is disputed by others (see Chamberlain and Moore, 1982, for a strong critique of the method). The long-lived families, the Phylloceratidae, Lytoceratidae, Desmoceratidae, and Tetragonitidae, generally possess thick siphuncles. The more common, short-lived families all possess thin siphuncles. If Westermann's interpretation is correct, it indicates that ammonites with greater depth tolerances (stronger siphuncles) had lower extinction rates than those ammonites with narrower (and shallower) depth ranges. D. Jablonski (pers. comm.) has told us that a similar pattern is also apparent in Cretaceous clams and snails. Wiedmann (1969, 1973) and Kennedy (1977) have shown that ammonite extinctions tend to occur during regressions, which should influence deep-water forms less than those dwelling on the continental shelves. Thus, habitat differences may explain much of the variation in evolutionary pattern among Jurassic and Cretaceous ammonites.

EXTINCTION OF THE AMMONITES

Wiedmann (1969, 1973), Kennedy (1977), and Birkelund (1979) have recently discussed the terminal Cretaceous extinction of the ammonites. They have shown that ammonite extinction at the end of the Maestrichtian involved only a small percentage of the total number of taxa that existed in the Campanian stage. Therefore, the reduction of ammonite genera was a gradual process from the Campanian through the Maestrichtian.

We have briefly examined this phenomenon in terms of changes in clade shape. Hancock (1967) has listed 11 families occurring in the Maestrichtian. The majority of genera in the Maestrichtian belonged to the long-lived families Tetragonitidae, Scaphitidae, Baculitidae, Phylloceratidae, and Diplomoceratidae. The clades of almost all of these families were long and thin, with high PRO values. A mean PRO value of 8.5 can be computed for the Maestrichtian ammonite families, as compared to 7.2 for the Upper Cretaceous as a whole, and 4.1 for the 61 Jurassic and Cretaceous families with 5 or more genera used in this study. Mean generic duration of the Maestrichtian families was 10.7 million years/genus, as compared to a mean value of 7 million years/genus for the entire sample.

The gradual reduction in diversity of the ammonites at the end of the Creta-ceous appears to have involved the removal of the short, fat clades, leaving mainly the long, thin clades. These remaining families produced few new taxa, but at the same time had very low extinction rates. This type of extinction pattern, showing removal of the high-turnover taxa first, followed by a sec-ondary removal of the low-turnover taxa, appears to have also characterized the graptolite extinction of the Paleozoic (W. Berry, pers. comm.). Stanley (1977, 1979) has suggested that rapidly evolving taxa should inherently be more susceptible to adverse conditions, and this is born out by the ammo-nites. By the end of the Cretaceous, however, the remaining ammonites could not be classified as "rapidly evolving" compared to their characteristic evolu-tionary tempo in the Jurassic. Whatever the cause, it is possible that the de-cline of the ammonites and their final extinction were separate events.

In our opinion, the most interesting aspect of this study was the obvious difference in evolutionary tempo between Jurassic and Cretaceous ammo-nites. A quick examination of Figure 1 points out this fact: the early and middle Jurassic clades are very wide, but short, and then give way to a greater number of longer, thinner clades in the late Jurassic and Cretaceous. This makes biological sense, since the Lower and Middle Jurassic ammonites were filling the void left by the nearly complete extinction of the ammonids at the end of the Triassic; following this catastrophe, the Lower and Middle Jurassic ammonites show the classic pattern of explosive, adaptive radiation. What is less clear, is why the pattern changes so markedly in the Cretaceous. The intu-itive answer is that by this time the ammonite niches were all filled, and that competition began to lessen the chance of new adaptive breakthroughs; in this sense the ammonite pattern is not very different from any period of reor-ganization following a major adaptive radiation. However, can the magnitude of the change be related solely to this? Also, although some of the difference in Jurassic and Cretaceous clade shapes may be monographic, it cannot all be attributed to a greater number of workers working on ever finer Jurassic sub-divisions.

The greatest single difference between the Jurassic and Cretaceous lies in the great radiation of heteromorphic forms in the latter. A second major differ-ence is in the increasing number of heavily ornamented ammonites which oc-cur in the Cretaceous (Ward, 1981). Finally, the last two stages of ammonite record, the Campanian and Maestrichtian, seem to show two directions of adaptation: either streamlining, as shown in the great radiations of spheno-dicids, pachydiscids, placenticeratids, engoniceratids; and a complete lack of streamlining, or even rapid swimming ability, as evidenced by the diverse het-eromorphs with ungainly, U-shaped body chambers. We cannot believe that this major divergence in form was due to chance factors. An ammonite in the Late Cretaceous had to face a very different set of ecological conditions than did the Jurassic forms. We feel that clade shape differences are perhaps related to this as well.

A second aspect of our analyses also warrants attention: why should longevity trade off with diversity? There is no inherent reason why the long-lived families were often of relatively low diversity, and that the longest lived, such as the Phylloceratidae and Lytoceratidae were *always* at extremely low generic diversity. This again, appears to stretch the realms of chance, and begs to be explained in causal, biological terms rather than stochastic ones. Although Phylloceratids and Lytoceratids are commonly considered as "deepwater" ammonites because of their occurrence in more offshore facies, it is our experience that they are found in a variety of facies, indicating that they were capable of a wide depth range. Can depth tolerance translate into characteristic clade shape?

CONCLUSIONS

Application of clade statistics to Jurassic and Cretaceous ammonite families has produced mixed results; some statistics seem to reflect evolutionary tempo (DUR, PRO), while others do not (UNI, MAX). Even those statistics which do correlate with evolutionary tempo leave considerable room for error. At the present level of development, little is gained from analysis of clade geometry that cannot be gained from more conventional tests.

Jurassic and Cretaceous ammonites exhibit a wide range of evolutionary rates. Clades composed of rapidly evolving genera are often diverse but short-lived, while clades composed of long-lived genera tend to persist for long periods at low diversities.

ACKNOWLEDGMENTS

We thank David Jablonski and Jennifer Kitchell for reviewing and, in the process, greatly improving this paper.

REFERENCES

Arnold, A. J., 1982. Species survivorship in the Cenozoic Globigerinida. 3rd North Am. Paleontol. Conv. Proc. 1:9-12.

Birkelund, T., 1979. The last Maastrichtian ammonites. In Christensen, W. K., and Birkelund, T. (eds.), Cretaceous-Tertiary boundary events, II Proceedings, Copenhagen: Univ. of Copenhagen, 51-60.

Chamberlain, J. A., Jr., and Moore, W. A., Jr., 1982. Rupture strength and flow rates of *Nautilus* siphuncular tubes. Paleobiology 8:408-425.

Gould, S. J., Raup, D. M., Sepkoski, J. J., Jr., Schopf, T. J. M., and Simberloff, D. S., 1977. The shape of evolution: A comparison of real and random clades. Paleobiology 3:23-40.

Hancock, J. M., 1967. Some Cretaceous-Tertiary marine faunal changes. In Harland, W. B., et al. (eds.), The fossil record, London: Geol. Soc. London, 91-104.

Kennedy, W. J., 1977. Ammonite evolution. In Hallam, A. (ed.), Patterns of evolution, Amsterdam: Elsevier Sci. Publ. Co., 251-304.

Obradovitch, J. D., and Cobban, W. A., 1975. A time scale for the Late Cretaceous of the western interior of North America. In Caldwell, W. E. G. (ed.), The Cretaceous system in the western interior of North America, Geol. Soc. Can. Spec. Pub. (13):31-54.

Raup, D. M., 1981. Extinction: Bad luck or bad genes? Acta Geol. Hispanica, 16:25-33.

Raup, D. M., and Gould, S. J., 1974. Stochastic simulations and evolution of morphology—towards a nomothetic paleontology. Syst. Zool. 23:305-322.

Raup, D. M., Gould, S. J., Schopf, T. J. M., and Simberloff, D. S., 1973. Stochastic models of phylogeny and the evolution of diversity. J. Geol. 81:525-543.

Schopf, T. J. M., 1979. Evolving paleontological views on deterministic and stochastic approaches. Paleobiology 5:337-352.

Schopf, T. J. M., Raup, D. M., Gould, S. J., and Simberloff, D. S., 1975. Genomic versus morphologic rates of evolution: Influence of morphologic complexity. Paleobiology 1:63-70.

Sepkoski, J. J., Jr., 1975. Stratigraphic biases in the analysis of taxonomic survivorship. Paleobiology 1:343-355.

Stanley, S. M., 1977. Trends, rates, and patterns of evolution in the Bivalvia. In Hallam, A. (ed.), Patterns of evolution, Amsterdam: Elsevier Sci. Pub. Co., 209-250.

Stanley, S. M., 1979. Macroevolution. San Francisco: Freeman and Co.; 332.

Stanley, S. M., Signor, P. W., III, Lidgard, S., and Karr, A., 1981. Natural clades differ from "random" ones: Simulations and analyses. Paleobiology 7:115-127.

Van Eysinga, F. W. B., 1975. Geological time table (3rd ed.). Amsterdam: Elsevier Sci. Pub. Co.

Van Hinte, J. E., 1976a. A Jurassic time scale. A.A.P.G. Bull. 60:489-497.

Van Hinte, J. E., 1976b. A Cretaceous time scale. A.A.P.G. Bull. 60:498-516.

Ward, P. D., 1981. Shell sculpture as a defensive adaptation in ammonoids. Paleobiology 7:96-100.

Ward, P. D., and Signor, P. W., III, 1983. Evolutionary tempo in Jurassic and Cretaceous ammonites. Paleobiology 9:183-198.

Westermann, G. E. G., 1971. Form, structure and function of shell and siphuncle in coiled Mesozoic ammonids. Life Sci. Cont., R. Ont. Mus.

(78); 39.

Wiedmann, J., 1969. The heteromorphs and ammonite extinction. Biol. Rev. 44:563-602.

Wiedmann, J., 1973. Evolution or revolution of ammonoids at Cretaceous system boundaries. Biol. Rev. 48:159-194.

Chapter 14

BIOTIC DIVERSITY
AND CLADE DIVERSITY

JAMES W. VALENTINE

Department of Geological Sciences, University of California, Santa Barbara

Many interesting themes recur in the preceding papers, and this is an attempt
to relate one of them to the body of data and interpretation available in the
literature at large. However, it is written from a certain point of view, and the
assumptions and inferences offered here would certainly not be granted by all
paleontologists, or even by all contributors to this volume.

BIOTIC RATES OF DIVERSIFICATION AND
EXTINCTION AND STANDING DIVERSITY

Some investigators who have considered the kinetics of diversity regulation
on an evolutionary scale (MacArthur, 1969; Rosenzweig, 1975) have assumed
a theoretical equilibrium value for species diversity which is controlled by a
balance of speciation and extinction rates. By analogy with the assumptions
of the island biogeographic model of diversity elaborated by MacArthur and
Wilson (1963, 1967), the speciation rate in these models is assumed to fall
with increasing diversity because opportunities are becoming more limited.
The per species extinction rate, on the other hand, is assumed to increase
with increasing diversity because of the multiplication of competition pressures attendant on the increased species numbers.

However, there is little empirical support for the idea that competition regulates strongly the diversity of ecosystems. On the contrary, the effects of competition in excluding species that immigrate into established ecosystems do not appear to be very significant. Simberloff (1981) reviewed data on over 800 cases of the successful establishment of invading species into ecosystems; the invaders rarely caused extinctions (less than 10% of the time) and only a few of the extinctions that did occur (three) could be attributed to classic competition. Usually the invading species was simply added to the ecosystem, which thus proved to be less than full.

The failure of this data to corroborate a model that considers ecosystems to be full at equilibrium and that regards competition as a powerful force in diversity regulation, renders more attractive the alternative hypothesis that per capita extinction rates are independent of diversity (Walker and Valentine, 1984). In this case extinction presumably depends upon intrinsic biological properties of the taxa and the pace of environmental change, and the per capita extinction rate is constant with respect to diversity. The expression for change in species numbers under this hypothesis is

$$\frac{dN}{dt} = [a(1 - \frac{N}{N_{max}}) - b] N \tag{1}$$

where N is the number of species present, N_{max} the maximum number of "niches" or potential places for species in the ecosystem, "a" the speciation rate and "b" the extinction rate inherent in the circumstances (Walker and Valentine, 1984). Then when an equilibrium condition is reached so that $dN/dt = 0$,

$$N = (1 - b/a) N_{max} \tag{2}$$

In this situation, ecosystems are never "full," with all potential niches occupied by species, so long as the extinction rate "b" is above zero. Competition does not affect niche availability disproportionately as the ecosystem acquires more species. When rates of "a" and "b" observed in the fossil record are substituted into equation (2) one can calculate the proportion of "empty niches" or unfilled opportunities that are available at equilibrium. For example, if N_{max} is arbitrarily set at 100, then N is the percentage of filled niches and the percentage of empty niches is $N_{max} - N$ or in this case $100 - N$. Because the observed extinction and speciation rates vary among taxa, there are more empty niches typically available for some taxa than others; there are more empty niches for ammonites, for example, than for corals. Chiefly because of uncertainties about speciation and extinction rates it is only possible to give a rough estimate of the usual or average percentage of open niches. At most times in most ecosystems the number should lie between 10% and 50% (Walker and Valentine, 1984); perhaps the average number in the average Phanerozoic benthic ecosystem is near 30%. In today's world, this would mean that there are over 150,000 openings for additional marine invertebrate species.

Why should adaptive space be so empty, assuming this model is roughly correct? Partly because extinction rates are fairly high, but also in large measure because observed speciation rates (which are rates of morphospeciation), even at times of the most extensive radiations ("unbridled speciation" of Stanley, 1979), do not begin to approach the maximum theoretical rates. Lande (1980) has estimated that morphospecies can be evolved in hundreds of thousands of generations under optimum circumstances; allowing for slightly less than optimum conditions we might say conservatively that such speciations can occur in about 10,000 years. Beginning with an ancestral species, then, and permitting only one speciation per lineage at this rate, more species could appear within a million years than have existed altogether since the origin of life. Observed rates of unbridled speciation, which are presumably not impeded by extinction and are supposed to represent radiation within rather empty parts of adaptive space, range around 0.25 to 0.60 new species per lineage per million years (Stanley,1979). Even when speciation is unbridled, something is holding it back.

It seems likely that the difference between the potential maximum speciation rate (which is certainly even higher than the estimates here) and the observed speciation rate, is a measure of the effectiveness of the barriers which separate niches (or the adaptive valleys between adaptive peaks, to use the common metaphor of an adaptive landscape). Presumably, opportunities remain unfulfilled because lineages which might produce daughter species capable of occupying an open adaptive space are restricted by stabilizing selection to their own adaptive peaks, unable to produce the less fit intermediates which can cross adaptive valleys leading to empty adaptive peaks. Such an explanation follows from the adaptive models of Wright (1931) and others. To the extent that this is true, evolution (via stabilizing selection) is acting to prevent the rapid filling of open adaptive space and thereby promoting rather open ecosystems.

The adaptive space actually occupied by lineages at one time or another may thus be kept partially empty by extinction and by the difficulty of speciation across adaptive barriers, even if the barriers have been previously broached and the adaptive space previously occupied. One can imagine also that some small fraction of empty adaptive space is fully as accessible to extant lineages as is the occupied space, but happens never to be occupied simply by chance. It is the density of occupation of the rather accessible previously occupied adaptive space that is estimated at about 30% on average.

If it is true that the adaptive barriers between niches are as strong as suggested above, then some of the puzzling patterns in the fossil record seem less enigmatic. This is particularly true of the abruptness with which many invertebrate taxa appear, at all levels from phylum to species. The famous punctuational pattern at the species level (Eldredge, 1971; Eldredge and Gould, 1972) can be viewed as representing episodic breakthroughs across adaptive barriers into "open niches." Once a niche is occupied, stabilizing selection will tend to prevent morphological change, except for improvements which should usually be accomplished early, and except for the tracking of new conditions as they arise.

However, there are other regions of adaptive space that are surrounded by a different, more formidable class of barriers, regions in which life can readily exist but to which entrée is particularly difficult for the extant lineages; whole new body plans or major modifications of existing ones, with all the physiological and behavioral changes implied, may be required. For example, most adaptive zones of the terrestrial environment represented such a space for animals of the earliest Paleozoic. Indeed, many of the adaptive regions which today contain rich communities of organisms were not occupied at the beginning of the Cambrian, while others were sparsely occupied but contained adaptive subregions or patches which were unoccupied (Bambach, herein; Ausich and Bottjer, herein). Many adaptive regions were occupied abruptly, sometimes in association with the origin of a new morphology distinctive enough to become classed as a taxon of ordinal or higher rank. This pattern can be reasonably interpreted as owing to the strong adaptive barriers surrounding the unoccupied regions and subregions, barriers broached only with difficulty. These barriers need not necessarily be higher than those surrounding some occupied regions; they may not have been penetrated owing in part to historical reasons or to chance. This pattern of zonal invasion is quite similar to that proposed long ago by Simpson (1944) from chiefly terrestrial vertebrate evidence, although evolution at the rapid rates that he termed "tachytely" may involve processes of genome change of which we were then unaware (and of which we are still largely uncertain). At any rate, the chances of penetrating barriers of a given magnitude would seem always to be better when there is much open space in the zones to be invaded, and when there are many such open zones available for invasion. Both these conditions were present near the Precambrian-Cambrian boundary, when so much evolutionary invention occurred. Furthermore, a weak but evidently real association of first appearances of higher taxa with the radiations that follow mass extinctions (Newell, 1967; Valentine, 1969) suggests that adaptive space formerly occupied by lineages removed by the extinctions became available for invasion and therefore played a role in the origin of some important clades.

CLADE RATES OF DIVERSIFICATION AND
EXTINCTION AND DIVERSITY HISTORIES

Aside from the requirement that they must expand to appear and contract to disappear, clades seem to obey no particular rules insofar as their diversity histories are concerned (for example, Hardy, herein, Ward and Signor, herein). On the contrary, at the family level, clades present such a rich variety of spindle shapes that they have been under suspicion as undergoing random fluctuations in response to stochastic processes (Raup et al., 1973). However, subsequent studies suggest that at least the largest and most abrupt changes in clade diversities are deterministic (Stanley et al., 1981; Raup, 1981). The relative volatility of clade diversity contrasts with the relative stability of family diversity in the marine biota as a whole (Kitchell and Carr, herein; Sepkoski and Hulver, herein; Sepkoski and Miller, herein). This disparity obviously reflects the differential representation of clades through time within their ecosystems.

Large clades are presumably extinguished because they share some trait which becomes disadvantageous owing to conditions arising. The trait may prevent these organisms from coping with new physical or biological conditions, or it may simply inhibit their speciation rates relative to other clades which may take over the adaptive space to which the slower-growing clades would have had access. In this way, extinctions owing to causes unrelated to any common characteristic of a clade will, nevertheless, not be replaced because of a common characteristic, and the clade will dwindle (see Stanley, 1979).

It seems likely that most clade expansions occur in unoccupied adaptive space, in which organisms have not previously been present or from which former occupants have been erased. It then seems plausible that since there have commonly been adaptive regions unoccupied for many tens of millions of years, and perhaps always many empty niches, the key to the differential success of clades has often been associated with their kinetics, particularly with the ability of their members to penetrate niche and zone boundaries, rather than with any superiority in physiological and behavioral tolerances or requirements *per se*.

REFERENCES

Eldredge, N., 1971. The allopatric model and phylogeny in Paleozoic invertebrates. Evolution 25:156-167.
Eldredge, N., and Gould, S. J., 1972. Punctuated equilibria: An alternative to phyletic gradualism. In Schopf, T. J. M. (ed.), Models in paleobiology, San

Francisco, Freeman and Cooper, 82-115.

Lande, R., 1980. Genetic variation and phenotype evolution during allopatric speciation. Am. Nat. 116:463-479.

MacArthur, R. H., 1969. Patterns of communities in the tropics. Biol. J. Linn. Soc. 1:19-30.

MacArthur, R. H., and Wilson, E. O., 1963. An equilibrium theory of insular zoogeography. Evolution 17:373-387.

MacArthur, R. H., and Wilson, E. O., 1967. The theory of island biography. Princeton, N.J.: Princeton Univ. Press; 203.

Newell, N. D., 1967. Revolutions in the history of life. Geol. Soc. Amer. Spec. Papers (89):63-91.

Raup, D. M., 1981. Extinction: Bad genes or bad luck? Acta. Geol. Hispanica 16:25-33.

Raup, D. M., Gould, S. J., Schopf, T. J. M., and Simberloff, D. S., 1973. Stochastic models of phylogeny on the evolution of diversity. J. Geol. 81: 525-542.

Rosenzweig, M. L., 1975. On continental steady states of species diversity. In Cody, M. L., and Diamond, J. M. (eds.), Ecology and evolution of communities, Cambridge, Mass.: Belknap Press, 121-140.

Simberloff, D. S., 1981. Community effects of introduced species. In Nitecki, M. H. (ed.), Biotic crises in ecological and evolutionary time, New York: Academic Press, 53-81.

Simpson, G. G., 1944. Tempo and mode in evolution. New York: Columbia Univ. Press; 237.

Stanley, S. M., 1979. Macroevolution: Pattern and process. San Francisco: Freeman; 332.

Stanley, S. M., Signor, P. W., III, Lidgard, S., and Karr, A. F., 1981. Natural clades differ from "random" clades: Simulations and analyses. Paleobiology 7:115-127.

Valentine, J. W., 1969. Patterns of taxonomic and ecological structure of the shelf benthos during Phanerozoic time. Palaeontology 12:684-709.

Walker, T. D., and Valentine, J. W., 1984. Equilibrium models of evolutionary species diversity and the number of empty niches. Am. Nat. 124: 887-899.

Wright, S., 1931. Evolution in Mendelian populations. Genetics 16:97-159.

CONCLUDING REMARK

The papers in this volume have described unexpected patterns of biotic diversity, of clade diversity, and of some of the events which contribute to each. Some patterns are unexpected because they indicate that widely held hypotheses are likely to be incorrect; some because they were simply unknown previously; and some because they provide new perspectives from which to analyze an established pattern. That many of the new findings and interpretations concern global patterns of the history of life involving whole eras of geologic time suggests that we have only begun to realize the potential of Phanerozoic diversity studies. The symposium contributors, though they have their differences, are certainly at one in their hope that this volume may represent at least a small step towards the successful exploitation of this rich scientific resource.

AUTHOR INDEX

Abbot, R.T.   373
Abbott, I.   258,273
Ahluwalia, D.S.   333
Allen, P.M.   284,307,308
Alvarez, L.W.   79,81,90
Alvarez, W.   80,90
Amsden, T.W.   185
Anderson, C.S.   278,307,361,371
Anderson, E.J.   170,177
Anderson, R.M.   308
Anderson, R.N.   344,348
Anderson, S.   278,307,346,348,358,
   360,361,371
Archibald, J.D.   75,76,78,90
Armstrong, R.   273
Arnold, A.J.   414,416
Arthur, M.A.   318,332
Asaro, F.   90
Ashworth, A.C.   15,37
Astrova, G.G.   210,249
Ausich, W.I.   145,149,153,177,188,
   255,256,257,259,260,261,263,266,
   267,268,271,272,422
Ayala, F.J.   117,118,126,284,305,308

Babloyantz, A.   308
Bakker, R.T.   42,43,49,90,92
Bambach, R.K.   6,7,8,11,35,95,97,126,
   128,131,132,145,149,150,153,159,
   177,179,180,185,191,194,217,241,
   242,243,247,249,251,255,273,274,
   343,345,346,348,422
Banks, H.P.   102,126
Barker, P.F.   339,349
Barnard, K.H.   373
Barnett, V.D.   281,307
Barron, E.J.   346,349
Barthel, K.W.   15,35
Bartlett, M.S.   281,307
Bartsch, P.   373
Basile, D.V.   123,126
Bassett, M.G.   102,126,178

Bassler, R.S.   210
Bateson, W.   360,371
Batten, R.L.   343,347,349
Bayer, T.N.   184
Beck, C.B.   106,126
Bedford, H.S.   128
Beerbower, J.R.   187
Behrensmeyer, A.K.   6,7,43,44,90,92
Bergquist, P.R.   200,250
Bergstrom, J.   170,178
Berry, W.B.N.   155,168,170,178,180,
   182,190,346,347,349,350,415
Beus, S.S.   189
Bierman, J.   308
Billingham, J.   252,396
Birkelund, T.   351,414,416
Blatt, H.   135,136,149
Bode, A.   15,35
Bolt, J.R.   55,90
Bolton, T.E.   184
Bonamo, P.M.   38
Bonham-Carter, G.F.   346,347,353
Bottjer, D.J.   145,149,153,175,177,
   178,249,255,256,257,259,261,263,
   266,267,268,271,272,273,348,422
Boucot, A.J.   32,35,149,158,168,169,
   170,172,178,185,186,190,264,273,
   274,346,349,350,352,369,371
Bowen, Z.P.   187
Bown, T.M.   60
Boyd, D.W.   236,250
Brabb, E.   185
Bradley, J.S.   190
Bramson, M.   329,332
Brasier, M.D.   269,273
Bray, R.E.   187
Bretsky, P.W.   155,170,171,177,178,
   183,259,261,273,347,349,369,371
Brew, D.C.   189
Bright, R.C.   181
Brink, A.S.   67,93
Brinkmann, R.   45,90

SUBJECT INDEX

Library of Congress Cataloging in Publication Data
Main entry under title:

Phanerozoic diversity patterns.

Includes bibliographies and indexes.
1. Paleontology—Congresses. 2. Evolution—Congresses
I. Valentine, James W.
QE701.P59   1985   560   84-42905
ISBN 0-691-08374-6
ISBN 0-691-08375-4 (pbk.)